PLANT CELL AND TISSUE CULTURE IN LIQUID SYSTEMS

Hanser Titles in Biotechnology

Series Advisory Editor
Michael L. Shuler

Plant Cell and Tissue Culture in Liquid Systems
Gregory F. Payne, Venkataraman Bringi, Christopher L. Prince, Michael L. Shuler

Related Hanser Titles

Bioprocess Monitoring and Control
Marie-Noëlle Pons

Protein Purification:
Design and Scale up of Downstream Processing
Scott M. Wheelwright

Diffusion In and Through Polymers:
Principles and Applications
Wolf R. Vieth

Membrane Systems: Analysis and Design
Wolf R. Vieth

Biotechnology Focus 1:
Fundamentals-Applications-Information
Edited by Robert K. Finn and Paul Präve

Biotechnology Focus 2:
Fundamentals-Applications-Information
Edited by Robert K. Finn and Paul Präve

Large Scale Cell Culture Technology
Edited by Bjorn K. Lydersen

PLANT CELL AND TISSUE CULTURE IN LIQUID SYSTEMS

G. F. PAYNE
V. BRINGI
C. L. PRINCE
M. L. SHULER

WILEY

Plant Cell and Tissue Culture in Liquid Systems

First Indian Reprint: 2013

Authorized reprint by Wiley India Pvt. Ltd., 4435-36/7, Ansari Road, Daryaganj, New Delhi – 110002.

Published by John Wiley & Sons Inc.

Printed at: *Syndicate Print-n-Pack, NOIDA*

ISBN: 978-81-265-4223-9

This book is dedicated to our family and friends who have provided the support and inspiration which have allowed us to pursue our interests.

Also GFP wishes to dedicate this book to the memory of Michael J. Schiavina.

CONTENTS

PREFACE

When we set out to write this book nearly three years ago, our intent was to provide a review of plant cell tissue culture for secondary metabolite production. Our goals were to write a book which provided a cohesive description of the field and was accessible to researchers from the various scientific and engineering disciplines. We envisioned that this would summarize the state of the art and provide insights into the generalities involved in cell and organ culture. Finally, we hoped that our initiative could be helpful someday when plant cell tissue culture technology was transferred from the laboratory to commercial practice (Recall that in 1988, the shikonin process was only three years old and there was no obvious second product).

With the sudden surge in interest in the production of plant products from cell culture, it is possible that that someday is today. For instance, academic and industrial researchers from several countries are actively pursuing strategies to exploit plant cultures for the production of taxol, a promising anticancer agent. It is possible that the dedicated study on this single product will provide the focused effort so necessary to overcome the technical barriers to commercialization of plant culture for secondary metabolite production. Perhaps history will repeat itself with the biosynthesis of taxol and a handful of other plant products ushering in a new commercial technology just as penicillin biosynthesis ushered in fermentation processing 50 years ago. We hope that our work can in some small way help to advance applications of commercial plant cell technology!

This book has become a reality due to the National Science Foundation, which has patiently supported research in our laboratories. We also wish to acknowledge Wolf G. W. Kurz of the Plant Biotechnology Institute of Canada who provided helpful suggestions to improve this book. Finally, we are grateful to Nadine Payne for drawing many of the figures which appear in these pages.

Gregory F. Payne
University of Maryland, Baltimore County Campus
Venkataraman Bringi
Phyton Catalytic Inc., Ithaca, New York
Christopher L. Prince
Phyton Catalytic Inc., Ithaca, New York
Michael L. Shuler
Cornell University, Ithaca, New York

June 25, 1991

BIOGRAPHICAL SKETCHES

Gregory F. Payne is a founding member of the Chemical and Biochemical Engineering Department, University of Maryland, Baltimore County (UMBC) Campus. His research focuses on how bioreactor operation can improve culture performance through interdisciplinary interaction related to cell physiology and gene expression. He also studies polymers for selective adsorption. He completed his Ph.D. degree at the University of Michigan with Henry Wang and post-doctoral study in secondary metabolite production at Cornell University with Michael Shuler. His Masters and Bachelors degrees are also in chemical engineering from Cornell.

Venkataraman (Bobby) Bringi is co-founder and Technical Director at Phyton Catalytic Inc. in Ithaca, New York. His research interests include the engineering and biological aspects of plant cultures, specifically mass transfer and reactor design, cell immobilization, and shoot cultures. He holds a Ph.D. degree from Cornell University, a M.S. from Colorado State University, and a B. Tech. from Banaras Hindu University, India, all in chemical engineering.

Christopher L. Prince is co-founder and Research Director at Phyton Catalytic, Inc. His research interest is in the area of primary and secondary metabolite production in plant cultures, focusing on physiological and reactor scale-up aspects of root organ cultures. He spent eight months as a visiting scientist at the Research Center for Cell and Tissue Culture in Kyoto, Japan. He holds a Ph.D. degree from Cornell University and a M.S. degree from Queen's University in Canada, both in chemical engineering; and a Bachelor's degree in chemistry and chemical engineering from Queen's University.

Michael L. Shuler is Professor of Chemical Engineering with a joint appointment in the Institute of Food Science at Cornell University. He is founding editor of the American Institute of Chemical Engineering (AIChE) journal "Biotechnology Progress" and Hanser Publisher's series-editor, for biotechnology. His research interests include: whole microbial cell immobilization using membrane culture; growth characteristics and product formation from plant cell tissue cultures; development and experimental verification of models for the growth of individual cells; response of natural eco-systems to toxic chemicals; genetic stability of recombinant bacteria; plasmid encoded protein excretion from E. coli; insect cell culture; cell culture analogs for pharmacokinetic models.

Dr. Shuler has five books published or in development, holds five patents, has written 16 book chapters and 94 technical papers. He has been a visiting professor at the Universities of Wisconsin, Washington (Seattle), Florida, Osaka, and Newcastle Upon Tyne. His honors include AIChE's Professional Progress Award for Outstanding Progress in Chemical Engineering, and AIChE's Food, Pharmaceutical & Bioengineering Division Award as well as the American Chemical Society's Marvin J. Johnson Award of the Microbial and Biochemical Technology Division. He is a member of the National Academy of Engineering. He completed his Ph.D. degree at the University of Minnesota and his B.S. degree at the University of Notre Dame, both in chemical engineering.

PLANT CELL AND TISSUE CULTURE IN LIQUID SYSTEMS

CHAPTER 1

THE QUEST FOR COMMERCIAL PRODUCTION OF CHEMICALS FROM PLANT CELL CULTURE

A. WHY PLANT CELL CULTURE?

The higher plants have been the key to human welfare from the earliest dawn of human existence. Food, materials, fragrances, dyes, and medicine have been derived from higher plants. The genetic diversity of plants has provided us not only survival, but a high degree of comfort and relief from illness.

Yet we have barely exploited this genetic diversity. The number of known chemical structures in plants is estimated to be nearly fourfold greater than in the microbial kingdom. In 1985, of 3,500 new chemical structures identified, 2,619 came from the higher plants. Worldwide, 121 clinically useful prescription drugs are derived from plants. Many of these compounds have been identified from folk or traditional medicine. Even today, 75% of the world's population relies on plants and traditional medicine. In the United States, where chemical synthesis dominates the pharmaceutical industry, 25% of the pharmaceutical market is based on plant-derived chemicals.

Yet only 5,000 of the estimated 250,000 to 300,000 plant species have been studied extensively for medical applications. Unfortunately, many plant species are becoming extinct even before it has been possible to identify them. The tropical rain forests are disappearing quickly. Even in the United States over 3,000 species are listed as endangered and 250 are predicted to become extinct by 1995. Our inherited natural genetic wealth is being lost.

Their genetic diversity is what makes plants potentially interesting as a source for production of chemicals. The obvious question is why might the "factory" production of chemicals using plant cell culture be preferable to simply recovering the product from whole plants.

There are several answers to such a question. The first is supply. In some cases, the compound of interest comes from a slow-growing woody species that may be rare or endangered. The supply of the compound cannot be expanded rapidly to meet the potential need for the chemical. A good example is the compound taxol, which comes from an endangered species, the Pacific Yew Tree. Taxol is in phase III trials as a potential chemotherapeutic agent for treatment of ovarian cancer - a cancer currently difficult to treat successfully. One 200-year-old tree must be cut and its bark extracted to provide a dose of taxol. To begin clinical trials, 12,000 trees were cut. Even if large plantations for cultivation of the Pacific Yew Tree were begun now, it would be many years before sufficient harvestable material would be available. Taxol is a chemical unique to the plant kingdom. Its *de novo* synthesis has not been accomplished, and genetically engineered microbes are not a realistic alternative source. Taxol is not a unique example; many compounds from slow-growing, rare plants are of potential interest as medicinals.

Other factors can also lead to supply problems that might be overcome using bioreactors and plant cell culture. Disease and drought can radically alter supply. War and politics can change availability of a raw material (e.g., natural rubber in World War II). Finally, fluctuations in growing conditions may alter the composition of plants, and variable yields complicate the processing of the raw material.

With standard plant-breeding techniques, it is difficult to rapidly increase yields of chemicals in slowly maturing plants. Cells in tissue culture can be manipulated much more rapidly to obtain higher yields.

A very intriguing possibility with "factory processes" using plant cell tissue culture is biotransformation processes making wholly novel compounds. Many plant enzymes will accept synthetic analogues of natural substrates and perform highly specific modifications on these analogues. Products that would be difficult or impossible to synthesize in the test tube can be created in high yield. This potential has been little explored.

Plant cells cultivated in bioreactors have the potential to make a significant impact on the production of chemicals for medicinals and other uses, such as flavors, fragrances, or pigments. How is this potential being converted into reality?

B. THEMES IN THE DEVELOPMENT OF CHEMICALS PRODUCTION FROM CELL CULTURE

The quest for practical commercial processes for the production of chemicals from plant cell tissue culture has been long and arduous. Success has come - but often only slowly. However, a sufficiently large body of experience has now developed which will support a rapid acceleration of these efforts.

Twenty years ago, the prospect for chemicals production from plant cell culture was only a dream. The low yields of secondary metabolites in suspension cultures clearly made commercialization problematic. In addition, consideration of engineering aspects of such cultures was virtually nonexistent. These early efforts treated plant cells in suspension culture by direct analogy to microbial systems, with little appreciation of the unique features of plant cell physiology and how bioreactor operation could influence the physiological state of such systems.

Since then, many useful themes have appeared. The potential usefulness of strategies to optimize growth and product formation separately began to develop in the period 1975-85. The development of production media and multistage or multistep cultivation were emerging themes. When these developments have been coupled to improved strategies for selection of cell lines with enhanced capability for product synthesis, yield improvements of 10- to 100-fold have often been obtainable. A combination of these three strategies gave sufficient yield improvements to make the establishment of the first commercial plant cell process possible. In 1984, Mitsui Petrochemical Industries began production of shikonin from tissue culture of *Lithospermum erythrorhizon*. Shikonin can be used both as a dye and as a medicinal (anti-inflammatory agent). The process has not been approved for production of shikonin as a medicinal, which has restricted the market to its production as a dye.

More recently, several promising themes have developed. The use of elicitors to activate secondary metabolism in tissue culture is highly promising. Immobilization of cells to alter intrinsic productivity has been demonstrated as well as the critical role mass transfer (e.g., diffusion of oxygen) may have on the physiological response of entrapped cells. Strategies to foster release of normally intracellular products from plant cells are being developed in conjunction with such efforts. Cell immobilization is probably an attractive technique for commercial use only if it can be coupled with secretion.

In some cases, none of these strategies have yielded product. Experiments have demonstrated that the gene(s) responsible for key steps in a pathway may be intact in suspension cultures but expressed only in differentiated and organized tissue. This observation has increased interest in the use of organ cultures (e.g., roots, shoots, or embryos). Recent advances in developing "hairy root" systems which grow rapidly has further accelerated interest in organ culture as a source for producing chemicals.

In related developments, bioreactors have proved useful in micropropagation. In some cases (e.g., artificial seeds), an understanding of control of developmental biology in tissue culture is critical, while in others the actual culture of organs (e.g., shoots) in submerged culture has become important.

Additional themes in the development of plant cell culture undoubtedly will emerge. A goal of this book is to provide a basis for wider development of an appreciation of the interaction between physical parameters and cellular physiology. This theme will lead to a clearer appreciation of how bioreactor design can alter cellular physiology. The interaction of mass transfer with cells and the level of association of cells with one another in a reactor will be key elements in achieving both optimal production and effective scale-up of plant cell cultures to commercial scale. We believe that this theme will be the most critical one over the next decade, and it is the focus of this book.

C. COMMENTS ON GENETIC ENGINEERING

Genetic engineering has yet to have an important impact on plant cell culture for chemicals production. We believe that the difficulties in metabolically engineering pathways in plant cells for predictable improvements in chemicals production are large. The lack of basic biochemical information on many of these pathways coupled with the sheer complexity of manipulating multiple pathways suggests that genetic engineering will not be a generally useful technique to enhance chemicals production for at least 10 more years. The only possible exception to this rule is the use of *Agrobacterium spp.*, as a vector for cloning key biosynthetic genes. Preliminary studies along these lines have been conducted by Hashimoto and Yamada at Kyoto University (personal communication).Consequently, we have chosen not to include specific information in this book on genetic engineering except for *Agrobacterium*.

A final speculation on the impact genetic engineering might have on plant cell tissue culture is the suggestion that plant cells could be hosts for the production of cloned protein products. Because of the chemically defined, protein-free media used for plant cells and the high cell density that can be achieved by plant cells, plant cell systems are indeed attractive when compared to systems in which animal cells are hosts. Plant cells do not compare well to bacterial or yeast cells in terms of production rates. Consequently, plant cells would be effective hosts for protein production *only* if they carry out post-translational processing of proteins in a manner identical or nearly identical to that in animal cells. The post-translational machinery in a variety of plant cells has not yet been well characterized. Thus, the bioreactor-based production of proteins from suspension cultures of plant cells is not likely in the short term and is not a subject of this book.

The use of transgenic plants to produce proteins such as insulin and antibodies has been demonstrated. Efficient production of such plants may require, at least initially, micropropagation. Bioreactor technology could easily be involved in the micropropagation phase, but the use of transgenic plants would not involve bioreactor issues different from those for normal plants. The primary theme of the book would still apply: "How does bioreactor design and operating strategy alter cellular physiology and development?"

D. COMMERCIALIZATION - SOME EXAMPLES AND A "RULE OF THUMB" FOR ECONOMIC FEASIBILITY

Most of the activity with commercial-scale plant cell culture has taken place in Japan. Plant cell culture has received more attention in Japan than in North America. Several reasons may be important. One is simply that natural products play a more prominent role in medicine and health-related products in Japan than in North America. Further, the commercialization of plant cell culture requires a long-term perspective, and most Japanese firms have been able to adopt such a viewpoint. Recently, the Japanese have targeted plant cell culture as an area for further development, and a substantial fund to support plant cell culture work has been established by the government and a consortium of interested corporations.

Three large-scale processes using plant cell culture have been established. Shikonin, sold as a dye, has been produced by Mitsui Petrochemical Industries for seven years (Tabata and Fujita 1985). Berberine, a medicinal that is sold almost exclusively in Asia, can be produced in high yield in a continuous-flow reactor in which cells are retained by a membrane and harvested on a periodic basis (Fujita et al. 1989). Apparently this system has been operated at a scale up to 4000 L. Production of ginseng as organized tissues (root or tuber-like structures) is being exploited commercially. The product is sold as a health food. It is cultured in vessels of, at least, 20,000 L (Ushiyama 1989). Currently, efforts are being made to exploit this system for production of ginsenosides.

In all three cases, regulatory barriers to the use of these products as medicinals have impeded further development. For example, both shikonin and chemicals from ginseng have medicinal uses as well as nonmedicinal applications. The full realization of the economic potential of these processes cannot be gained until full regulatory approval is achieved.

In North America no commercial-size facilities are in operation. However, two processes are actively being developed for commercialization. The production of a natural vanilla flavor from plant cell culture is being developed by Escagenetics. The second process is the production of sanguinarine. Elicitation is apparently a key component of the production strategy. Sanguinarine appears effective in treating gum disease (gingivitis). A 300-L airlift system has been constructed by Vipont Research Laboratories for the production of sanguinarine from *Papaver somniferum* (Park et al. 1990). Colgate has developed a formal arrangement with Vipont which suggests strong interest in completing the development of commercialization of this project. A new start-up company, Phyton Catalytic, has been founded to make commercial scale levels of taxol and other plant products.

In Europe, several processes have been seriously considered for commercialization but ultimately put on hold or abandoned. However, a new 75,000-L facility has been built near Hamburg, Germany. Among the cells cultured in this facility is *Echinacea purpurea* which produces immunoactive polysaccharides.

In addition to these efforts to produce chemicals, whole plants are being produced. Levin et al. (1988) report that Plant Biotechnology Industries (Israel) uses a bioreactor as an important component of a micropropagation process that has been applied successfully to 14 species. A collaborative venture between Kirin Brewery and the U.S. firm Plant Genetics has produced a large-scale lettuce and celery crop from synthetic seeds. Based on previous reports (e.g., Walker and Sluis 1985), it is likely that bioreactor technology was employed to generate the embryos upon which the artificial seeds were made. Although the focus of this book is on chemicals production, these examples are still relevant to bioreactor considerations.

Thus, plant cell tissue culture in bioreactors for production of chemicals and plant propagules is a commercial reality. It is clearly an infant technology, and its full commercial potential is unclear. However, the recent advances discussed as themes in plant cell development are providing the intellectual and experimental foundation for the more predictable exploitation of plant cell culture.

The key factor in many of these decisions on commercialization is volumetric productivity (e.g.,milligrams of product made per liter per day). Cooney (1989) has observed that the revenue generated per unit volume of reactor capacity from a wide variety of fermentation products is surprisingly constant. That is, the wholesale price of a product times its volumetric productivity is nearly constant. For compounds from ethanol to lysine and penicillin to vitamin B_{12}, the revenue is about 15¢/(L × day) with a range of values from 7¢/(L × day) to 23¢/(L × day).

If we assume plant cell culture would follow the same "rule of thumb," we can estimate what productivity a plant cell tissue culture must achieve to be economic at a given price, or, given a particular productivity, what price it must command to be potentially economically attractive. Table 1.1 lists some compounds with their reported productivities, the required price to achieve a revenue of 15¢/(L × day) and, in some cases, the wholesale price for the compound of interest. Most of these products have productivities between 0.025 and 1.0 g/(L × day) which corresponds to prices of $150/kg to $5,000/kg. On the other hand, we can use the analysis to set productivity goals. For example, the wholesale prices for codeine and rose oil are 90¢/g and

750¢/g, respectively. Plant cell processes for these compounds would become economically interesting at productivities of 0.17 g/(L × day) and 0.02 g/(L × day), respectively.

The compounds listed in Table 1.1 are ones that have been subjects of fairly intense investigation and optimization. Not all products examined in detail have achieved such productivities; the indole alkaloids from *Catharanthus roseus* are a case in which some products were never found and others, such as ajmalicine, have values less than 0.01 g/

Table 1.1. Volumetric productivities for some products made in plant cell culture and their corresponding value

Product	Cell line	Productivity [g/(L×day)]	Investigators	Bulk price to give 15¢/(L×day) revenue ¢/g	Bulkprice (Fall,1989) ¢/g
Anthocyanin	Grape (Bailey Alicant A, *Vitis* hybrid)	0.06	Yamakawa et al. (1983)	250	-
		0.15	Hirasuna (1990)	100	-
Berberine	*Thalictrum minus*[a]	0.05	Kobayashi et al. (1988)	300	325[b]
	Coptis japonica[c]	0.60	Fujita et al. (1989)	25	325
Diosgenin	*Dioscorea* spp.	0.75	Based on table in Fowler and Scragg (1988)	20	105[d]
Podoverine	*Podophyllum versipelle*	0.15	Ulbrich et al. (1988)	100	-
Rosmarinic acid	*Coleus blumei*	0.91	Ulbrich et al. (1985)	16	-
Sanguinarine	*Papaver somniferum*	0.034	Park et al. (1990)	440	-
Shikonin	*Lithospermum erythrorhizon*	0.15	Tabata and Fujita (1985)	100	400[e]

[a] Execreted into the medium, where it crystallizes

[b] Special quote from Sigma Chemical Co. (St. Louis, MO) for 1 kg of berberine hydrochloride

[c] Continuous flow system with cell retention; intracellular product.

[d] Sigma Chemical Co. (St. Louis, MO) $105/100 g; $2,420/kg from ICN Biochemicals (Cleveland, OH). Price dependent on purity.

[e] From Curtin 1983, for 1983 wholesale prices.

(L × day). Increases in productivity of 10- to 100-fold or even more from initial levels are not unusual, and such potential improvements should be considered in applying this "rule of thumb" to prospective products. Such a productivity rule may be a helpful guide to assessing the economic feasibility of bioreactor-produced products from plant cell culture.

Some cautions in using such a rule must be observed. First, the analysis does not consider alternative supply sources. The critical factor is the ratio of productivity in cell culture to that in field-grown plants. The analysis by Drapeau et al. (1987) concluded that commercial production of shikonin was economically attractive when compared to recovery from whole plants because the ratio of product formation (g product/g cells per day) for bioreactor-grown material was 870 times larger than for field-grown material. They also concluded that, at a ratio of 24:1, ajmalicine was uneconomic. Their calculations suggest that a 40-fold increase in ajmalicine productivity would give a ratio of 960 and would make recovery of the product from bioreactor-grown material more economic than from field-grown plants. So, as a second rule of thumb, we suggest that the productivity should be high enough to give not only a revenue of 15¢/(L × day) but also a productivity advantage over field-grown material of 800 to 1000. Additionally, if chemical or microbial syntheses are well-established processes, they would not be displaced by plant cell culture at a price to yield 15¢/(L × day). The manufacturing cost from plant cell culture would have to be significantly less (e.g., able to give >30¢/ (L × day) at current prices).

The second factor the analysis does not consider is market demand. The investment in new technology will probably only be attractive if the annual sales are sufficiently high (i.e., price × quantity sold per year = annual sales) to recover research and development costs. "Sufficiently high" will depend on the culture of a corporation, range of products already sold, and whether the basic technology could be applied to several different products. As a very rough rule of thumb, a total market of $10,000,000/ year may be near the threshold for development of a single product.

When considering these factors, it is apparent that plant cell tissue culture is most likely to be amenable to commercialization when the product commands a moderate to high price and comes from a slow-growing woody or rare species and where no well-established alternative route exists. If plant cells can make a novel product, such as through feeding a synthetic analogue of a natural substrate, then the opportunities for commercialization are enhanced, depending almost solely on the projected need for the product (price and volume).

References

C. L. Cooney. 1989. Personal communication.

M. E. Curtin. 1983. Harvesting Profitable Products from Plant Tissue Culture. *Bio/Technology 1*, 649-657.

D. Drapeau, H. W. Blanch, and C.R. Wilke. 1987. Economic Assessment of Plant Cell Culture for the Production of Ajmalicine. *Biotechnol. Bioeng. 30*, 946-953.

M. W. Fowler and A. H. Scragg. 1988. Natural Products from Higher Plants and Plant Cell Culture. In "Plant Cell Biotechnology," M. S. S. Pais, F. Mavituna, and J. M. Novais (eds.), NATO ASI Series Vol. H18, Springer-Verlag, Berlin, p. 165.

Y. Fujita, T. Yoshioka, and Y. Hara. 1989. Production of Shikonin and Berberine by Large Scale Cultures of Plant Cells. Paper read at the 1989 International Chemical Congress of Pacific Basin Societies, Honolulu,Hawaii, Dec. 17-22, 1989.

T. Hirasuna. 1991. Ph.D. Thesis, Cornell University, Ithaca, New York.

Y. Kobayashi, H. Fukui, and M. Tabata. 1988. Berberine Production by Batch and Semi-Continuous Cultures of Immobilized *Thalictrum* Cells in an Improved Bioreactor. *Plant Cell Rep.* 7, 249-252.

R. Levin, V. Gaba, B. Tal, S. Hirsch, D. DeMola, and I.K. Vasil.. 1988. Automated Plant Tissue Culture for Mass Propagation. *Bio/Technology 6*, 1035-1040.

J. M. Park, K. L. Giles, and D. D. Songstad. 1990. Industrial Production of Sanguinarine from Cell Suspension Cultures of *Papaver somniferum*. Proceedings of Asia-Pacific Biochemical Engineering Conference '90, Kyungju, Korea, April 22-25, 1990.

M. Tabata and Y. Fujita. 1985. Production of Shikonin by Plant Cell Cultures. In "Biotechnology in Plant Science," M. Zatlin, P. Day, and A. Hollaender (eds.), Academic Press, Orlando, Florida, p. 207-218.

B. Ulbrich, W. Wiesner, and H. Arens. 1985. Large-Scale Production of Rosmarinic Acid from Plant Cell Cultures of *Coleus blumei*. In "Primary and Secondary Metabolism of Plant Cell Cultures," K. H. Neumann, W. Barz, and E. Reinhard, (eds.). Springer-Verlag, Berlin, p. 293-303.

B. Ulbrich, H. Osthoff, and W. Wiesner. 1988. Aspects of Screening Plant Cell Cultures for New Pharmacologically Active Compounds. In "Plant Cell Biotechnology," M. S. S. Pais, F. Mavituna, and J. M. Novais, (eds.). NATO ASI Series Vol. H18, Springer-Verlag, Berlin, pp. 461-474.

K. Ushiyama. 1989. Production of Saponins by Large Scale Tissue Culture of *Panax* ginseng. Paper read at the 1989 International Chemical Congress of Pacific Basin Societies, Honolulu, Hawaii, Dec. 17-22, 1989.

K. A. Walker and C. Sluis. 1985. *In Vitro* Plant Production - Costs and Methodologies. In "Research Needs in Non-Conventional Bioprocesses," D. J. Fink, L. M. Curran, and B. R. Allen (eds.). Battelle Press, Columbus, Ohio, pp. 123-126.

T. Yamakawa, S. Kato, K. Ishida, T. Kodama, and Y. Minoda. 1983. Production of Anthocyanins by *Vitis* Cells in Suspension Culture. *Agric. Biol. Chem.* 47, 2185-2191.

CHAPTER 2

STRUCTURE AND METABOLISM OF PLANT CELLS

A. INTRODUCTION

In this chapter we have attempted to summmarize several structural and metabolic features of plant cells. We have chosen to focus on those features which are especially important for subsequent discussions and have particularly concentrated on issues related to secondary metabolism. We have targeted our discussion to readers familiar with microbial or animal cell culture technology who wish to gain an understanding of plant cells. Obviously, it is impossible to cover all the important and interesting issues in such a chapter, and we have therefore provided the reader with many references on the various topics.

In our discussion of metabolism, we describe how cells obtain usable energy from the carbohydrate source and assimilate nitrogen and phosphorus from the medium. The goal here is to provide the background for understanding how cells respond to their nutritional environment. With respect to secondary metabolism, we provide a brief general discussion followed by a more extensive description of two well-studied systems. Although this approach is somewhat unconventional, we felt that a detailed description of two secondary metabolic systems would provide a deeper insight into the subject than a superficial review of many systems.

B. STRUCTURE

In this section, the structure of the plant cell will be briefly described. The goal here is to discuss cell structure in relation to function - especially with respect to the function of organelles as sites of biosynthesis and metabolite storage. More rigorous discussions of plant cell structure can be found in plant biochemistry texts (e.g., Tolbert 1980; Goodwin and Mercer 1983).

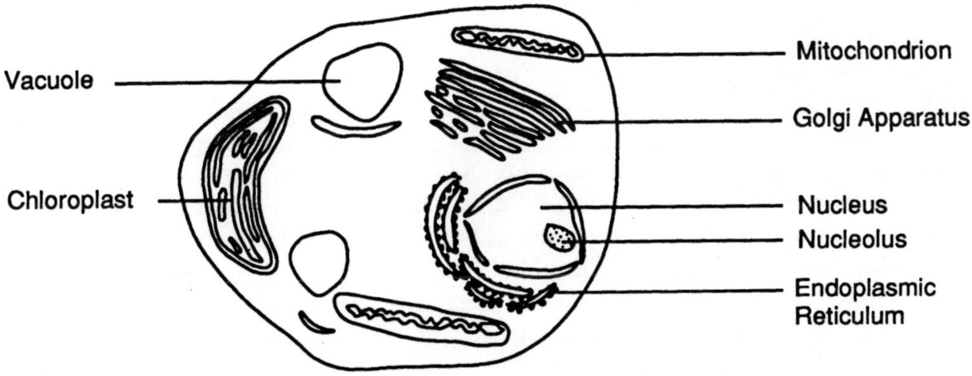

Figure 2.1. Illustration of the plant cell, showing key structural features. Adapted from Rosevear and Lambe (1983).

Figure 2.1 is a sketch of a plant cell. Plant cells vary in size from 20 to 200 μm. In suspension, they usually appear as aggregates of a few to a few hundred cells. Cells are surrounded by a primary cell wall which consists of polysaccharides (e.g., cellulose, hemicellulose, and pectin) and proteins (e.g., hydroxyproline-rich glycoproteins). The wall confers rigidity to the cell and provides some resistance to shear damage. Thus, plant cells are considerably more shear resistant than animal cells, which lack a cell wall. As some cells age, they develop a secondary cell wall between the primary wall and the plasmalemma. Secondary wall formation is also associated with lignification. In addition to containing hydroxyproline-rich glycoproteins, which appear to play a structural role, the cell wall also has enzymatic activity. Various hydrolases (e.g., invertase), oxidases, and phosphatases, are present in the cell wall.

The plasmalemma, or membrane surrounding the cells, offers a barrier to the transport of solutes from the medium into the cells. Plasmodesmata are tubular channnels between cells which connect the cytoplasm of adjacent cells. Plasmodesmata permit the transfer of low-molecular-weight (less than 1,000) metabolites and signals between contiguous cells. In addition, plasmodesmata may also connect the endoplasmic reticulum of adjacent cells. It is currently believed that plasmodesmata may play an important role in the communication between cells, and Meiners et al. (1988) drew an analogy between plasmodesmata in plant cells and gap junctions in animal cells.

Like other eucaryotic cells, plant cells contain a membrane- enclosed nucleus. Continuous with the nuclear region is the endoplasmic reticular membrane. The mitochondrion is the site of oxidative phosphorylation. Since most plant cells are cultured under heterotrophic conditions, mitochondria, and not chloroplasts, are the major energy-converting organelles in the cell. Plastids are a family of related organelles. In addition to being sites of photosynthesis, plastids can also be important sites of biosynthesis [e.g., for aromatic amino acid biosynthesis (Jensen 1986)].

The vacuole, which is surrounded by the tonoplast membrane, is a storage organelle which plays important physiological roles (see Marty et al. 1980). Since solute concentrations within vacuoles are approximately 0.5 M, the vacuole is involved in maintaining cell turgor and in cell expansion. As the cell ages, the vacuole generally expands and occupies a larger portion of the cell (up to 90% of the cell volume). In addition to its role in cell expansion, the vacuole is the storage site for inorganic salts such as phosphate (Brodelius and Vogel 1985), neutral sugars (Wagner 1979), Kreb's cycle acids (Wagner 1979), amino acids (Sasse et al. 1979; Dietz et al. 1990), and secondary metabolites (see Zenk et al. 1985). The vacuole is considerably more acidic than the cytosol. Although the vacuolar pH can vary, it can be several units below the near neutral pH of the cytosol. This pH difference has been suggested to be important in "trapping" basic compounds in the vacuole (Renaudin et al. 1985). The active processes involved in the transport of hydrogen ions and other solutes across the tonoplast are currently being studied.

C. CATABOLISM

Carbohydrates, especially sucrose and glucose, are typically used as the carbon and energy source in plant cell tissue culture (Verma and Dougall 1977). Although there appear to be some differences, it is believed that sugar catabolism by plant cells is similar to catabolism used by other heterotrophic cells. Figure 2.2, which was adapted from Ashihara et al. (1988), summarizes the glycolytic pathway believed to operate in plant cells. Sugar is often supplied to the cultures as the disaccharide sucrose, and several groups have observed the rapid hydrolysis of this disaccharide with the accumulation of the hexoses glucose and fructose in the extracellular medium (Kanabus et al. 1986; Amino and Tazawa 1988; Ashihara et al. 1988; Masuda et al. 1988). This hydrolysis appears to be catalyzed by a wall-bound or extracellular acid invertase (Masuda et al. 1988). An alkaline invertase has also been observed in the cytoplasm of sugar beet cells. There is some discrepancy as to whether there is preferential utilization of the individual hexoses, as Kanabus et al. (1986) observed that carrot cells took up fructose only after most of the glucose had been consumed. Others have observed either no (Amino and Tazawa 1988) or small (Ashihara et al. 1988; McDonald and Jackman 1989) preferential utilization of glucose over fructose.

Whether most of the sugar is taken up through active or passive transport is currently unclear (Fowler and Stepan-Sarkissian 1985), and this ambiguity is illustrated in Figure 2.2 by the dotted arrows for this step(s). The remainder of the glycolytic pathway is shown in an abbreviated form. The pathway enzymes appear to be located in the cytosol of the cells, although it is possible that glycolysis occurs in chloroplasts (Stitt and ap Rees 1979) if these organelles are present. Regulation of this pathway may occur at three steps: the phosphorylation of the hexoses; the phosphorylation of fructose-6-phosphate; and the dephosphorylation of phosphoenolpyruvate (Goodwin and Mercer 1983; Turner and Turner 1980). Evidence that enzymatic steps may be regulatory is provided by observations that they catalyze reactions which are far from equilibrium under physiological conditions, and that activities of these enzymes are low compared to those of other pathway enzymes. As illustrated by the arrows in Figure 2.2, the postulated regulatory reactions are irreversible under physiological conditions. Also of interest are recent observations that fructose-6-phosphate may be phosphorylated by two enzymes: phosphofructokinase, which is common in other biological systems, and a PP_i:fructose-6-phosphate 1-phosphotransferase. Black et al. (1987) have postulated that this latter enzyme system plays a critical role in regulating glycolysis and gluconeogenesis in the plant cell cytosol (see also Taiz 1986). Ashihara et al. (1988) observed that the activity of this phosphotransferase enzyme was always three to five times higher than that of phosphofructokinase for suspension cultures of *Catharanthus roseus*.

If glucose is catabolized to pyruvate by the glycolytic pathway, then the overall reaction can be represented by the equation

$$C_6H_{12}O_6 + 2\,NAD^+ + 2\,(ADP + P_i) \rightarrow 2\,C_3H_4O_3 + 2\,(NADH + H^+) + 2ATP$$

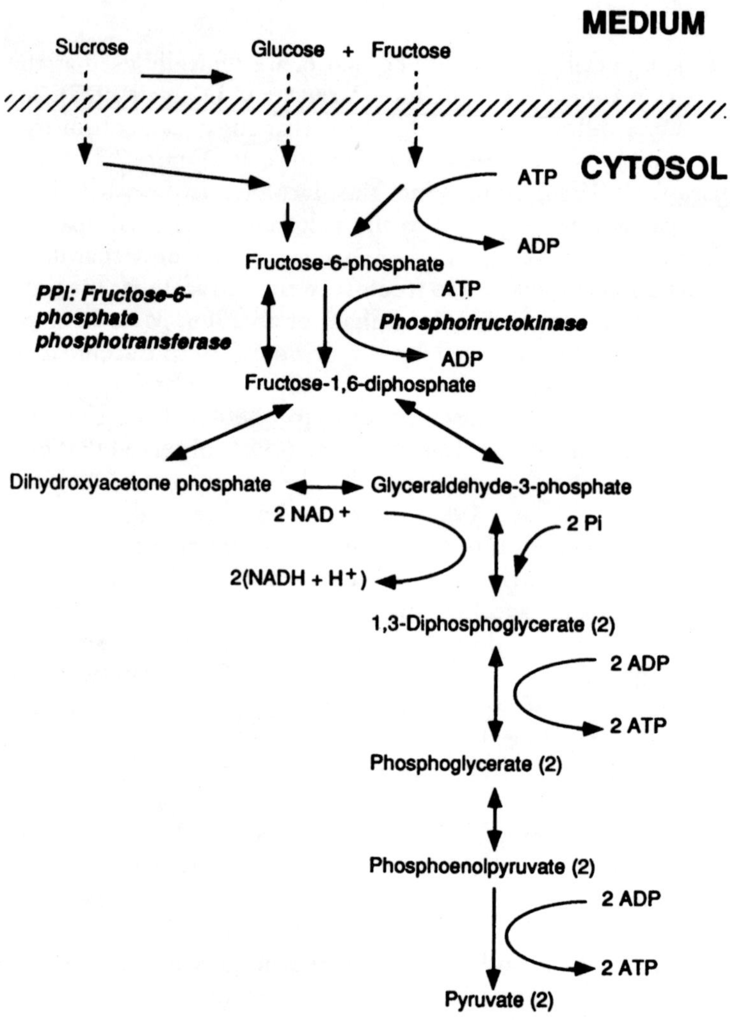

Figure 2.2. Sugar catabolism by plant cells. Adapted from Ashihara et al. (1988).

Under anaerobic conditions (i.e., conditions in which oxygen is unavailable to oxidize the reduced NADH), living systems must develop alternative strategies to oxidize these cofactors and make NAD^+ available for further glycolysis. Under anaerobic conditions, ethanol formation appears to be the favored route for plant cells to reoxidize NADH:

$$2\ C_3H_4O_3 + 2\ (NADH + H^+) \rightarrow 2\ CH_3CH_2OH + 2\ CO_2 + 2\ NAD^+$$

Due to its relevance to oxygen transfer, anaerobic metabolism will be further discussed in Chapter 5.

Under aerobic conditions, the NADH is reoxidized by oxidative phosphorylation, and thus the accumulation of this reduced cofactor does not present a bottleneck to further catabolic activity. Thus, the pyruvate generated by glycolysis can be further oxidized through the tricarboxylic acid (TCA) cycle (for a review, see Wiskich 1980). For this to occur, the pyruvate from the cytosol must first enter the mitochondria and then undergo the following irreversible reaction which is catalyzed by pyruvate dehydrogenase:

$$C_3H_4O_3 + CoASH + NAD^+ \rightarrow CH_3CO\text{-}S\text{-}CoA + CO_2 + NADH + H^+$$

As shown in Figure 2.3, the acetyl-CoA from the above reaction enters the TCA cycle by reacting with oxaloacetate to form citrate. This six-carbon compound is then oxidized and decarboxylated by an NAD-dependent dehydrogenase to form α-ketoglutarate. The subsequent oxidation and decarboxylation of α-ketoglutarate is catalyzed by an NAD-dependent dehydrogenase and is the only irreversible step in the cycle (see Goodwin and Mercer 1983). The next step of the cycle is catalyzed by succinate thiokinase and leads to the formation of ATP by substrate level phosphorylation. Succinate is dehydrogenated by an FAD-dependent enzyme which is embedded in the mitochondrial membrane. The final step in the pathway is catalyzed by an NAD-dependent malate dehydrogenase which catalyzes a reaction which has a large positive standard free energy change, and thus equilibrium favors malate formation. Cells apparently overcome this problem by removing the product of this reaction, oxaloacetate.

In addition to the traditional view of the TCA cycle discussed above, Day and Hanson (1977) have reported that under conditions of high energy demand, pyruvate transport into the mitochondria may become limiting. Under such conditions, these authors proposed that the reactions shown on the left portion of Figure 2.3 may become important. In this scheme, phosphoenolpyruvate (PEP) is carboxylated to oxaloacetate (OAA) by the enzyme PEP carboxylase. Cytosolic OAA is then converted to malate by an NADP-dependent malate dehydrogenase. Malate, which is more rapidly transported than pyruvate, then enters the mitochondria. Once inside the mitochondria, the malate can enter into the TCA cycle directly or be converted to pyruvate by the NAD-dependent malic enzyme.

The oxidation of NADH occurs at the inner mitochondrial membrane through the process of oxidative phosphorylation. This is illustrated in Figure 2.4, which shows that the electrons from NADH are transferred through a series of oxidation/reduction reactions to molecular oxygen. Stoichiometrically, this reaction can be represented as

$$NADH + H^+ + \frac{1}{2}O_2 \rightarrow NAD^+ + H_2O$$

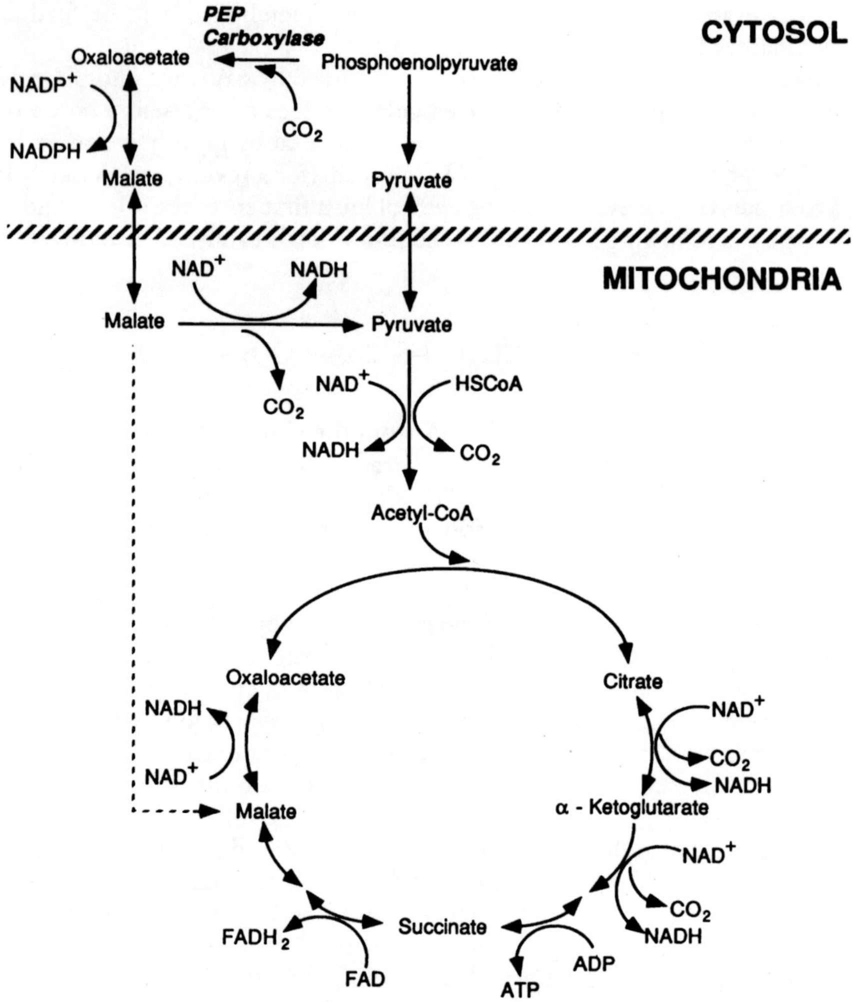

Figure 2.3. Tricarboxylic acid cycle. Adapted from Goodwin and Mercer (1983).

A considerable amount of energy is released from this reaction, and, according to the chemiosmotic theory, the cells conserve this energy by coupling NADH oxidation to the asymmetric transport of hydrogen ions across the inner mitochondrial membrane. This expulsion of hydrogen ions from the mitochondrial matrix results in the formation of an electrochemical gradient across the membrane. Since the hydrogen ion cannot freely diffuse through the inner mitochondrial membrane, this gradient cannot be freely dissipated. The cells couple the dissipation of this gradient to the synthesis of ATP. The indirect coupling of NADH oxidation to ATP synthesis

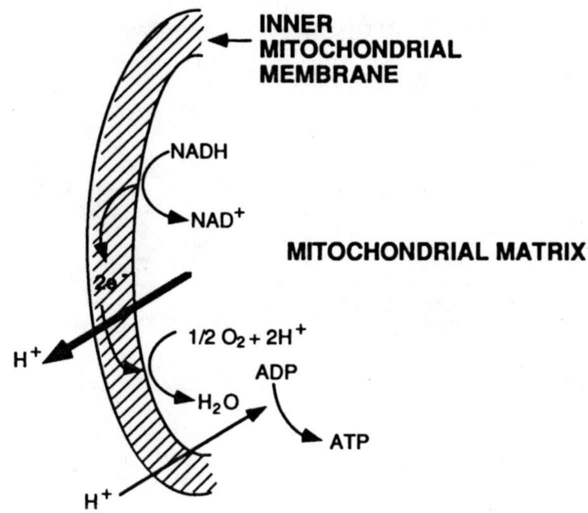

Figure 2.4. Oxidative Phosphorylation

through an intermediate electrochemical gradient is substantially different from ATP-generating steps in glycolysis, in which dephosphorylation of high-energy intermediates is directly coupled to ATP synthesis. It is generally believed that the oxidation of one mole of mitochondrial NADH is coupled to the formation of three moles of ATP, whereas each mole of $FADH_2$ oxidized results in the formation of two moles of ATP. If each mole of NADH generated in the cytosol by the glycolytic path results in only two moles of ATP, then the overall stoichiometry for the complete oxidation of glucose through glycolysis, the TCA cycle and oxidative phosphorylation is

$$C_6H_{12}O_6 + 6\ O_2 + 36\ (ADP + P_i) \rightarrow 6\ CO_2 + 36\ ATP + 42\ H_2O$$

In addition to the traditional oxidative phosphorylation mechanisms described above, it should also be mentioned that plants appear to have an alternative pathway for respiration (often referred to as cyanide-resistant respiration). This alternative pathway is also located on the inner mitochondrial membrane; however, this pathway is only weakly coupled to ATP synthesis. Current understanding of the role of this pathway is limited, and the interested reader is referred to reviews by Solomos (1977), Laties (1982), and Siedow and Berthold (1986).

An alternative pathway for sugar metabolism is the pentose phosphate pathway. The enzymes for the pentose phosphate pathway, like those for glycolysis, appear to be located in the cytosol (see Kelly and Latzko 1980). However, it is likely that this

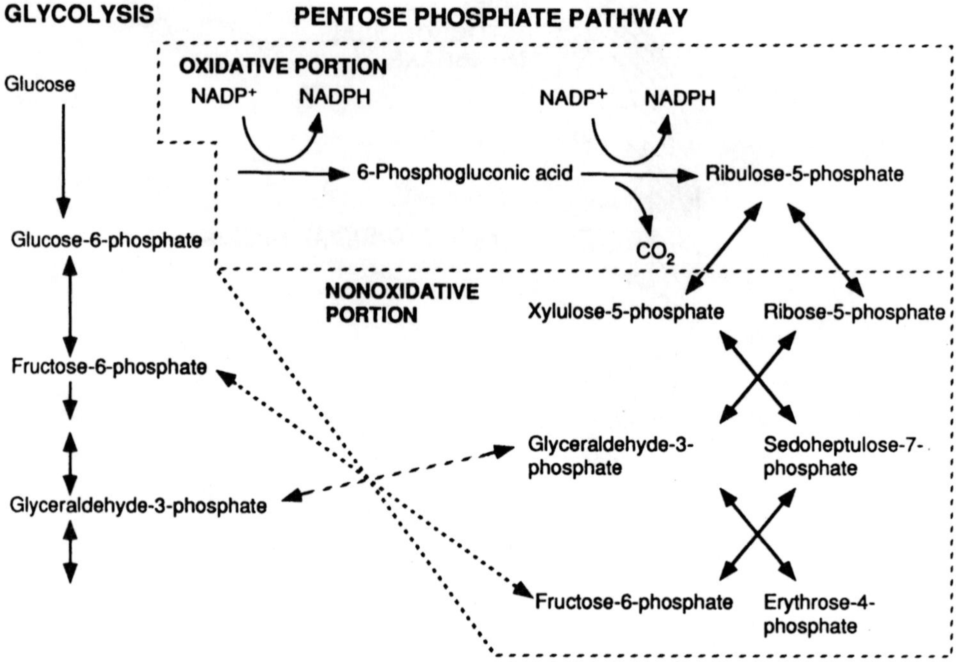

Figure 2.5. The pentose phosphate pathway

pathway serves more of a biosynthetic rather than a catabolic role (Turner and Turner 1980). As shown in Figure 2.5, the pentose phosphate pathway can be considered in two parts. The irreversible, oxidative portion involving the NADP-dependent oxidation steps from glucose-6-phosphate to ribulose-5-phosphate appears to provide reduced pyridine nucleotides (NADPH) for biosynthetic operations. The reversible, nonoxidative portion involves several interconversions between sugar phosphates including fructose-6-phosphate and glyceraldehyde-3-phosphate (intermediates of glycolysis), ribose-5-phosphate (the precursor for nucleic acid and tryptophan biosynthesis), and erythrose-4-phosphate (the precursor for aromatic amino acid biosynthesis). The activity of the pentose phosphate pathway appears to vary depending on the demand of the cells for various biosynthetic operations (Fowler and Clifton 1974). Pryke and ap Rees (1976) reported high activities of the pentose phosphate pathway during periods of lignification. These authors suggested that these high activities were required to meet the NADPH requirements for lignification. Thorpe and Laishley (1973) reported increased activities of the pentose phosphate pathway during shoot initiation in tobacco cultures. These increased activities may be required to meet the precursor (erythrose-4-phosphate) demands for the high levels of the shikimic acid pathway-derived metabolites required during shoot initiation (Beaudoin-Eagan and Thorpe 1983, 1985). In contrast, Hirose and Ashihara (1984) reported that during periods of rapid nucleotide synthesis (i.e. high demands for ribose-5-phosphate), there was not an increase in activity of the oxidative portion of the pentose phosphate pathway. These authors suggested that the ribose-5-phosphate demand was met by increases in the carbon flow through the nonoxidative portion of the pentose phosphate pathway.

D. ASSIMILATION AND PRIMARY METABOLISM

D.1. Carbon

Carbon, which constitutes approximately 40% of the dry cell mass, is the major element of plant cells. As shown in Figure 2.6, the carbonaceous precursors for biosynthesis are obtained from intermediates of the above-mentioned catabolic pathways (i.e., glycolysis, the pentose phosphate shunt, and the TCA cycle). Plants and some plant cells in culture also obtain carbon from carbon dioxide and energy from light. However, since most plant cell cultivation studies involve heterotrophic, and not autotrophic, growth, photosynthetic carbon assimilation will not be considered here.

In addition to carbon, plant cells also contain significant amounts of nitrogen [van Gulik et al. (1989) reported *C. roseus* cells to contain 7% nitrogen on a dry weight basis] and phosphorus (about 0.3% of dry weight). The assimilation of these elements is considered below.

D.2. Nitrogen

Nitrogen is supplied to most plant cells as a combination of ammonium and nitrate salts. The ratio of these salts appears to be important, because growth on these salts results in changes in the medium pH, and typical plant cell media are poorly buffered. After inoculation, the medium pH usually decreases during ammonia assimilation and then increases during nitrate uptake (McDonald and Jackman 1989). Bayley et al. (1972) grew soybean and wheat cells on media containing nitrate as the sole nitrogen source, while Martin et al. (1977) observed that ammonium could be used as the sole nitrogen source for plant cells, provided the pH was controlled by external means. Wetherell and Dougall (1976) observed that other reduced nitrogen forms (e.g., glutamine and alanine) could also be used as a sole nitrogen source. Interestingly, Veliky and Rose (1973) suggested that although the use of simple inorganic nitrogen salts is beneficial for examining how individual nutritional components affect cell metabolism, the use of inorganic nitrogen may result in significant reductions in the efficiency of plant cell metabolism. In contrast, most industrial media for microbial secondary metabolite production contain organic nitrogen sources (e.g., amino acids and peptides). It is possible that significant improvements in plant secondary metabolite formation could be achieved if alternate, richer media were used.

The biochemistry of nitrogen assimilation is reviewed elsewhere (see Miflin and Lea 1976; Dougall 1980) and will only briefly be discussed here. As with bacteria,

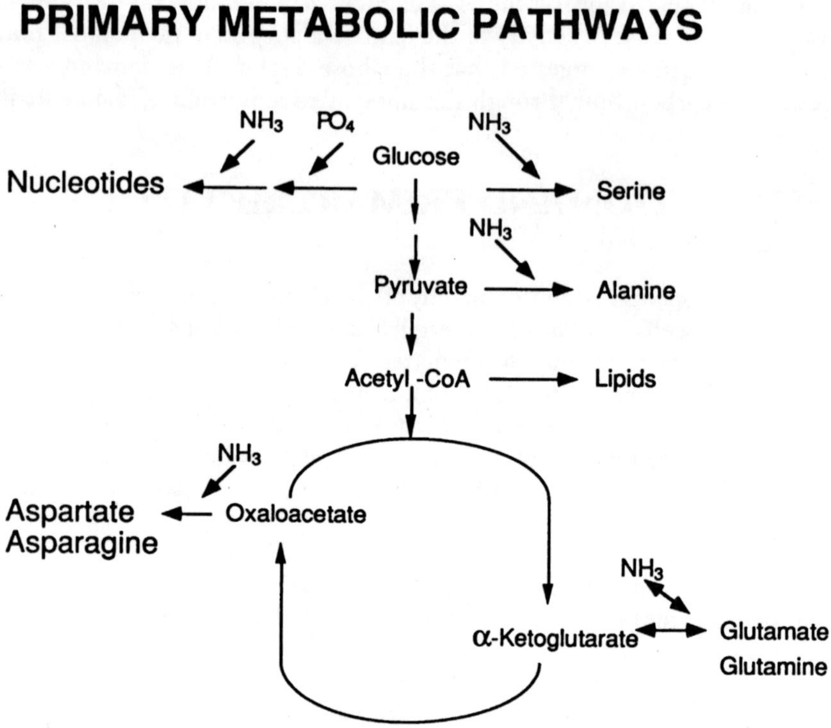

Figure 2.6. Schematic of biosynthetic pathways for primary metabolites

plant cells possess both glutamate dehydrogenase (GDH) and a glutamine synthetase-glutamate synthase (GS-GOGAT) system which can be involved in ammonia assimilation. The GS-GOGAT path (and not the GDH reaction) is commonly believed to be the major route for inorganic ammonia assimilation (see Miflin and Lea 1976, 1977; Yamaya and Oaks 1987). Arguments in favor of this belief are based on comparisons of the K_m for ammonia for the GS and GDH enzymes, results from studies involving radioactively labeled inorganic nitrogen, and the use of specific inhibitors (Skokut et al. 1978). The reader should be cautioned that under conditions where high levels of ammonia are present, arguments based on K_m are probably not valid. The GS-GOGAT pathway involves the following two steps:

Glutamine Synthetase (GS)

$$\text{Glutamate} + NH_4^+ + \text{ATP} \rightarrow \text{Glutamine} + \text{ADP} + P_i + H^+$$

Glutamate Synthase (GOGAT)

$$\text{Glutamine} + \alpha\text{-Ketoglutarate} + \text{NAD(P)H} + H^+ \rightarrow 2\ \text{Glutamate} + \text{NAD(P)}^+$$

NAD(P)-dependent GOGAT is most commonly observed with plant cells (Dougall 1974; Dougall and Bloch 1976); however, a ferredoxin-dependent GOGAT which may be related to chloroplast development has also been reported (Suzuki et al. 1984; Lenee and Chupeau 1989).

Glutamate dehydrogenase catalyzes the following reaction:

Glutamate Dehydrogenase (GDH)

$$\alpha\text{-Ketoglutarate} + NAD(P)H + NH_4^+ \rightarrow \text{Glutamate} + NAD(P)^+ + H_2O$$

The role of GDH is unclear although it has been suggested that it may play a catabolic role or be used for detoxifying high levels of ammonia (Givan 1979).

To assimilate nitrate, plant cells must first reduce it to ammonia. The reduction of nitrate to ammonia proceeds through the following steps:

Nitrate Reductase

$$NO_3^- + NAD(P)H + H^+ \rightarrow NO_2^- + NAD(P)^+ + H_2O$$

Nitrite Reductase

$$NO_2^- + 6e^- + 6H^+ \rightarrow NH_3 + OH^- + H_2O$$

As shown above, the reduction of nitrite involves multiple electrons and leads to the formation of ammonia. McDonald and Jackman (1989) reported that alfalfa cells preferentially utilized ammonia when both ammonia and nitrate were available.

It has been observed that during periods of active ammmonia assimilation, a significant amount of carbon is drained from the TCA cycle. From Figure 2.6, it can be seen that for each ammonia group assimilated through GS-GOGAT, a single α-ketoglutarate molecule is removed from the TCA cycle. To ensure that sufficient oxaloacetate is available for subsequent "turns," the carbon drained from the TCA cycle must be replenished. It appears that one means of replenishing the TCA cycle is through the nonphotosynthetic fixation of carbon dioxide by phosphoenolpyruvate carboxylase (PEPCase). The significance of this anaplerotic reaction is considered further in our dissussion of carbon dioxide (see Chapter 5).

Once assimilated into glutamine, the nitrogen groups can be transferred to other carbon skeletons via transamination reactions. These transamination reactions provide the amine groups for the other amino acids. Further discussion of amino acid biosynthesis can be found in Miflin and Lea (1977).

D.3. Phosphorus

In plant cell media, phosphorus is supplied in the form of inorganic phosphate. This phosphate is required for the biosynthesis of nucleic acids, phospholipids, nucleotides, and sugar phosphates (see Ashihara and Tokoro 1985). Also, phosphate, through its effects on ATP synthesis, plays a significant role in the energetics of the cell (Ashihara and Ukaji 1986).

From phosphorus-31 NMR studies it was observed that *Nicotiana tabacum* (Wray et al. 1983) and *Catharanthus roseus* (Brodelius and Vogel 1985) cultures rapidly took up inorganic phosphate from the medium, and this P_i was stored in the vacuole. As P_i was consumed for biosynthesis, the vacuolar pool decreased. During this time, however, cytoplasmic inorganic phosphate levels remained unchanged. In contrast, phosphate is taken up slowly from the medium by *Daucus carota* (Brodelius and Vogel 1985), *Capsicum frutescens* (Lindsey and Yeoman 1984), and *Thalictrum minus* (Kobayashi et al. 1989).

In addition to its role in the synthesis of phosphorus-containing organics, P_i also exerts dramatic effects on cellular metabolism. Ashihara and coworkers observed that the addition of P_i to phosphate-depleted *C. roseus* cultures resulted in a marked increase in intracellular ATP (Ashihara and Ukaji 1986) and reductions in the intracellular levels of most amino acids (Ukaji and Ashihara 1987). These reductions in free amino acids were believed to be due to increased rates of protein synthesis accompanied by reductions in protease levels. Also, phosphate has been observed to reduce the production of several secondary metabolites (Knobloch and Berlin 1981; Knobloch 1982; Knobloch and Berlin 1983).

E. PHYTOHORMONES

The term hormone is borrowed from work with higher animals. Hormones refer to chemicals which are produced in small amounts by cells in one tissue, are transported through the blood to cells in the "target" tissue, and result in a change in structure or function of the target tissue. Although this definition can only loosely be applied to plants, there are five recognized groups of phytohormones. Auxins, cytokinins, and gibberellins are plant growth promoters, while abscisic acid and ethylene are plant growth inhibitors. A more thorough discussion of plant hormones can be found in Takahashi (1986). Although all of these types of phytohormones have been used with cultured cells, most culture media employ only auxins and/or cytokinins. Table 2.1 lists several commonly used phytohormones.

Table 2.1. Typical auxins and cytokinins used in plant cell tissue culture media.

Auxins:	2,4-Dichlorophenoxyacetic acid (2,4-D)
	Indole-3-acetic acid (IAA)
	Naphthalene acetic acid (NAA)
	Indolebutyric acid (IBA)
Cytokinins:	Kinetin
	Benzyladenine (BA)
	Isopentenyl adenosine (IPA)
	Zeatin

The synthetic auxin 2,4-dichlorophenoxyacetic acid (2,4-D) has been traditionally used for culturing cells. However, because of 2,4-D's inhibitory effect on secondary metabolite production and morphogenesis, there has been some effort to use "weaker" auxins such as the natural auxin indoleacetic acid (IAA) and naphthalene acetic acid (NAA) (see Morris 1986a).

Habituated cells can be cultivated in the absence of specific phytohormones (see Jackson and Lyndon 1990). Plant cells which are transformed by *Agrobacteria* commonly gain the ability to produce the auxin IAA, and therefore these cells no longer require the exogenous addition of auxins. A somewhat more detailed discussion of hormonal effects can be found in Chapter 8.

F. SECONDARY METABOLISM

The study of secondary metabolism is diverse, and entire volumes have been written on various aspects of the field. In this section, we will not attempt an exhaustive summary of secondary metabolism, but rather we will provide an overview of the area and then consider, in detail, two well-studied examples. The emphasis of this discussion will be to describe important biochemical aspects of secondary metabolism and show how these features impact on bioprocess design. More general reviews of plant secondary metabolite production from cell culture are readily available (e.g., Misawa 1985; Staba 1985; Rokem and Goldberg 1985; Stafford et al. 1986; Parr 1988; Hay et al. 1988; Herbert 1989).

F.1. Key Branch Points between Primary and Secondary Metabolism

Despite the diversity of secondary metabolites, a few key intermediates in primary metabolism supply the precursors for most secondary products. As shown in Figure 2.7, the key intermediates from which primary and secondary metabolism diverge are sugars, acetyl-CoA, nucleotides, and amino acids. Cyanogenic glycosides and glucosinolates are derived from sugars. Terpenes and steroids are produced from isoprene units which are derived from acetyl-CoA. The nucleotide bases are the precursors to purine and pyrimidine alkaloids (e.g., nicotine, anabasine, and caffeine). Many different types of aromatic compounds are derived from shikimic acid pathway intermediates (e.g., morphine, codeine, shikonin, berberine, and rosmarinic acid). The nonaromatic amino acid arginine is the precursor to polyamines (e.g., spermine and putrescine) and the tropane alkaloids (hyoscyamine and scopolamine).

Not shown in Figure 2.7 is that many natural products are derived from pathways involving more than one of these intermediates. The phenylpropanoids are derived from the amino acid phenylalanine, with acetyl-CoA and sugar units being added later in the biosynthetic pathway. The indole (e.g., ajmalicine and vincristine) and the quinoline (e.g., quinine and quinidine) alkaloids are derived from the amino acid tryptophan and from monoterpenes. The aglycon moieties of cyanogenic glycosides and glucosinolates are derived from amino acids.

A more thorough presentation of secondary metabolite biosynthesis can be found in Herbert (1989).

F.2. Phenylpropanoid Metabolism.

F.2.1. Biosynthesis of Phenylpropanoids. The phenylpropanoids are probably the most thoroughly studied group of plant secondary metabolites. Within this group are the lignins (Grisebach 1981) and the flavonoids (Hahlbrock and Grisebach 1979; Hahlbrock 1981). Functionally, the flavonoids appear to serve several roles: animal attraction; protection from UV light; and protection against infection (Hahlbrock and Grisebach 1979).

As shown in Figure 2.8, the three initial steps in the phenylpropanoid pathway are catalyzed by the enzymes phenylalanine ammonia lyase (PAL), cinnamate 4-hydroxylase, and 4-coumarate:CoA ligase (4-CL). These three steps are often distinguished from the remainder of the biosynthetic pathways because of their generality to phenylpropanoid metabolism (i.e., these are the initial steps for the biosynthesis of both flavonoids and lignins) and because it appears that the enzymes which catalyze

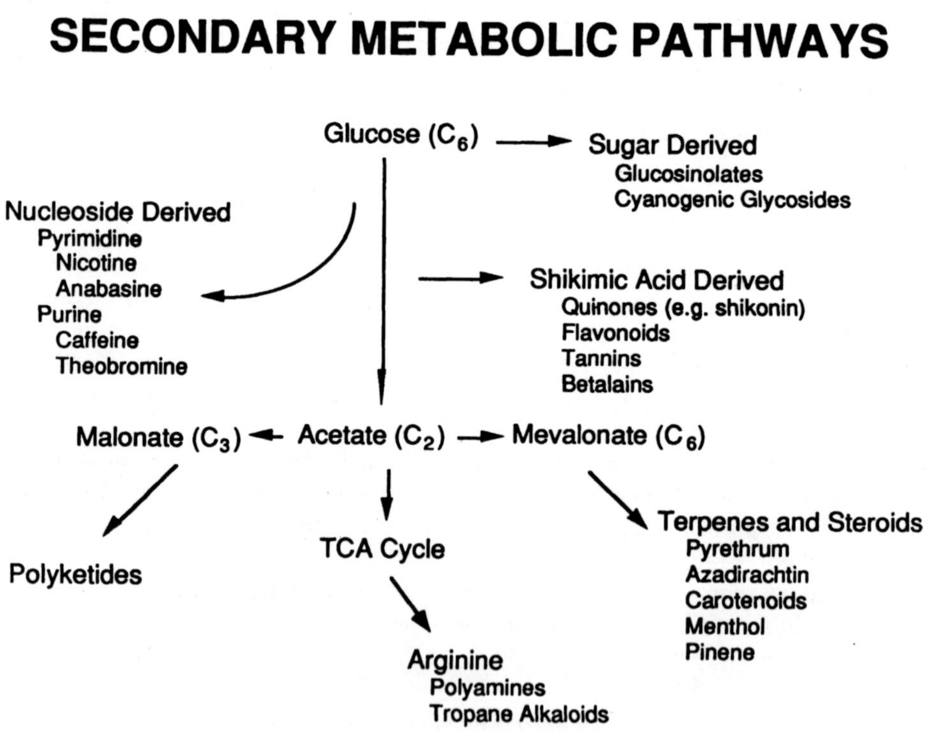

Figure 2.7. Schematic of biosynthetic pathways for secondary metabolites

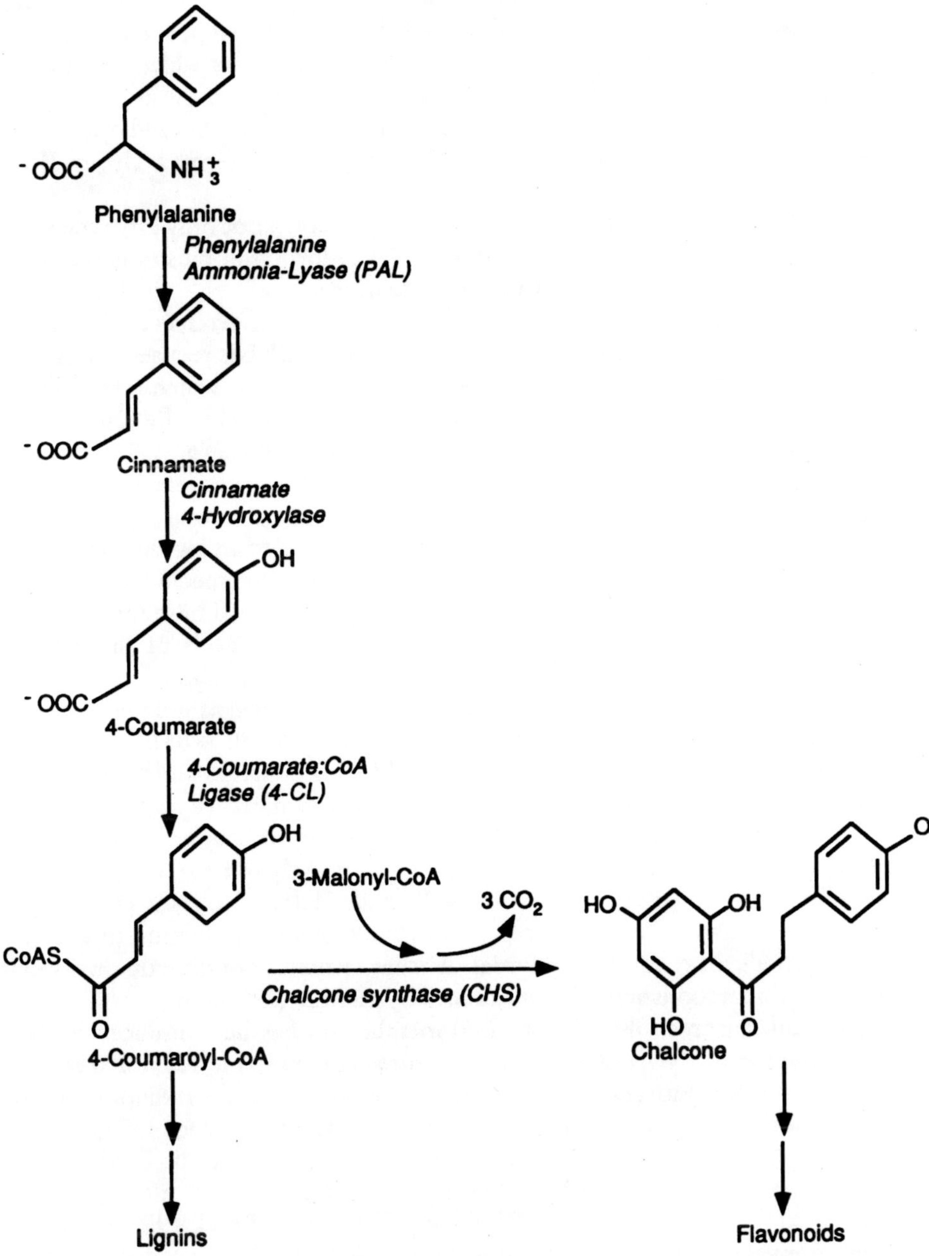

Figure 2.8. Biosynthetic pathway for phenylpropanoids. Adapted from Hahlbrock (1981).

these steps are regulated separately from those which catalyze later steps in the biosynthetic paths. These three enzymes are often referred to as Group I enzymes.

The first step specific for the flavonoid branch is catalyzed by chalcone synthase (CHS), which was formerly known as flavone synthase. At this step, acetate units are added to the aromatic structure. These two-carbon acetate units are added via malonyl-CoA in a process which is considered to be similar to fatty acid synthesis. A later step in the pathway is the conversion of chalcone to flavonones and is catalyzed by chalcone isomerase (CHI). Often these flavonoid-specific enzymes are termed Group II enzymes.

The biosynthesis of lignins differs from that of flavonoids not only after the 4-CL step in Figure 2.8, but also in that substrates for 4-CL for lignin biosynthesis can be hydroxylated or methoxylated derivatives of coumarate.

The production of phenylpropanoid metabolites has been extensively studied in three systems: bean, parsley, and soybean. Much of this work has focused on the role of phenylpropanoids as plant defense compounds, phytoalexins, which are produced in response to an elicitor. In the case of bean suspension cells, the isoflavonoid phytoalexin phaseollin was observed to be produced in response to to elicitor treatment (Dixon and Lamb 1979). Typically, the elicitor used in these studies is a heat-released extract from the mycelial cell walls of a pathogenic fungus. The onset of phytoalexin synthesis (Dixon and Lamb 1979) was correlated to an increased synthesis of PAL and CHS, the first enzymes of the general, and the specific isoflavonoid branch of the phenylpropanoid pathway (Lawton et al. 1983a). These increased enzyme activities were observed to be preceded by increased levels of the mRNAs coding for the PAL and CHS enzymes (Lawton et al. 1983b). Later work demonstrated that increased mRNA levels for these isoflavonoid biosynthetic enzymes resulted from *de novo* synthesis (Edwards et al. 1985; Cramer et al. 1985).

The timing of appearance of the PAL, CHS, and CHI enzymes and their mRNAs suggests that these pathway enzymes are coordinately induced (Ryder et al. 1984; Edwards et al. 1985; Cramer et al. 1985). Also, induction of transcription has been reported to be very rapid (Lawton and Lamb 1987) and likely to be an early event in a defense response (Robbins et al. 1985; Bolwell et al. 1985a). Robbins et al. (1985) observed that the elicitor treatments resulted in the selective induction of enzymes involved in phytoalexin synthesis. Activities of other enzymes not directly involved in phenylpropanoid metabolism were not induced by elicitor treatment.

In parsley cell cultures, phenylpropanoid metabolism has been induced by both elicitor treatment and UV irradiation. The cultures responded to elicitor treatment by producing fungitoxic furanocoumarins and responded to UV irradiation by producing UV-absorbing flavonoid compounds (Chappell and Hahlbrock 1984). As with bean cultures, phenylpropanoid biosynthetic enzymes appeared to be induced, with mRNA levels and rates of mRNA synthesis observed to increase after treatment (Hahlbrock et al. 1981; Kuhn et al. 1984; Chappell and Hahlbrock 1984). Although the timing of appearance of PAL (a Group I enzyme) was somewhat different from that of CHS (a Group II enzyme) (Hahlbrock et al. 1976; Hahlbrock 1981), both enzymes were induced in response to UV irradiation (Schroder et al. 1979; Chappell and Hahlbrock 1984). On the other hand, PAL, but not CHS, was induced after elicitor treatment. According to Chappell and Hahlbrock (1984), the synthesis of furanocoumarins does not require CHS, but rather involves a less clearly defined pathway (catalyzed by Group III enzymes) which branches from the general phenylpropanoid path. Also, induction by elicitor was more rapid than induction resulting from UV irradiation (Hahlbrock et al. 1981; Chappell and Hahlbrock 1984).

Early work with soybean cell cultures has been reviewed by Hahlbrock and Grisebach (1979). These cultures have the ability to produce both flavonoids and lignin. As with bean and parsley, the enzymes and mRNAs for flavonoid biosynthesis can be induced by treating soybean cells with elicitors (Habereder et al. 1988). Recently, Parker et al. (1988) examined different cell wall components of a phytopathogenic fungus to better characterize the elicitor signals. Suprisingly, these authors observed that a proteinaceous fraction was active in the parsley cells, whereas carbohydrate components were active in soybean.

F.2.2. Regulation of Phenylpropanoid Metabolism. In the previous discussion, we considered the role of induction on the production of phenylpropanoid metabolites. Although transcriptional regulation governs the timing of appearance of phenylpropanoid metabolites (Chappell and Hahlbrock 1984), and enzyme levels may in some cases be controlling the pathway flux (Hahlbrock et al. 1976; Schroder et al. 1979), other mechanisms may also be important in regulating phenylpropanoid biosynthesis. Early work reviewed by Hahlbrock and Grisebach (1979) and Hahlbrock (1981) has shown that the enzyme 4-CL exists as isoenzymes, with isoenzyme 1 having substrate specificities consistent with its proposed role in lignin biosynthesis and isoenzyme 2 having a different substrate specificity and possibly playing a role more specific to flavonoid production (Knobloch and Hahlbrock 1975). Thus, these isoenzymes may play a role in directing metabolites toward either flavonoid or lignin pathways in soybean cultures.

Bolwell et al. (1985b) reported that PAL exists as isoenzymes and that elicitor treatment leads to the rapid appearance of the isoenzyme having the lowest K_m. A low K_m would permit phenylalanine pools to be more efficiently directed - presumably toward phytoalexin biosynthesis. This argument is similar to older suggestions that primary and secondary metabolism compete for a common pool of precursors and that the rate limitation to phenylpropanoid production is the precursor supply. Phillips and Henshaw (1977) emphasized the competition for precursor phenylalanine between protein synthesis and phenylpropanoid production. Margna (1977) supported this hypothesis by reviewing examples in which phenylpropanoid production could not be correlated to PAL activities; *in vitro* measured PAL activities were significantly greater than observed phenylpropanoid production rates; the addition of phenylalanine stimulated phenylpropanoid production; and the observed intracellular phenylalanine levels were low relative to observed K_ms of PAL.

Later work has identified significant post-translational control of phenylpropanoid metabolism. Consistently it has been observed that after induction, the activities of phenylpropanoid pathway enzymes decay over time. Thus, enzymatic activities represent a balance between synthesis and inactivation/degradation (Betz at al. 1978). For parsley, Schroder et al. (1979) estimated the half-life of PAL and CHS to be between 7-10 hours and 5-7 hours respectively. Later work with bean cells suggested that in addition to enhancing enzyme synthesis, high concentrations of elicitors also appeared to stabilize the PAL enzyme (Lawton et al. 1980). This observation suggests that PAL inactivation, the irreversible loss of activity, may be different from a nonspecific proteolysis, referred to as degradation (Creasy 1987). Subsequent studies have suggested that this inactivation may be mediated by a specific protein (Dixon et al. 1980) and that inactivation may result from the modification of a specific site on the PAL enzyme (Bolwell et al 1986). The possibility that irreversible post-translational enzyme modification may be a general mechanism by which plants control phenylpropanoid metabolism was further suggested by Robbins and Dixon (1984), who showed that immunodetectable CHI appears earlier than CHI activity in bean cultures. Thus, some of the CHI is apparently produced in an inactive form, which is immunologically recognized as CHI, yet enzymatically inactive. Subsequent activation steps are then needed to convert the inactive to the active CHI.

In addition to the previously mentioned mechanisms, it has been further suggested that other, intracellular physiological signals may play a role in regulating the rate of phenylpropanoid metabolism. Since the 4-CL enzyme of parsley shows sigmoidal kinetics with respect to the ATP substrate, and this enzyme is inhibited by AMP (Knobloch and Hahlbrock 1977), it is possible that adenylates and/or energy charge may be regulating. Also, it has been observed that cinnamic acid has a significant effect on phenylpropanoid pathway activities. Cinnamic acid appears to act as a feedback regulator of PAL activity both by reducing the induced synthesis of PAL and by enhancing PAL inactivation reactions (Bolwell et al. 1986). Cinnamic acid also appears to act as a feedforward signal by inducing CHI activities (Gerrish et al. 1985).

F.2.3. Implications for Bioprocess Development. Most of the work on phenyl–propanoid metabolism in plant cells has been aimed at elucidating the plant's defense response (Templeton and Lamb 1988). However, several findings with these cultures have implications with respect to culturing plant cells for the commercial production of secondary metabolites.

From the previous discussion, it is clear that biosynthesis is regulated in a complex manner. As this complexity becomes more apparent, it becomes easier to understand why empirical approaches to strain improvement and to bioprocess development have been of limited success.

The observation that secondary metabolism can be induced has led several groups to add elicitors to cultures in the hope that production will be induced or enhanced. Inducing agents have included extracts from pathogens, ethylene-generating compounds, light, and abiotic compounds.

A final point, which was not mentioned in the previous discussion, is that several groups reported that the level of phenylpropanoid enzymes obtained after induction was dependent on the aggregate size of the culture (Kuboi and Yamada 1976; Knobloch and Hahlbrock 1977). In fact, Kinnersley and Dougall (1980a) obtained high anthocyanin-producing carrot cell lines by selecting for cells in small aggregates. The difficulties of reproducing clump or aggregate sizes in various types of bioreactors have been noted (Knobloch and Hahlbrock 1977).

F.3. Indole Alkaloid Production

In contrast to investigations of the phenylpropanoids, studies of indole alkaloid production by *Catharanthus roseus* are often aimed toward the commercial production of the high-value alkaloids - especially vincristine and vinblastine. In initial approaches, techniques from industrial microbiology were transferred directly in an effort to obtain production from undifferentiated suspended or immobilized cells. Although undifferentiated *C. roseus* cells are able to produce the monomeric intermediates, it has generally been observed that they are unable to produce the dimeric alkaloids, which are products from later steps in the alkaloid biosynthetic pathway. More recent studies have considered using differentiated organs for the production of these dimeric compounds.

F.3.1. Biosynthesis of Indole Alkaloids. Indole alkaloid biosynthesis has been reviewed several times (Madyastha and Coscia 1978; Zenk 1980; Stockigt 1980; Scott et al. 1981; Kutney at al. 1983). More recently, van der Heijden et al. (1989) published an extensive review on the cultivation of *C. roseus* cells and the production of indole alkaloids by these cultures.

Figure 2.9 shows key early steps in the biosynthetic pathway. Tryptophan decarboxylase (TDC) catalyzes the first step committing the amino acid tryptophan to secondary metabolism (Noe et al. 1984). Geraniol-10-hydroxylase (G-10-H) appears to be a key enzyme in the monoterpene portion of the pathway (McFarlane et al. 1975). Strictosidine synthase (SS) catalyzes the condensation of tryptamine and secologanin (Treimer and Zenk 1979; Mizukami et al. 1979). Reactions between strictosidine and ajmalicine have also been studied (Treimer and Zenk 1978). Later steps in the pathway leading to the formation of the dimeric alkaloids will be discussed in a subsequent section.

Alkaloid biosynthetic enzymes appear to be in the cytosol, while the alkaloids accumulate in the vacuole (Deus-Neumann and Zenk 1984). Neumann et al. (1983) reported that alkaloid "storage" cells have a vacuolar pH of 3 compared to normal cells, which have a vacuolar pH of 5. This has led to the suggestion that an ion-trapping mechanism is involved in alkaloid accumulation in the vacuole. Ion- trapping mechanisms suggest that the neutral form of the alkaloid base can pass through the tonoplast but the protonated form, which will predominate under the low-pH conditions of the vacuole, cannot passively diffuse across the membrane. Thus, once inside

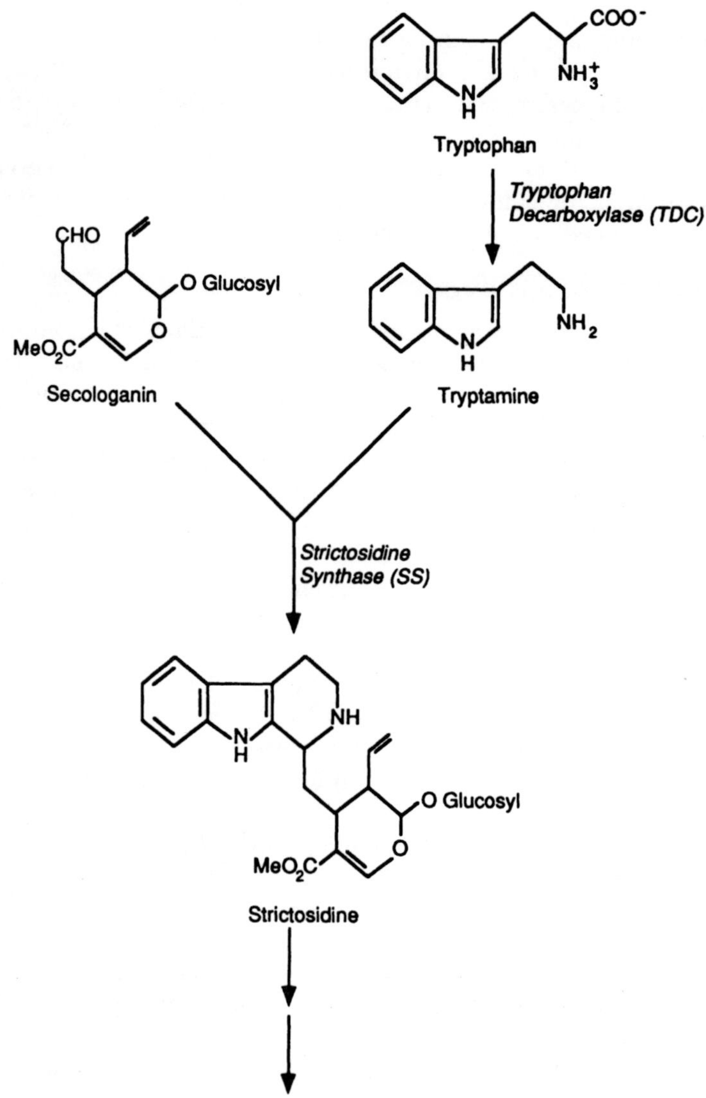

Figure 2.9. Early portion of the indole alkaloid pathway.

the vacuole, the alkaloid is trapped. There was considerable early work in support of the ion-trapping mechanism (Courtois et al. 1980; Renaudin 1981; Renaudin and Guern 1982), and it is possible that ion trapping may play a role in the vacuolar accumulation of some alkaloids (Renaudin et al. 1985). However, more recent work has suggested that alkaloid uptake into the vacuole can be energy dependent and highly selective (Deus-Neumann and Zenk 1984, 1986).

F.3.2. Strategies for Improving Production. Since a major goal of studying alkaloid biosynthesis has been to provide the knowledge to permit the economic production of these valuable compounds from cell culture systems, a considerable amount of applied research has been conducted to improve productivities of the culture. Table 2.2 illustrates the approaches used to increase alkaloid production. Many of these approaches have been transferred directly from the field of industrial microbiology.

Table 2.2. Approaches used to improve indole alkaloid production.

1. Strain improvement
2. Medium optimization
    - Nutrients
    - Precursors
    - Hormones
3. Cultivation strategies
    - Suspension culture
    - Immobilization
    - Organ culture
4. Specialized techniques
    - Elicitor treatment
    - *In situ* product removal

Strain Improvement. Industrial microbiologists have successfully used several approaches to obtain secondary metabolite over-production from bacterial and fungal strains (Demain 1973). Many of these approaches have been attempted to enhance plant cell secondary metabolite production [for reviews on selecting for high- producing plant cells, see Yamada (1984) and Berlin and Sasse (1985)]. Basically, these approaches require that: (1) high variabilities are present or can be induced (i.e., by mutation) in the culture; (2) the resulting overproducers can be identified with a rapid and sensitive screen; and (3) single high-producing cells can be propagated (Rokem and Goldberg 1985).

The development of selective and rapid screens to detect plant products is essential if high-producing cell lines are to be identified. However, efficient screening procedures may be difficult to develop if the desired activities are not easily assayed. For instance, the detection of high-alkaloid-producing cell lines requires specific chemical techniques (e.g., radioimmune or fluorescence techniques) to estimate alkaloid levels (Deus and Zenk 1982). This is in contrast to the selection of shikonin-overproducing cells, which can be readily screened by visual inspection.

To obtain high-producing cell lines, Zenk and co-workers (Zenk et al. 1976; Deus and Zenk 1982) reported that statistically, high- alkaloid-producing plants give rise to high-producing cell lines. In other systems, production levels by cells have shown variable correlations to levels observed in the whole plant (Kinnersley and Dougall 1980b; Dougall 1985). Thus, selection of high producers generally involves labor-intensive and empirical approaches. Kurz and co-workers screened large numbers of cell lines to identify high producers (Kurz et al. 1980, 1985).

Mutation strategies have also been employed to obtain alkaloid-overproducing cell lines. Rational strategies have generally involved the selection of cultures which are resistant to toxic analogues of tryptophan. One such strategy has been to use 5-methyltryptophan in the hope that cultures which are resistant to this analogue will be deregulated such that the shikimic acid pathway enzymes will no longer be feedback inhibited by tryptophan. Thus, these deregulated cultures should be capable of overproducing the precursors for alkaloid biosynthesis. Scott et al. (1979) and Schallenberg and Berlin (1979) obtained *C. roseus* cultures which were resistant to 5-methyltryptophan, and Scott et al. (1979) reported that shikimic acid enzymes of their resistant line were insensitive to tryptophan feedback inhibition. Both groups observed significant (e.g., 30-fold) increases in the accumulation of tryptophan by the resistant cultures. However, in both cases the mutations did not improve the production of either tryptamine or indole alkaloids. These results suggest that the supply of the indole precursors was not limiting the biosynthesis of alkaloids.

An alternative strategy involved the use of 4-methyltryptophan. This analogue can be detoxified by TDC, and thus resistant mutants were identified which contained high levels of the TDC (Sasse et al. 1983a,b). These mutants were observed to accumulate higher levels of tryptamine, and in some cases, higher alkaloid levels.

A major problem with plant cell cultures is their inherent instability. Although variability is essential if high producers are to be identified from a population, variability often leads to reductions in culture productivity with subculturing. Variability observed in cultures has been attributed to: (1) genetic changes which are reversed only at low frequency; (2) epigenetic changes which are due to physiological conditions and can be reversed by changes in the culture environment; and (3) selection for a given cell population from the heterogeneous population typically present in cell cultures (Petiard et al. 1985). This latter cause can be quite important when cell lines are maintained by subculture and thus selective pressure is exerted over a long period of time (Morris et al. 1989).

Medium Optimization. In contrast to the biosynthesis of phenylpropanoids, which is induced by adding elicitors or by light treatment, indole alkaloid production has typically been "induced" by changing medium from a growth medium to a production/induction medium which permits slower growth and alkaloid biosynthesis. Although the stimulus involved in alkaloid induction is not as well characterized as with the phenylpropanoids, this "induction" does appear to be at the transcriptional level - at least with respect to the TDC (Noe and Berlin 1985). After the cultures are transferred to an induction medium, the activity changes of the three enzymes (TDC, G-10-H, and SS) follow different patterns. Knobloch et al. (1981) observed TDC activities to increase 12-fold while SS activities remained relatively constant. Eilert et al. (1987b) observed both TDC and SS activities to rise after transferring *C. roseus* cultures into an induction medium; however, SS activities increased several days before the TDC activities. The timing of activity increases for TDC and G-10-H was also observed to differ (Schiel and Berlin 1986; Schiel et al. 1987). These differences in the timing of appearance of these enzymes suggest that the indole alkaloid pathway is not coordinately regulated.

Although enzyme induction is required for indole alkaloid formation, it appears that, once induced, alkaloid biosynthesis is not limited by the enzyme activities. Merillon et al. (1986) used three different media and observed that alkaloid production could not be correlated to TDC levels. In agreement with the suggestion that alkaloid production is not limited by the tryptophan branch of the pathway, these authors also observed that tryptamine additions did not stimulate alkaloid production. G-10-H appears to correlate a bit better to alkaloid production, although high levels of this enzyme are also no guarantee of enhanced alkaloid production (Schiel et al. 1987). Further support for the idea that the monoterpene branch may limit the rate of alkaloid formation was obtained by Merillon et al. (1986), who observed that secologanin additions could stimulate alkaloid production, and by Schiel et al. (1987), who reported that tryptamine accumulated shortly after cells were transferred into the induction medium.

As previously mentioned, indole alkaloid production has traditionally been induced by transferring the cultures from a growth to a production medium. The medium component which has gained the most attention is the auxin. It has generally been observed that the synthetic auxin 2,4-D suppresses alkaloid production (Knobloch and Berlin 1980; Roustan et al. 1982; Eilert et al. 1987a), and, in fact, Morris (1986a) has shown that if an alternative auxin (naphthalene acetic acid) is used, then a single medium can be used for both growth and production.

In addition to 2,4-D, the phosphate and nitrogen salts also appear to suppress alkaloid production. Phosphate additions to the phosphate-deficient induction medium have been observed to suppress both alkaloid production and the induction of G-10-H (Schiel et al. 1987). However, phosphate additions had no effect on the induction of TDC (Knobloch and Berlin 1983). Regulation of secondary metabolism by such nutritional factors is well established in microbial systems (Martin 1977; Martin and Demain 1980).

Alkaloid production appears to be stimulated by high sugar levels, and, in fact, Knobloch and Berlin (1980) reported that dilution of a culture into an 8% sucrose solution was sufficient for indole alkaloid production. Also, Smith et al. (1987c) observed that lactose was superior to glucose, fructose, or sucrose for catharanthine production.

Another condition which is frequently optimized in microbial secondary metabolite-producing systems is temperature (Weinberg 1974). Courtois and Guern (1980) and Morris (1986c) observed that the optimal temperature for growth is higher than the optimal temperature for alkaloid biosynthesis.

When production rates are limited by precursor availability, precursor feeding can enhance production. Precursor feeding to enhance alkaloid production has led to variable results. Krueger and Carew (1978) observed that alkaloid production was enhanced when precursors of the indole pathway (e.g., tryptamine) were fed, but production was not enhanced by the addition of intermediates in the monoterpene

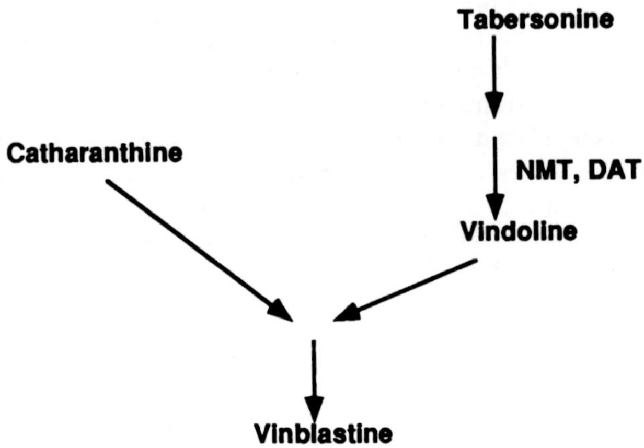

Figure 2.10. Later portion of the indole alkaloid pathway. Adapted from DeLuca et al. (1987)

branch (e.g., geraniol and secologanin). In contrast, Knobloch and Berlin (1980) observed that tryptophan additions reduced alkaloid accumulation, while Merillon et al. (1986) reported that secologanin additions stimulated production.

Cultivation strategies to improve secondary metabolism which are more specific to plant cell systems have also been examined. Light, although unable to induce alkaloid synthesis, had a modifying effect such that serpentine, rather than ajmalicine, was produced by irradiated cultures (Knobloch et al. 1982). Elicitor treatment resulted in rapid increases in TDC and SS activities and alkaloid production in some suspension cultures (Eilert et al. 1986, 1987b), whereas elicitor treatment of habituated or tumorous cell suspension cultures had no effect on alkaloid production (Eilert et al. 1987a). Chemical inducing agents have also been added to stimulate biosynthesis. Chemicals known to induce monoterpene biosynthesis (Lee et al. 1981; Kutney et al. 1985) and cytochrome P-450 enzymes (e.g., G-10-H) (Simpson and Kelly 1989) were observed to enhance alkaloid production. Vanadyl sulfate, an alternative to biologically derived elicitors (Smith et al. 1987a), and abscisic acid (Smith et al. 1987b) were also observed to stimulate catharanthine and ajmalicine accumulation.

Cultivation Strategies. In addition to manipulating the cell line and the cultivation medium, efforts to enhance alkaloid production have included the use of novel or improved bioreactors and operating strategies. Since suspension, immobilized, and organ culture strategies will be discussed in greater detail in subsequent chapters, our discussion here is very brief. Despite extensive efforts, unorganized cells - either in suspension or immobilized - have been observed to be unable to produce the dimeric indole alkaloids. The biosynthetic pathway for these dimeric alkaloids is shown in Figure 2.10. Although ajmalicine, serpentine, and one of the monomeric precursors (i.e., catharanthine) to dimeric alkaloid synthesis have been observed, the vindoline precursor does not appear to be produced by unorganized cultues. Fahn et al. (1985) and DeLuca et al. (1987) have related this failure to produce vindoline to the lack of two key enzymes, *N*-methyl transferase (NMT) and deacetylvindoline acetyltransferase (DAT), in cell suspension cultures. These enzymes were present in vindoline-containing plant tissue. Efforts to induce activity of these enzymes with fungal elicitors were unsuccessful - despite the induction of the early pathway enzymes TDC and SS (DeLuca et al. 1987). Interestingly, Misawa et al. (1988) and Endo et al. (1988) reported that suspension-cultured cells do contain the necessary enzymes to catalyze the reaction between vindoline and catharanthine to form dimeric alkaloids.

The most recent efforts to produce the dimeric alkaloids have focused on the use of differentiated tissue. Constabel et al. (1982) observed that although vindoline was present in shoots regenerated from callus, callus-derived suspension cultures were unable to accumulate vindoline. Subsequent work by Morris (1986b) and Endo et al. (1987) showed that although the major root alkaloids serpentine and ajmalicine can be produced in suspension cultures, vindoline and catharanine - alkaloids associated with aerial plant parts - are either not present or only occasionally present in suspension cultures. Recently, *C. roseus* shoot cultures have been reported to accumulate dimeric alkaloids - including vinblastine (Endo et al. 1987; Miura et al. 1988). In contrast, transformed root cultures were observed to produce alkaloids common in the root (Parr et al. 1988).

G. CONCLUSION

In comparison to microbes and mammalian cells, plant cells have many similarities but also some important differences. Due to the size of plant cells (20-200 μm), they are more shear sensitive than bacteria. However, because of their cell walls, plant cells are more shear tolerant than mammalian cells. Since plant cells elongate with age, their shear sensitivity may also increase with age. The ability or tendency of plant cells to grow as aggregates and as organized tissue makes it possible for intercellular communication to play an important role in determining how plant cells respond to their environment. In fact, many secondary metabolites do not accumulate unless tissue organization/differentiation occurs. A final point with respect to the structure of plant cells is that the vacuole is a site of storage. Thus, in contrast to microbial cultures, plant cells often retain their secondary metabolites within the vacuole.

Metabolically, plant cells are capable of growing in very simple media. Most cultures grow well with a sugar as the sole carbon source, ammonia and nitrate as the nitrogen source, and phosphate as the phosphorus source. Amino acids are generally not required for plant cell growth. Vitamins (e.g., thiamine) are often incorporated into typical plant cell media. In contrast to microbes, plant cells often require phytohormones (e.g., auxins and cytokinins) for growth. These hormonal requirements are considerably less demanding than those for mammalian cell cultivation. Unfortunately, our limited understanding of phytohormone action prevents a truly rational approach in choosing the types and levels of these compounds used.

Finally, for commercial production of plant secondary metabolites, it will probably be essential that improved strains be selected and optimal media developed. It has been difficult to achieve these two objectives due in part to our limited understanding of the complexities of secondary metabolism.

References

S. Amino and M. Tazawa. 1988. Uptake and Utilization of Sugars in Cultured Rice Cells. *Plant Cell Physiol. 29*, 483-487.

H. Ashihara and T. Tokoro. 1985. Metabolic Fate of Inorganic Phosphate Absorbed by Suspension Cultured Cells of *Catharanthus roseus. J. Plant Physiol. 118*, 227-235.

H. Ashihara and T. Ukaji. 1986. Inorganic Phosphate Absorption and Its Effect on the Adenosine 5"-Triphosphate Level in Suspension Cultured Cells of *Catharanthus roseus . J. Plant Physiol. 124*, 77-85.

H. Ashihara, T. Horikosi, X.-N. Li, K. Sagishima, and Y. Yamashita. 1988. Profiles of Enzymes Involved in Glycolysis in *Catharanthus roseus* Cells in Batch Suspension Culture. *J. Plant Physiol. 133*, 38-45.

J. M. Bayley, J. King, and O. L. Gamborg. 1972. The Effect of the Source of Inorganic Nitrogen on Growth and Enzymes of Nitrogen Assimilation in Soybean and Wheat Cells in Suspension Cultures. *Planta 105*, 15-24.

L. D. Beaudoin-Eagan and T. A. Thorpe. 1983. Shikimate Pathway Activity during Shoot Initiation in Tobacco Callus Cultures. *Plant Physiol. 73*, 228-232.

L. D. Beaudoin-Eagan and T. A. Thorpe. 1985. Shikimate Pathway Enhancement in the Lower Half of Shoot -Forming Tobacco Callus. *J. Plant Physiol. 120*, 87-90.

J. Berlin and F. Sasse. 1985. Selection and Screening Techniques for Plant Cell Cultures. *Adv. Biochem. Eng. Biotechnol. 31*, 99-132.

B. Betz, E. Schafer, and K. Hahlbrock. 1978. Light Induced Phenylalanine-Ammonia Lyase in Cell Suspension Cultures of *Petroselinum hortense. Arch. Biochem. Biophys. 190*, 126-135.

C. C. Black, L. Mustardy, S. S. Pang, S. S. Sung, P. P. Kormanik, D.-P. Xu, and N. Paz. 1987. Regulation and Roles for Alternative Pathways of Hexose Metabolism in Plants. *Physiol. Plant. 69*, 387-394.

G. P. Bolwell, M. P. Robbins, and R. A. Dixon. 1985a. Metabolic Changes in Elicitor-Treated Bean Cells: Enzymic Responses Associated with Rapid Changes in Cell Wall Components. *Eur. J. Biochem. 148*, 571-578.

G. P. Bolwell, J. N. Bell, C. L. Cramer, W. Schuch, C. J. Lamb and R. A. Dixon. 1985b. L-Phenylalanine Ammonia-Lyase from *Phaseolus vulgaris*: Characterization and Differential Induction of Multiple Forms from Elicitor-Treated Cell Suspension Cultures. *Eur. J. Biochem. 149*, 411-419.

G. P. Bolwell, C. L. Cramer, C. J. Lamb, W. Schuch, and R. A. Dixon. 1986. L-Phenylalanine Ammonia-Lyase from *Phaseolus vulgaris*: Modulation of the Levels of Active Enzyme by *trans*-Cinnamic Acid. *Planta 169*, 97-107.

P. Brodelius and H. J. Vogel. 1985. A Phosphorus-31 Nuclear Magnetic Resonance Study of Phosphate Uptake and Storage in Cultured *Catharanthus roseus* and *Daucus carota* Plant Cells. *J. Biol. Chem. 260*, 3556-3560.

J. Chappell and K. Hahlbrock. 1984. Transcription of Plant Defense Genes in Response to UV Light or Fungal Elicitor. *Nature 311*,76-78.

F. Constabel, P. Gaudet-LaPrairie, W. G. W. Kurz, and J. P. Kutney. 1982. Alkaloid Production in *Catharanthus roseus* Cell Cultures. XII. Biosynthetic Capacity of Callus from Original Explants and Regenerated Shoots. *Plant Cell Rep. 1*, 139-142.

D. Courtois and J. Guern. 1980. Temperature Response of *Catharanthus roseus* Cells Cultivated in Liquid Medium. *Plant Sci. Lett. 17*, 473-482.

D. Courtois, A. Kurkdjian, and J. Guern. 1980. Tryptamine Uptake and Accumulation by *Catharanthus roseus* Cells Cultivated in Liquid Medium. *Plant Sci. Lett. 18*, 85-96.

C. L. Cramer, T. B. Ryder, J. N. Bell, and C. J. Lamb. 1985. Rapid Switching of Plant Gene Expression Induced by Fungal Elicitor. *Science 227*, 1240-1243.

L. L. Creasy. 1987. The Role of Enzyme Inactivation in the Regulation of Synthetic Pathways: A Case History. *Physiol. Plant. 71*, 389-392.

D. A. Day and J. B. Hanson. 1977. Pyruvate and Malate Transport and Oxidation in Corn Mitochondria. *Plant Physiol. 59*, 630-635.

V. DeLuca, J. Balsevich, R. T. Tyler, and W. G. W. Kurz. 1987. Characterization of a Novel N-Methyltransferase (NMT) from *Catharanthus roseus* plants. *Plant Cell Rep. 6*, 458-461.

A. L. Demain. 1973. Mutation and the Production of Secondary Metabolites. *Adv. Appl. Microbiol. 16*, 177-202.

B. Deus and M. H. Zenk. 1982. Exploitation of Plant Cells for the Production of Natural Products. *Biotechnol. Bioeng. 24*, 1965-1974.

B. Deus-Neumann and M. H. Zenk. 1984. A Highly Selective Alkaloid Uptake System in Vacuoles of Higher Plants. *Planta 162*, 250-260.

B. Deus-Neumann and M. H. Zenk. 1986. Accumulation of Alkaloids in Plant Vacuoles Does Not Involve an Ion-Trap Mechanism. *Planta 167*, 44-53.

K.-J. Dietz, R. Jager, G. Kaiser, and E. Martinoia. 1990. Amino Acid Transport across the Tonoplast of Vacuoles Isolated from Barley Mesophyll Protoplasts. *Plant Physiol. 92*, 123-129.

R. A. Dixon and C. J. Lamb. 1979. Stimulation of *de novo* Synthesis of L-Phenylalanine Ammonia-Lyase in Relation to Phytoalexin Accumulation in *Colletotrichum lindemuthianum* Elicitor-Treated Cell Suspension Cultures of French Bean (*Phaseolus vulgaris*). *Biochim. Biophys. Acta 586*, 453-463.

R. A. Dixon, T. Browne, and M. Ward. 1980. Modulation of L-Phenylalanine Ammonia-Lyase by Pathway Intermediates in Cell Suspension Cultures of Dwarf French Bean (*Phaseolus vulgaris* L.). *Planta 150*, 279-285.

D. K. Dougall. 1974. Evidence for the Presence of Glutamate Synthase in Extracts of Carrot Cell Cultures. *Can. J. Bot.. 54*, 2924-2927.

D. K. Dougall. 1980. The Use of Tissue Cultures in Studies of Metabolism. In "The Biochemistry of Plants," Vol. 2, D. D. Davies (ed.), Academic Press, New York, pp. 627-642.

D. K. Dougall. 1985. Chemicals from Plant Cell Cultures: Yields and Variation. In "Biotechnology in Plant Science: Relevance to Agriculture in the Eighties," M. Zaitlin, P. Day, and A. Hollaender (eds.), Academic Press, New York, pp. 179-190.

D. K. Dougall and J. Bloch. 1976. A Survey of the Presence of Glutamate Synthase in Plant Cell Suspension Cultures. *Biochem. Biophys. Res. Commun. 58*, 639-646.

K. Edwards, C. L. Cramer, G. P. Bolwell, R. A. Dixon, W. Schuch, and C. J. Lamb. 1985. Rapid Transient Induction of Phenylalanine Ammonium-Lyase mRNA in Elicitor-Treated Bean Cells. *Proc. Natl. Acad. Sci. USA 82*, 6741-6744.

U. Eilert, F. Constabel, and W. G. W. Kurz. 1986. Elicitor-Stimulation of Monoterpene Indole Alkaloid Formation in Suspension Cultures of *Catharanthus roseus. J. Plant Physiol. 126*, 11-22.

U. Eilert, V. DeLuca, W. G. W. Kurz, and F. Constabel. 1987a. Alkaloid Formation by Habituated and Tumorous Cell Suspension Cultures of *Catharanthus roseus. Plant Cell Rep. 6*, 271-274.

U. Eilert, V. DeLuca, F. Constabel, and W. G. W. Kurz. 1987b. Elicitor-Mediated Induction of Tryptophan Decarboxylase and Strictosidine Synthase Activities in Cell Suspension Cultures of *Catharanthus roseus*. *Arch. Biochem. Biophys. 254*, 491-497.

T. Endo, A. Goodbody, and M. Misawa. 1987. Alkaloid Production in Root and Shoot Cultures of *Catharanthus roseus*. *Planta Med. 53*, 479-482.

T. Endo, A. Goodbody, J. Vukovic, and M. Misawa. 1988. Enzymes from *Catharanthus roseus* Cell Suspension Cultures That Couple Vindoline and Catharanthine to Form 3',4'-Anhydrovinblastine. *Phytochemistry* 27, 2147-2149.

W. Fahn, H. Gundlach, B. Deus-Neumann, and J. Stockigt. 1985. Late Enzymes of Vindoline Biosynthesis. Acetyl-CoA:17-O-Deacetylvindoline 17-O-Acetyltransferase. *Plant Cell Rep. 4*, 333-336.

M. W. Fowler and A. Clifton. 1974. Activities of Carbohydrate Metabolism in Cells of *Acer pseudoplatanus* L. Maintained in Continuous (Chemostat) Culture. *Eur. J. Biochem. 45*, 445-450.

M. W. Fowler and G. Stepan-Sarkissian. 1985. Carbohydrate Source, Biomass Productivity and Natural Product Yield in Cell Suspension Cultures. In "Primary and Secondary Metabolism of Plant Cell Cultures," K.-H. Neumann, W. Barz, and E. Reinhard (eds.), Springer-Verlag, New York, pp. 66-73.

C. Gerrish, M. P. Robbins, and R. A. Dixon. 1985. Trans-Cinnamic Acid as a Modulator of Chalcone Isomerase in Bean Cell Suspension Cultures. *Plant Sci. 38*, 23-27

C. V. Givan. 1979. Metabolic Detoxification of Ammonia in Tissues of Higher Plants. *Phytochemistry 18*, 375-382.

T. W. Goodwin and E. I. Mercer. 1983. "Introduction to Plant Biochemistry," Second ed., Pergamon Press, New York.

H. Grisebach. 1981. Lignins. In "The Biochemistry of Plants," Vol. 7, E. E. Conn (ed.), Academic Press, New York, pp. 457-478.

H. Habereder, G. Schroder, and J. Ebel. 1988. Rapid Induction of Phenylalanine Ammonia-Lyase and Chalcone Synthase mRNA's during Fungus Infection of Soybean (Glycine max L.) Roots or Elicitor Treatment of Soybean Cell Cultures at the Onset of Phytoalexin Synthesis. *Planta* 177, 58-65.

K. Hahlbrock. 1981. Flavonoids. In "The Biochemistry of Plants," Vol. 7, E. E. Conn (ed.), Academic Press, New York, pp. 425-456.

K. Hahlbrock and H. Grisebach. 1979. Enzymic Controls in the Biosynthesis of Lignins and Flavanoids. *Annu. Rev. Plant Physiol. 30*, 105-130.

K. Hahlbrock, K.-H. Knobloch, F. Kreuzaler, J. R. M. Potts, and E. Wellman. 1976. Coordinated Induction and Subsequent Activity Changes of Two Groups of Metabolically Interrelated Enzymes: Light Induced Synthesis of Flavonoid Glycosides in Cell Suspension Cultures of Petroselinum hortense. *Eur. J. Biochem. 61*, 199-206.

K. Hahlbrock, C. J. Lamb, C. Purwin, J. Ebel, E. Fautz, and E. Schafer. 1981. Rapid Response of Suspension-Cultured Parsley Cells to the Elicitor from *Phytophthora megasperma* var *sojae*. *Plant Physiol. 67*, 768-773.

C. A. Hay, L. A. Anderson, M. F. Roberts, and J. D. Phillipson. 1988. Alkaloid Production by Plant Cell Cultures. *Adv. Biotechnol. Proc. 9*, 99-140.

R. B. Herbert. 1989. "The Biosynthesis of Secondary Metabolites," Second ed., Chapman and Hall, London.

F. Hirose and H. Ashihara. 1984. Interrelationship between the Pentose Phosphate Pathway and Nucleotide Synthesis in Cultured Cells of *Catharanthus roseus*. *Plant Sci. Lett. 35*, 123-126.

J. A. Jackson and R. F. Lyndon. 1990. Habituation: Cultural Curiosity or Developmental Determinant? *Physiol. Plant. 79*, 579-583.

R. A. Jensen. 1986.Tyrosine and Phenylalanine Biosynthesis: Relationship between Alternative Pathways, Regulation and Subcellular Location. *Recent Adv. Phytochem. 20*, 57-81.

J. Kanabus, R. A Bressan, and N. C. Carpita. 1986. Carbon Assimilation in Carrot Cells in Liquid Culture. *Plant Physiol. 82*, 363-368.

G. J. Kelly and E. Latzko. 1980. The Cytosol. In "The Biochemistry of Plants," Vol. 1, N. E. Tolbert (ed.), Academic Press, New York, pp. 183-208.

A. M. Kinnersley and D. K. Dougall. 1980a. Increase in Anthocyanin Yield from Wild-Carrot Cell Cultures by a Selection System Based on Cell-Aggregate Size. *Planta 149*, 200-204.

A. M. Kinnersley and D. K. Dougall. 1980b. Correlation between the Nicotine Content of Tobacco Plants and Callus Cultures. *Planta 149*, 205-206.

K.-H. Knobloch. 1982. Uptake of Phosphate and Its Effect on Phenylalanine Ammonia-Lyase Activity and Cinnamoyl Putrescine Accumulation in Cell Suspension Cultures of *Nicotiana tabacum*. *Plant Cell Rep. 1*, 128-130.

K.-H. Knobloch and J. Berlin. 1980. Influence of Medium on the Formation of Secondary Compounds in Cell Suspension Cultures of *Catharanthus roseus*.. *Z. Naturforsch. 35c*, 551-556.

K.-H. Knobloch and J. Berlin. 1981.Phosphate Mediated Regulation of Cinnamoyl Putrescine Biosynthesis in Cell Suspension Cultures of *Nicotiana tabacum*. *Planta Med. 42*, 167-172.

K.-H. Knobloch and J. Berlin. 1983. Influence of Phosphate on the Formation of Indole Alkaloids and Phenolic Compounds in Cell Suspension Cultures of *Catharanthus roseus*..I Comparison of Enzyme Activities and Product Accumulation *Plant Cell Tissue Organ Culture 2, 333-340.*

K.-H. Knobloch and K. Hahlbrock. 1975. Isoenzymes of p-Coumarate:CoA Ligase from Cell Suspension Cultures of *Glycine max*. *Eur. J. Biochem. 52*, 311-320.

K.-H. Knobloch and K. Hahlbrock. 1977. 4-Coumarate:CoA Ligase from Cell Suspension Cultures of *Petroselinum hortense* Hoffm. *Arch. Biochem. Biophys. 184*, 237-248.

K.-H. Knobloch, B. Hansen, and J. Berlin. 1981. Medium-Induced Formation of Indole Alkaloids and Concomitant Changes of Interrelated Enzyme Activities in Cell Suspension Cultures of *Catharanthus roseus*. *Z. Naturforsch. 36c*, 40-43.

K.-H. Knobloch, G. Bast, and J. Berlin. 1982. Medium- and Light-Induced Formation of Serpentine and Anthocyanins in Cell Suspension Cultures of *Catharanthus roseus*. *Phytochemistry 21*, 591-594.

R. J. Krueger and D. P. Carew. 1978. *Catharanthus roseus* Tissue Culture: The Effects of Precursors on Growth and Alkaloid Production. *Lloydia 41*, 327-331.

T. Kuboi and Y. Yamada. 1976. Caffeic Acid-O-Methyltransferase in a Suspension of Cell Aggregates of Tobacco. *Phytochemistry 15*, 397-400.

D. N. Kuhn, J. Chappell, A. Boudet, and K. Hahlbrock. 1984. Induction of Phenylalanine Ammonia-Lyase and 4-Coumarate:CoA Ligase mRNA's in Cultured Cells by UV Light or Fungal Elicitor. *Proc. Natl. Acad. Sci. USA 81*, 1102-1106.

W. G. W. Kurz, K. B. Chatson, F. Constabel, J. P. Kutney, L.S. L. Choi, P. Kolodziejczyk, S. K. Sleigh, K. L. Stuart, and B. R. Worth. 1980. Alkaloid Production in *Catharanthus roseus* Cell Cultures: Initial Studies on Cell Lines and Their Alkaloid Content. *Phytochemistry 19*, 2583-2587.

W. G. W. Kurz, K. B. Chatson, and F. Constabel. 1985. Biosynthesis and Accumulation of Indole Alkaloids in Cell Suspension Cultures of *Catharanthus roseus* Cultivars. In "Primary and Secondary Metabolism of Plant Cell Cultures," K.-H. Neumann, W. Barz and E. Reinhard (eds.), Springer-Verlag, New York, pp. 143-153.

J. P. Kutney, B. Aweryn, L. S. L. Choi, T. Honda, P. Kolodziejczyk, N. G. Lewis, T. Sato, S. K. Sleigh, K. L. Stuart, and B. R. Worth. 1983. Studies in Plant Tissue Culture: The Synthesis and Biosynthesis of Indole Alkaloids. *Tetrahedron 39*, 3781-3795.

J. P. Kutney, B. Aweryn, K. B. Chatson, L. S. L. Choi, and W. G. W. Kurz. 1985. Alkaloid Production in *Catharanthus roseus* (L.) G. Don Cell Cultures. XIII. Effects of Bioregulators on Indole Alkaloid Biosynthesis. *Plant Cell Rep. 4*, 259-262.

G. G. Laties. 1982. The Cyanide-Resistant Alternative Path in Higher Plant Respiration. *Annu. Rev. Plant Physiol. 33*, 519-555.

M. A. Lawton and C. J. Lamb. 1987. Transcriptional Activation of Plant Defense Genes by Fungal Elicitor, Wounding and Infection. *Mol. Cell. Biol.* 7, *335-341.*

M. A. Lawton, R. A. Dixon, and C. J. Lamb. 1980. Elicitor Modulation of the Turnover of L-Phenylalanine Ammonium-Lyase in French Bean Cell Suspension Cultures. *Biochim. Biophys. Acta 633*, 162-175.

M. A. Lawton, R. A. Dixon, K. Hahlbrock, and C. J. Lamb. 1983a. Rapid Induction of the Synthesis of Phenylalanine Ammonium-Lyase and of Chalcone Synthase in Elicitor-Treated Plant Cells. *Eur. J. Biochem. 129*, 593-601.

M. A. Lawton, R. A. Dixon, K. Hahlbrock, and C. J. Lamb. 1983b. Elicitor Induction of mRNA Activity: Rapid Effects of Elicitor on Phenylalanine Ammonium-Lyase and Chalcone Synthase mRNA Activities in Bean Cells. *Eur. J. Biochem. 129*, 593-601.

S.-L. Lee, K.-D. Cheng and A. I. Scott. 1981. Effects of Bioregulators on Indole Alkaloid Biosynthesis in *Catharanthus roseus* Cell Culture. *Phytochemistry 20*, 1841-1843.

P. Lenee and Y. Chupeau. 1989. Development of Nitrogen Assimilating Enzymes during Growth of Cells Derived from Protoplasts of Sunflower and Tobacco. *Plant Sci. 59*, 109-117.

K. Lindsey and M. M. Yeoman. 1984. The Viability and Biosynthetic Activity of Cells of *Capsicum frutescens* Mill. cv. annuum Imobilized in Reticulate Polyurethane. *J. Exp. Bot. 35*, 1684-1696.

K. M. Madyastha and C. J. Coscia. 1978. Enzymology of Indole Alkaloid Biosynthesis. *Recent Adv. Phytochem. 13*, 85-129.

U. Margna. 1977. Control at the Level of Substrate Supply - An Alternative in the Regulation of Phenylpropanoid Accumulation in Plant Cells. *Phytochemistry 16*, 419-426.

J. F. Martin. 1977. Control of Antibiotic Biosynthesis by Phosphate. *Adv. Biochem. Eng. 6*, 105-127.

J. F. Martin and A. L. Demain. 1980. Control of Antibiotic Biosynthesis. *Microbiol. Rev. 44*, 230-251.

S. M. Martin, D. Rose, and V. Hui. 1977. Growth of Plant Cell Suspension Cultures with Ammonium as the Sole Source of Nitrogen. *Can. J. Bot. 55*, 2838-2843.

F. Marty, D. Branton and R. A. Leigh. 1980. Plant Vacuoles. In "The Biochemistry of Plants," Vol. 1, N. E. Tolbert (ed.), Academic Press, New York, pp. 625-658.

H. Masuda, T. Takahashi, and S. Sugawara. 1988. Acid and Alkaline Invertases in Suspension Cultures of Sugar Beet Cells. *Plant Physiol. 86*, 312-317.

K. A. McDonald and A. P. Jackman. 1989. Bioreactor Studies of Growth and Nutrient Utilization in Alfalfa Suspension Cultures. *Plant Cell Rep. 8*, 455-458.

J. McFarlane, K. M. Madyastha, and C. J. Coscia. 1975. Regulation of Secondary Metabolism in Higher Plants. Effect of Alkaloids on a Cytochrome P-450 Dependent Monooxygenase. *Biochem. Biophys. Res. Comm. 66*, 1263-1269.

S. Meiners, O. Baron-Epel, and M. Schindler. 1988. Intercellular Communication - Filling in the Gaps. *Plant Physiol. 88*, 791-793.

J. M. Merillon, P. Doireau, A. Guillot, J. C. Chenieux, and M. Rideau. 1986. Indole Alkaloid Accumulation and Tryptophan Decarboxylase Activity in *Catharanthus roseus* Cells Cultured in Three Different Media. *Plant Cell Rep. 5*, 23-26.

B. J. Miflin and P. J. Lea. 1976. The Pathway of Nitrogen Assimilation in Plants. *Phytochemistry 15*, 873-885.

B. J. Miflin and P. J. Lea. 1977. Amino Acid Mretabolism. *Annu. Rev. Plant Physiol. 28*, 299-329.

M. Misawa. 1985. Production of Useful Plant Metabolites. *Adv. Biochem. Eng. Biotechnol. 31*, 59-88.

M. Misawa, T. Endo, A. Goodbody, J. Vukovic, C. Chapple, L. Choi and J. P. Kutney. 1988. Synthesis of Dimeric Alkaloids by Cell Free Extracts from Cell Suspension Cultures of *Catharanthus roseus*. *Phytochemistry* 27, 1355-1359.

Y. Miura, K. Hirata, N. Kurano, K. Miyamoto, and K. Uchida. 1988. Formation of Vinblastine in Multuple Shoot Culture of *Catharanthus roseus*. *Planta Med. 54*, 18-20.

H. Mizukami, H. Nordlov, S.-L. Lee, and A. I. Scott. 1979. Purification and Properties of Strictosidine Synthase (an Enzyme Condensing Tryptamine and Secologanin) from *Catharanthus roseus* Cultured Cells. *Biochemistry 18*, 3760-3763.

P. Morris. 1986a. Regulation of Product Synthesis in Cell Cultures of *Catharanthus roseus*. II. Comparison of Production Media. *Planta Med. 52*, 121-126.

P. Morris. 1986b. Regulation of Product Synthesis in Cell Cultures of *Catharanthus roseus*. III. Alkaloid Metabolism in Cultured Leaf Tissue and Primary Callus. *Planta Med. 52*, 127-132.

P. Morris. 1986c. Regulation of Product Synthesis in Cell Cultures of *Catharanthus roseus*. Effect of Culture Temperature. *Plant Cell Rep. 5*, 427-429.

P. Morris, K. Rudge, R. Cresswell, and M. W. Fowler. 1989. Regulation of Product Synthesis in Cell Cultures of *Catharanthus roseus*. V. Long Term Maintenance of Cells on a Production Medium. *Plant Cell Tissue Organ Culture* 17, 79-90.

D. Neumann, G. Krauss, M. Hieke, and D. Groger. 1983. Indole Alkaloid Formation and Storage in Cell Suspension Cultures of *Catharanthus roseus. Planta Med. 48*, 20-23.

W. Noe and J. Berlin. 1985. Induction of *de-novo* Synthesis of Tryptophan Decarboxylase in Cell Suspensions of *Catharanthus roseus. Planta 166*, 500-504.

W. Noe, C. Mollenschott, and J. Berlin. 1984. Tryptophan Decarboxylase from Catharan*thus roseus* Cell Suspension Cultures: Purification, Molecular and Kinetic Data of the Homogenous Protein. *Plant Mol. Biol. 3*, 281-288.

J. A. Parker, K. Hahlbrock, and D. Scheel. 1988. Different Cell-Wall Components from Phytophthora megasperma f. sp. *glycinea* Elicit Phytoalexin Production in Soybean and Parsley. *Planta 176*, 75-82.

A. J. Parr. 1988. Secondary Products from Plant Cell Culture. *Adv. Biotechnol. Proc. 9*, 1-34.

A. J. Parr, A. C. J. Peerless, J. D. Hamill, N. J. Walton, R. J. Robbins, and M. J. C. Rhodes. 1988. Alkaloid Production by Transformed Root Cultures of *Catharanthus roseus*. *Plant Cell Rep.* 7, 309-312.

V. Petiard, C. Baubault, A. Bariaud, M. Hutin, and D. Courtois. 1985. Studies on Variability of Plant Tissue Cultures for Alkaloid Production *Catharanthus roseus* and *Papaver somniferum* Callus Cultures. In "Primary and Secondary Metabolism of Plant Cell Cultures", K.-H. Neumann, W. Barz, and E. Reinhard (eds.), Springer-Verlag, New York, pp. 133-142.

R. Phillips and G. G. Henshaw. 1977. The Regulation of Synthesis of Phenolics in Stationary Phase Cell Cultures of *Acer pseudoplatanus* L. *J. Exp. Bot. 28*, 785-794.

J. A. Pryke and T. ap Rees. 1976. Activity of the Pentose Phosphate Pathway during Lignification. *Planta 132*, 279-284.

J.-P. Renaudin. 1981. Uptake and Accumulation of an Indole Alkaloid, [14-C] Tabernanthine, by Cell Suspension Cultures of *Catharanthus roseus* (L.) G. Don and *Acer Pseudoplatanus* L. *Plant Sci. Lett. 22*, 59-69.

J.-P. Renaudin and J. Guern. 1982. Compartmentation Mechanisms of Indole Alkaloids in Cell Suspension Cultures of *Catharanthus roseus* . *Physiol. Veg. 20*, 533-547.

J.-P. Renaudin, S. C. Brown, and J. Guern. 1985. Compartmentation of Alkaloids in Cell Suspension of *Catharanthus roseus* : A Reappraisal of the Role of pH Gradients. In "Primary and Secondary Metabolism of Plant Cell Cultures," K.-H. Neumann, W. Barz, and E. Reinhard (eds.), Springer-Verlag, New York, pp. 124-132.

M. P. Robbins and R. A. Dixon. 1984. Induction of Chalcone Isomerase in Elicitor-Treated Bean Cells: Comparison of Rates of Synthesis and Appearance of Immunodetectable Enzyme. *Eur. J. Biochem. 145*, 195-202.

M. P. Robbins, G. P. Bolwell, and R. A. Dixon. 1985. Metabolic Changes in Elicitor-Treated Bean Cells: Selectivity of Enzyme Induction in Relation to Phytoalexin Accumulation. *Eur. J. Biochem. 148*, 563-569.

J. S. Rokem and I. Goldberg. 1985. Secondary Metabolites from Plant Cell Suspension Cultures: Methods for Yield Improvement. *Adv. Biotechnol. Proc. 4*, 241-274.

A. Rosevear and C. A. Lambe. 1983. Immobilized Plant and Animal Cells. In "Topics in Enzyme and Fermentation Biotechnology," A. Wiseman (ed.), John Wiley, New York, pp. 13-37.

J.-P. Roustan, C. Ambid, and J. Fallot. 1982. Influence de l'acide 2,4-dichlorophenoxyacetique sur l'accumulation de certains alcaloides indoliques dans les celles quiescentes de *Catharanthus roseus* cultivees *in vitro. Physiol. Veg. 20*, 523-532.

T. B. Ryder, C. L. Cramer, J. N. Bell, M. P. Robbins, R. A. Dixon, and C. J. Lamb. 1984. Elicitor Rapidly Induces Chalcone Synthase mRNA in *Phaseolus vulgaris* Cells at the Onset of the Phytoalexin Defense Response. *Proc. Natl. Acad. Sci. USA 81*, 5724-5728.

F. Sasse, D. Backs-Husemann, and W. Barz. 1979. Isolation and Characterization of Vacuoles from Cell Suspension Cultures of *Daucus carota*. *Z. Naturforsch. 34c*, 848-853.

F. Sasse, M. Buchholz, and J. Berlin. 1983a. Site of Action of Growth Inhibitory Tryptophan Analogues in *Catharanthus roseus* Cell Suspension Cultures. *Z. Naturforsch. 38c*, 910-915.

F. Sasse, M. Buchholz, and J. Berlin. 1983b. Selection of Cell Lines of *Catharanthus roseus* with Increased Trytptophan Decarboxylase Activity. *Z. Naturforsch. 38c*, 916-922.

J. Schallenberg and J. Berlin. 1979. 5-Methyltryptophan Resistant Cells of *Catharanthus roseus*. *Z. Naturforsch. 34c*, 541-545.

O. Schiel and J. Berlin. 1986. Geraniol-10-Hydroxylase of *Catharanthus roseus* and Its Correlation with Indole Alkaloid Biosynthesis. *Planta Med. 52*, 422.

O. Schiel, L. Witte, and J. Berlin. 1987. Geraniol-10-Hydroxylase Activity and Its Relation to Monoterpene Indole Alkaloid Accumulation in Cell Suspension Cultures of *Catharanthus roseus*. *Z. Naturforsch. 42c*, 1075-1081.

J. Schroder, F. Kreuzaler, E. Schafer, and K. Hahlbrock. 1979. Concomitant Induction of Phenylalanine Ammonia-Lyase and Flavanone Synthase mRNA's in Irradiated Plant Cells. *J. Biol. Chem. 254*, 57-65.

A. I. Scott, H. Mizukami, and S.-L. Lee. 1979. Characterization of a 5-Methyltryptophan Resistant Strain of *Catharanthus roseus* Cultured Cells. *Phytochemistry 18*, 795-798.

A. I. Scott, S.-L. Lee, M. G. Culver, W. Wan, T. Hirata, F. Gueritte, R. L. Baxter, H. Nordlov, C. A. Dorschel, H. Mizukami, and N. E. MacKenzie. 1981. Indole Alkaloid Biosynthesis. *Heterocycles 15*, 1257-1274.

J. N. Siedow and D. A. Berthold. 1986. The Alternative Oxidase: A Cyanide-Resistant Respiratory Pathway in Higher Plants. *Physiol. Plant. 66*, 569-573.

A. P. Simpson and S. L. Kelly. 1989. Cytochrome P-450 Inducer/Inhibitor Effects on Cell Cultures of *Catharanthus roseus*. *Plant Sci. 60*, 231-236.

T. A. Skokut, C. P. Wolk, J. Thomas, J. C. Meeks, and P.W. Shaffer. 1978. Initial Organic Products of Assimilation of [13-N] Ammonium and [13-N] Nitrate by Tobacco Cells Cultured on Different Sources of Nitrogen. *Plant Physiol. 62*, 299-304.

J. I. Smith, N. J. Smart, M. Misawa, W. G. W. Kurz, S. G. Tallevi, and F. DiCosmo. 1987a. Increased Accumulation of Indole Alkaloids by Some Cell Lines of *Catharanthus roseus* in Response to Addition of Vanadyl Sulphate. *Plant Cell Rep. 6*, 142-145.

J. I. Smith, N. J. Smart, W. G. W. Kurz, and M. Misawa. 1987b. Stimulation of Indole Alkaloid Production in Cell Suspension Cultures of *Catharanthus roseus* by Abscisic Acid. *Plant Med. 53*, 470-474.

J. I. Smith, A. A. Quesnel, N. J. Smart, and M. Misawa. 1987c. The Development of a Single-Stage Growth and Indole Alkaloid Production Medium for *Catharanthus roseus* (L.) G. Don Suspension Cultures. *Enzyme Microb. Technol. 9*, 466-469.

T. Solomos. 1977. Cyanide-Resistant Respiration in Higher Plants. *Annu. Rev. Plant Physiol. 28*, 279-297.

E. J. Staba. 1985. Milestones in Plant Tissue Culture Systems for the Production of Secondary Products. *J. Nat. Prod. 48*, 203-209.

A. Stafford, P. Morris, and M. W. Fowler. 1986. Plant Cell Biotechnology: A Perspective. *Enzyme Microb. Technol. 8*, 578-587.

M. Stitt and T. ap Rees. 1979. Capacities of Pea Chloroplasts to Catalyze the Oxidative Pentose Phosphate Pathway and Glycolysis. *Phytochemistry 18*, 1905-1911.

J. Stockigt. 1980. The Biosynthesis of Heteroyohimbine-Type Alkaloids. In "Indole and Biogenetically Related Alkaloids," J. D. Phillipson and M. H. Zenk (eds.), Academic Press, New York, pp. 113-141.

A. Suzuki, A. Nato, and P. Gadal. 1984. Glutamate Synthase Isoforms in Tobacco Cultured Cells. *Plant Sci. Lett. 33, 93-101.*

L. Taiz. 1986. Are Biosynthetic Reactions in Plant Cells Thermodynamically Coupled to Glycolysis and the Tonoplast Proton Motive Force? *J. Theor. Biol. 123*, 231-238.

N. Takahashi. 1986. "Chemistry of Plant Hormones," CRC Press, Boca Raton, Florida.

M. D. Templeton and C. J. Lamb. 1988. Elicitors and Defense Gene Activation. *Plant Cell Environ. 11*, 395-401.

T. A. Thorpe and E. J. Laishley. 1973. Glucose Oxidation during Shoot Initiation in Tobacco Callus Cultures. *J. Exp. Bot. 24*, 1082-1089.

N. E. Tolbert. 1980. "The Biochemistry of Plants," Vol. 1, Academic Press, New York.

J. F. Treimer and M. H. Zenk. 1978. Enzymatic Synthesis of Corynanthe-Type Alkaloids in Cell Cultures of *Catharanthus roseus*: Quantitation by Radioimmunoassay. *Phytochemistry 17*, 227-231.

J. F. Treimer and M. H. Zenk. 1979. Purification and Properties of Strictosidine Synthase, the Key Enzyme in Indole Alkaloid Formation. *Eur. J. Biochem. 101*, 225-233.

J. F. Turner and D. H. Turner. 1980. The Regulation of Glycolysis and the Pentose Phosphate Pathway. In "The Biochemistry of Plants," Vol. 2, D. D. Davies (ed.), Academic Press, New York. pp, 279-316.

T. Ukaji and H. Ashihara. 1987. Effect of Inorganic Phosphate on the Levels of Amino Acids in Suspension-Cultured Cells of *Catharanthus roseus. Ann. Bot. 60*, 109-114.

R. van der Heijden, R. Verpoorte, and H. J. G. Ten Hoopen. 1989. Cell and Tissue Cultures of *Catharanthus roseus* (L.) G. Don: A Literature Survey. *Plant Cell Tissue Organ Culture 18*, 231-280.

W. M. van Gulik, J. J. Meijer, H. J. G. ten Hoopen, K. Ch. A. M. Luyben, and K. R. Libbenga. 1989. Growth of a *Catharanthus roseus* Cell Suspension Culture in a Modified Chemostat under Glucose-Limiting Conditions. *Appl. Microbiol. Biotechnol. 30*, 270-275.

I. A. Veliky and D. Rose. 1973. Nitrate and Ammonium as Nitrogen Nutrients for Plant Cells. *Can. J. Bot. 51*, 1837-1844.

D. C. Verma and D. K. Dougall. 1977. Influence of Carbohydrates on Quantitative Aspects of Growth and Embryo Formation in Wild Carrot Suspension Cultures. *Plant Physiol. 59*, 81-85.

G. J. Wagner. 1979. Content and Vacuole/Extravacuole Distribution of Neutral Sugars, Free Amino Acids, and Anthocyanin in Protoplasts. *Plant Physiol. 64*, 88-93.

E. D. Weinberg. 1974. Secondary Metabolism. Control by Temperature and Inorganic Phosphate. *Dev. Ind. Microbiol. 15*, 70-81.

D. F. Wetherell and D. K. Dougall. 1976. Sources of Nitrogen Supporting Growth and Embryogenesis in Cultured Wild Carrot Tissue. *Physiol. Plant. 37*, 97-103.

T. Wiskich. 1980. Control of the Krebs Cycle. In "The Biochemistry of Plants," Vol. 2, D. D. Davies (ed.), Academic Press, New York, pp. 243-278.

V. Wray, O. Schiel, and J. Berlin. 1983. High Field Phosphorus-31 Nuclear Magnetic Resonance Investigation of the Phosphate Metabolites in Cell Suspension Cultures of *Nicotiana tabacum. Z. Pflanzenphysiol. 112*, 215-220.

Y. Yamada. 1984. Selection of Cell Lines for High Yields of Secondary Metabolites. In "Cell Culture and Somatic Cell Genetics of Plants," Vol 1, I. K. Vasil (ed.), Academic Press, New York, pp. 629-636.

T. Yamaya and A. Oaks. 1987. Synthesis of Glutamate by Mitochondria - An Anaplerotic Function for Glutamate Dehydrogenase. *Physiol. Plant. 70*, 749-756.

M. H. Zenk. 1980. Enzymatic Synthesis of Ajmalicine and Related Indole Alkaloids. *J. Nat. Prod. 43*, 438-451.

M. H. Zenk, H. El-Shagi, T. Arens, J. Stockigt, E. W. Weiler, and B. Deus. 1976. Formation of Indole Alkaloids Serpentine and Ajmalicine in Cell Suspension Cultures of *Catharanthus roseus*. In "Plant Tissue Culture and Its Biotechnological Applications," W. Barz, E. Reinhard, and M. H. Zenk (eds.), Springer-Verlag, New York, pp. 3-16.

M. H. Zenk, M. Rueffer, M. Amann, and B. Deus-Neumann. 1985. Benzylisoquinoline Biosynthesis by Cultivated Cells and Isolated Enzymes. *J. Nat. Prod. 48*, 725-738.

CHAPTER 3

QUANTIFYING GROWTH AND PRODUCT SYNTHESIS: KINETICS AND STOICHIOMETRY

A. INTRODUCTION

In Chapter 2, we considered plant cell metabolism from a mechanistic and biochemical perspective. By integrating these various reactions in a coordinated manner, cells are able to grow and differentiate. Mathematically, it is a formidable task to attempt to describe overall culture growth in terms of the individual biochemical reactions. Thus, empirical models have been developed to quantitatively describe growth and biosynthesis in terms of measurable parameters. In this chapter, we present some of the quantitative approaches which have been used to describe plant cells. Our goal is to provide the kinetic and stoichiometric framework which we believe is helpful in analyzing culture performance.

Typically, kinetic and stoichiometric approaches consider a simple reaction to represent substrate consumption, cell growth, and product synthesis:

$$\text{Substrates} \rightarrow \text{Cells} + \text{Products} \qquad [3.1]$$

In our discussion, a substrate is any component taken up by the cells and converted into different compounds (i.e., cell components or products).

Kinetics describes the rate at which (i.e., how fast) reaction [3.1] occurs. Thus, kinetic models consider the reaction in terms of time.

In contrast, stoichiometry does not consider the rate of reaction [3.1], but rather considers only the relative amounts of substrates, cells, and products which are consumed or produced from the reaction. A stoichiometric quantity which is commonly used to describe plant cell performance is the product concentration expressed as a fraction of the cell mass (this is the yield commonly used by plant biologists). It is important to recognize the inherent limitations of this measurement. As a stoichiometric quantity, the number of milligrams of product per of gram cell provides no information on the time required to produce the product nor can the absolute amount of product be determined unless the cell mass is also known. This is illustrated by the fact that the amount of product per gram of cell can increase in the absence of biosynthesis if cell lysis occurs (i.e., the cell mass, or the denominator in this relationship, decreases). Also, a constant amount of product per gram of cell could result if there was no growth and no production or if growth and production occurred at the same relative rate. Thus, although the amount of product per gram of cell is commonly used as a measure of production - especially for intracellular products - great care must be used in interpreting results based on this measurement.

B. ENGINEERING CONSIDERATIONS

B.1. Why Use Quantitative Descriptions?

The simplest reason for quantifying data is to obtain a concise description of the culture's behavior. In this sense, modeling is a means of organizing data and answering questions such as: does production follow growth, or do these processes occur simultaneously? At another level of complexity, modeling is used to provide a quantitative measure of culture behavior. Using such measures as doubling times or production rates, it is possible to quantitatively compare the performance of different cultures. Quantitative measures are also important for design. Knowledge of a culture's oxygen requirements is important if a bioreactor is to be designed to match the culture's oxygen demand. At an even greater level of complexity, models attempt to incorporate knowledge of the underlying physical, chemical, and biological phenomena. Currently, however, our understanding of plant cell physiology is rather limited, and therefore models relating key intracellular events to growth or product synthesis will not be discussed.

B.2. Measures of Bioprocess Performance

Engineers typically use three measures for assessing a biological process: productivity, maximum product concentration, and product yield. When bioreactor capital and operating costs are most significant (as is likely to be the case with plant cell processes), productivity is the most important measure of process performance. The productivity is a measure of the product synthesized per unit time. Productivity, or production rates, can be measured in a few ways. A scientist comparing the performance of two strains may consider the specific productivity (mg product/g cell-day) while a process engineer measures volumetric productivity as the amount of product obtained per unit reactor volume per total cultivation time (mg product/L-day). Since productivity describes the rate of production, productivities are kinetic parameters.

The second measure of a biological process is the maximum product concentration. Since product recovery is facilitated by increasing product concentrations, the final product concentration is the critical measure when product recovery costs dominate the economics of the process.

The third measure of a culture's performance is the product yield. Here the engineers differ from plant biologists in the definition of yield. While the plant biologist typically thinks of the yield as an amount of product per unit of cell mass, the engineer typically uses the yield to represent the amount of product produced per unit of substrate consumed. Although in both cases the yield is a stoichiometric parameter, the yield used by the engineer is a measure of the conversion efficiency of the culture with respect to substrate utilization. This yield is important when the nutrient costs dominate the process economics. It is unlikely that nutrient costs will dominate for plant cell processes.

B.3. Material Balance

To analyze bioreactors, engineers use material balances to relate changes in concentrations to feed and effluent flow rates and to utilization and production rates. In a general sense, a material balance can be written as:

$$\text{Accumulation} = \text{Inlet Flow} - \text{Exit Flow} + \text{Production} \qquad [3.2]$$

where the accumulation term accounts for changes with time. In this chapter, we will consider only batch operations (i.e., operations where there are no inlet and exit flow streams). For batch operations, the first two terms on the right-hand side of the above equation are equal to zero and the material balance simplifies to:

$$\text{Accumulation} = \text{Production} \quad [3.3]$$

A material balance can be written for each chemical species and for cells. When the total liquid volume does not change, then the material balances for substrate (S), product (P), and cells (X) are:

$$\frac{dS}{dt} = -r_S \quad [3.4a]$$

$$\frac{dP}{dt} = r_P \quad [3.4b]$$

$$\frac{dX}{dt} = r_X \quad [3.4c]$$

where t is time. As shown in the above equations, for batch operations, the net rate of product synthesis (r_P), cell growth (r_X), and substrate utilization (r_S) can be directly related to the change in concentration of these species. In all of the above equations, concentrations are expressed as a mass or molar quantity per unit volume. It should be noted that the rate terms refer to net rates. If, for instance, significant product degradation occurs, it may be useful to express r_P as the net effect of a biosynthetic rate (r_B) and a degradation rate (r_D):

$$r_P = r_B - r_D \quad [3.5]$$

C. QUANTIFYING GROWTH AND GROWTH RELATED PROCESSES

C.1. Kinetics

In this chapter, we will consider only growth of unorganized cells. Kinetic descriptions of organ growth require knowledge of organ anatomy and will therefore be discussed in later organ culture chapters. Figure 3.1 shows a typical pattern for the growth of unorganized plant cells. Characteristically, a lag phase is followed by a phase of rapid cell growth and then a stationary phase (Wilson et al. 1971; Nash and Davies 1972). When cultures are frequently transferred, and sufficient inoculum concentrations are used, the lag phase may be short or indistinguishable. From a

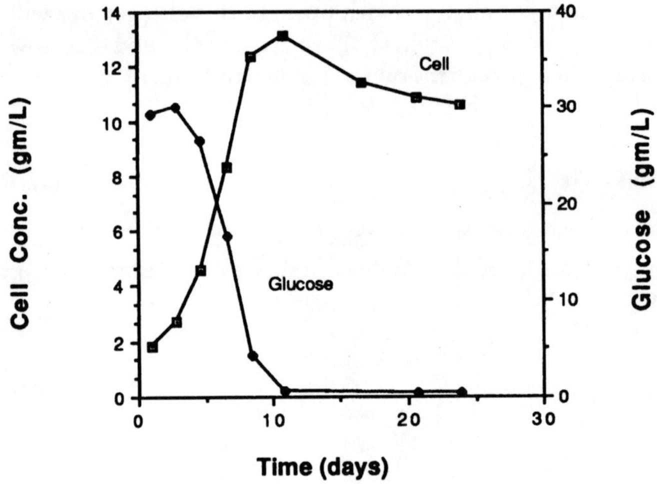

Figure 3.1. Cell and glucose concentrations as a function of time for the growth of an undifferentiated culture of *Catharanthus roseus*. Data from Payne et al. (1988).

mathematical modeling point of view, the lag phase is not tractable because changes in readily measurable parameters (i.e., cell and substrate concentrations) are not observed, while the critical metabolic events which are occurring during the lag phase are not understood well enough to permit relationships to be established between observable phenomena (e.g., secretion and uptake of solutes) and subsequent growth. Thus, mathematical descriptions of growth do not generally include consideration of the lag phase. Similarly, the stationary phase is often not considered in mathematical descriptions of growth. However, since the stationary phase is often the period in which significant amounts of secondary metabolites are produced, mathematical models describing metabolite synthesis during the stationary phase are important and will be considered in a subsequent section.

Most mathematical models describing culture growth are confined to the period of active growth. The most familiar model is the exponential growth model. Fundamentally, exponential growth is expected when cell division results in the formation of daughter cells which are each equally capable of further division. Although exponential growth will result when each cell further divides, exponential growth will also result if only a fraction (but a constant fraction) of the cells divide. Many plant cell cultures (including the one shown in Figure 3.1) have been observed to grow exponentially. Exponential growth is modeled by the kinetic equation

$$r_X = \mu X \qquad [3.6]$$

where X is cell concentration and μ is the specific growth rate in units of day^{-1}. As shown in equation [3.6], the rate at which the culture grows can be expressed in two ways. Scientific studies often focus on μ, which is the specific growth rate and provides information on how rapidly growth occurs on a per cell basis (e.g., sometimes the units of μ are given as grams of cells produced per gram of cell per day). In contrast, process engineers are more interested in r_x, or the volumetric rate of cell growth, which expresses the amount of cell mass generated per reactor volume per time. As can be seen from equation [3.6], r_x is directly proportional to X, and therefore to increase the volumetric rate of cell growth, it is desirable to operate at higher cell concentrations.

Combining equation [3.6] with the material balance (equation [3.4c]) and integrating leads to the familiar semilogarithmic expression

$$\ln \frac{X}{X_o} = \mu\,(t - t_o) \qquad [3.7]$$

where X_o is the cell concentration at the time t_o when exponential growth begins. As suggested by equation [3.7], a semilogarithmic plot of X versus t is linear for an exponentially growing culture, and the slope of this semilogarithmic plot is the specific growth rate, μ. When the cell concentration data from Figure 3.1 are replotted on a semilogarithmic graph (Figure 3.2), it can be seen that the culture grew exponentially until the eighth day, after which growth deviated from this pattern. The specific growth rate for this culture over the eight-day period was calculated from the slope of Figure 3.2 to be 0.26 day-1. The specific growth rate can be used to calculate the culture's doubling time (td) by the equation

$$t_d = \frac{\ln 2}{\mu} = \frac{0.693}{\mu} \qquad [3.8]$$

The specific growth rate of 0.26 day_{-1} corresponds to a doubling time of 2.7 days. Typical doubling times for plant cell cultures range from 1.5 to 4 days (Nash and Davies 1972; Wilson 1976).

In plant cell systems, growth is often reported in terms of the growth index, which is the ratio of the final to the initial cell concentration:

$$\text{Growth Index} = \frac{X}{X_o} \qquad [3.9]$$

Average doubling times have also been used to describe plant cell growth. These average doubling times are calculated based on some initial and some final cell concentration and may include lag and stationary phases.

Growth rates as well as other metabolic rates generally depend on the nutritional environment which the cells experience. It is common that the substrate dependence of cell metabolism is approximately described by mathematical models originally derived for enzyme kinetics. Rates of growth, nutrient uptake, respiration, and me-

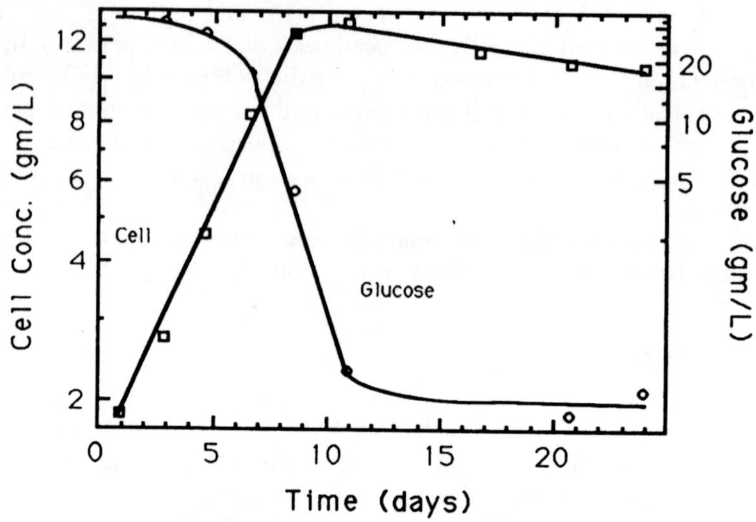

Figure 3.2. A semilogarithmic plot of *C. roseus* growth and sugar consumption as a function of time. The linearity of this curve between days 0 and 8 indicates that the culture was growing exponentially. Data from Payne et al. (1988).

tabolite formation are often described by simple Michaelis-Menten kinetics. While V is used to denote the reaction rate in Michaelis-Menten kinetics, μ is used to denote specific growth rates and q is used to denote nutrient uptake rates (e.g., see equation [5.1] for oxygen uptake). Clearly K_s has a somewhat different meaning depending on which process is being described. Here we consider the kinetics of growth as described by:

$$\mu = \frac{\mu_{max} S}{K_S + S} \qquad [3.10]$$

where μ and μ_{max} are the specific and maximum specific growth rates in units of day^{-1}, S is the substrate concentration, and K_s is a constant with units of g/L or mol/L. As shown in Figure 3.3, there are two extremes for the Michaelis Menten form. At high substrate concentrations, equation [3.10] reduces to

$$\mu \approx \mu_{max} \quad \text{when } S >> K_S \qquad [3.11]$$

If the analogy with enzyme kinetics is valid, equation [3.11] corresponds to the condition that the cells are "saturated" with substrate. Under this condition, the culture's growth rate is not limited by the availability of nutrients, but rather is limited by the biosynthetic abilities of the cell. Typically, cells experience such nutrient-nonlimiting conditions during the exponential growth phase in a batch culture. Thus, the specific growth rate determined from Figure 3.2 corresponds to the μ_{max} of the culture for the conditions studied. The maximum specific growth rate, μ_{max}, may change with changes in temperature, pH, and hormone levels.

The second extreme predicted by the Michaelis-Menten equation is the low-substrate regime where the culture's growth is reduced due to a limitation in the substrate. Under this condition, equation [3.10] reduces to:

$$\mu \approx \frac{\mu_{max}}{K_S} S \quad \text{when } S << K_S \tag{3.12}$$

This low concentration extreme becomes relevant for batch cultures as the cells approach the stationary phase. As substrates are depleted in batch culture, the growth rate decreases below μ_{max}. However, since the substrate concentration does not remain constant but rather decreases in batch operation, the growth rate μ often decreases rapidly as the culture leaves the exponential growth phase and approaches the stationary phase.

In addition to μ_{max}, the other parameter in equation [3.10] is K_S. K_s is a measure of the affinity of the culture for substrate S and is the substrate concentration at which the growth rate is half of μ_{max}. Although equation [3.10] can be used to describe the substrate dependencies for several nutrients, K_s is specific to the specific nutrient in question.

If mass transfer limitations are important, then equation [3.10] is often modified such that an apparent K_s is used. This apparent K_s accounts both for the affinity of the cells for substrate and for mass transfer limitations. When mass transfer limitations occur, the apparent K_s is larger than the intrinsic K_s of the cells (i.e., the apparent affinity of the culture is reduced). As will be shown in Chapter 4, it is possible to account for mass transfer effects in a more rigorous manner.

Another measure of the affinity of a culture for a nutrient is the critical concentration. As shown in Figure 3.3, above the critical concentration, cell growth is unaf-

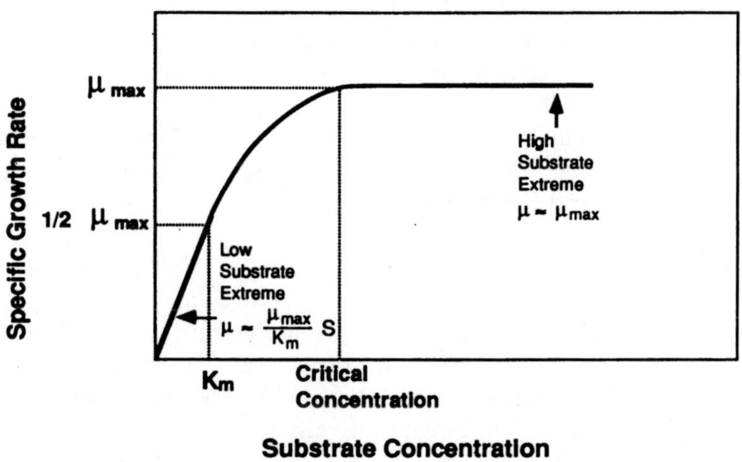

Figure 3.3. The Michaelis-Menten form used to characterize the substrate dependence of growth.

fected by the nutrient level (i.e., the cells are "saturated"). The critical concentration is typically two to four times K_S. The critical concentration will be further discussed in relation to oxygen (see Chapter 5).

It should be noted that describing the kinetics of growth using enzyme kinetic models is not rigorously valid. However, enzyme kinetic models have been observed to perform at least as well as other empirical models and thus they are commonly used. In addition to Michaelis-Menten, other enzyme models have been used to describe the effects of medium components on cell growth. For instance, an inhibition model may be appropriate for describing the adverse effects of auxins on secondary metabolite formation

Although Figure 3.1 shows a typical growth curve for plant cells, it should be recognized that deviations from this pattern are not unusual. In addition to exponential growth, linear growth can be observed for the growth of a single rootlet (see Chapter 8) and when growth is limited by the rate of transfer of a critical nutrient such as oxygen (Kato et al. 1975). Quantitatively, linear growth is described by the equation

$$r_X = K \qquad [3.13]$$

where K is a constant and has units of g cells/L-day. Also, a gradual transition between the rapid growth and stationary phases is often observed (Wilson et al. 1971). Once growth ceases, it is possible that the cell concentration will not remain constant, but rather the cell concentration may decrease after rapid growth has ceased. Such a decrease in cell concentration is often termed a death phase.

Finally, it should be noted that the dynamics of growth will depend on how growth is measured (e.g., dry weight, wet weight, or cellular nitrogen) (Wilson et al. 1971). In particular, a phase of cell division in which cell size remains constant may be followed by a phase of cellular elongation in which the cell number may remain constant and the cell size increases. Thus, what are called the exponential and the stationary phase would depend on whether cell number or total cell mass is measured. Also, the dynamics of growth may change over time depending on the subculture routine (Morris et al. 1989).

When cultivation modes other than batch are used, the material balance becomes more complicated since the flow rate terms in equation [3.2] cannot be neglected (for example, see Westgate and Emery 1990).

C.2. Stoichiometry

Stoichiometry provides a framework to quantitatively relate substrate consumption to growth and product formation. For example, when a culture is growing exponentially, it is logical to expect the substrate concentration to be decreasing exponentially. This behavior is illustrated in Figures 3.1 and 3.2. Stoichiometry provides a framework for expressing this intuitive knowledge quantitatively. In the fermentation and animal cell culture industries, stoichiometric analysis has been useful for gaining insights into metabolism and for bioprocess monitoring and control. We believe that stoichiometric analysis could be usefully applied to understand plant cell systems.

C.2.1. Empirical Description of Carbohydrate Conversion. Most frequently, stoichiometric analysis is limited to describing the conversion in reaction [3.1] in terms of a single substrate (i.e., the carbon and energy source, glucose) and a single product (i.e., cells) during a single phase of the growth curve (i.e., the period of rapid growth). Despite limiting consideration to a single reactant and a single product, reaction [3.1] suggests that there is a quantitative relationship between growth and sugar consumption. This relationship is often expressed by engineers as a growth yield ($Y_{X/S}$) by:

$$Y_{X/S} = -\frac{r_X}{r_S} \qquad [3.14]$$

For batch operations, equation [3.14] can be expressed in terms of the changes in cell concentration (ΔX) and changes in substrate concentration (ΔS) over a finite time period by:

$$Y_{X/S} = -\frac{\Delta X}{\Delta S} \qquad [3.15]$$

The negative sign in the above two equations is used because although the cell concentration is increasing (ΔX is positive), the substrate concentration is decreasing (ΔS is negative).

For the exponential growth phase of the culture shown in Figure 3.1, the growth yield can be calculated in two ways. First, the yield can be calculated from the ratio of the change in cell concentration to the change in substrate concentration at any given time. For these calculations, it is convenient to choose the time when exponential growth ceases (e.g., 8 days). The amount of glucose consumed at that time was 25 g/L while 10.5 g cell/L were generated up to that time (note that the 1.8 g cell/L in the inoculum is subtracted from the cell concentration on day 8). From these values, the growth yield is calculated to be 0.4 g cell/g glucose consumed. An alternate approach for calculating the cell yield is to plot the cell concentration as a function of the substrate concentration as shown in Figure 3.4. The slope of this plot is the growth

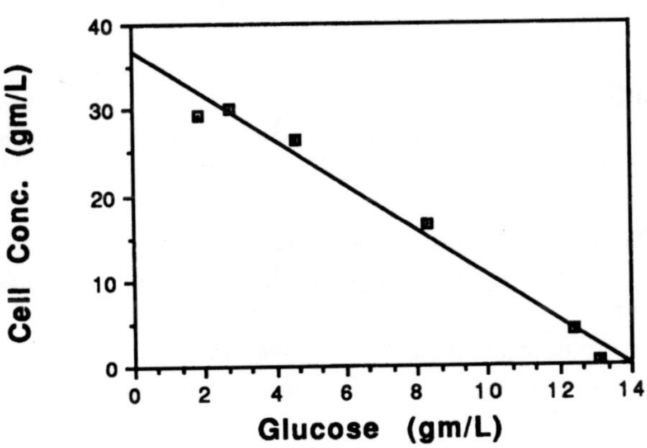

Figure 3.4. Plot of cell versus substrate concentration for determination of the growth yield.

yield and is observed to be 0.4 g cell/ g glucose consumed. It can be seen that yields calculated from either method are in good agreement.

Growth yields for plant cells have been observed within a range of about 0.3 - 0.7 g cell/g sugar with a value of 0.5 g cell/g sugar being common (Kato et al. 1975; Pareilleux and Vinas 1983; Hashimoto and Azecki 1988; van Gulik et al. 1989). These values are also similar to those observed in microbial systems. Higher yields have been reported when photosynthetic processes contribute to the energetics of the system (Dalton 1984).

Calculations of growth yields are useful for comparing the metabolic efficiency with which a culture converts substrates to cell mass. For instance, growth yields will be considered in Chapter 5 when we discuss the effects of anaerobic conditions on growth. Also, if the growth yield is known, and this yield does not change over time, then cell growth can be monitored simply by monitoring substrate consumption. Park et al. (1989) used this approach to monitor growth in shoot cultures, while Toivonen et al. (1990) used this approach to measure the growth of hairy roots. Because of the difficulty in obtaining representative measurements of shoot mass, the yield and equation [3.15] were used to relate substrate consumption to shoot growth.

C.2.2. Relating Growth to Nitrogen and Phosphorus Uptake. In the above discussion, we have focused on growth yields based on the carbohydrate. Accounting for how the cell uses the carbohydrate to form cell mass is complicated by the fact that some of the carbon from this substrate is incorporated into cell mass while some is evolved as carbon dioxide by the catabolic processes required to generate ATP. For substrates like the nitrogen and phosphorus sources, relating measured growth yields to cell physiology is simplified since these substrates are taken up and incorporated directly into cell mass (i.e., none is lost in respiration). Thus, the cell yield for the nitrogen source expressed as grams of cells produced per gram of nitrogen consumed is the inverse of the cellular nitrogen content. Similarly, the phosphorus taken up by the cell becomes cellular phosphorus, and therefore the growth yield on a gram of phosphorus basis is the inverse of the cellular phosphorus content.

Although it seems reasonable to expect the cellular composition of nitrogen and phosphorus (and therefore their respective growth yields) to be constant, Dougall et al. (1983) observed that these yields can vary depending on the culture conditions. With plant cells, it is common for some nutrients (e.g., phosphate) to be rapidly taken up from the medium, stored within the cell, and slowly incorporated into organic cell mass. In these cases, stoichiometric and kinetic models which attempt to relate growth and metabolism to extracellular nutrient concentrations will not be useful.

C.2.3. Potential Pitfalls to Stoichiometric Analysis. When using growth yields, there are a few potential problems which must be considered. Theoretically, a single stoichiometric equation may not be valid for describing the overall metabolic processes for various phases of the growth cycle. For instance, when cultures are growing exponentially, limited amounts of nutrients are diverted to the production of secondary metabolites and thus the growth yield is typically high. When cultures enter a stationary phase and begin devoting metabolic precursors and energy to secondary metabolite biosynthesis, then the efficiency with which substrates are converted into cell mass is reduced. The growth yield is also observed to be reduced during this time.

Experimental determination of growth yields can be prone to error when determination of either ΔX or ΔS is imprecise. Since cell concentration is often difficult to measure reproducibly, it is important that significant differences in the cell concentration are measured. Also, calculations are often confined to the rapid growth phase to eliminate problems associated with the loss of cells due to lysis. Measuring ΔS can be a problem especially when calculations are based on a sugar like sucrose which is rapidly hydrolyzed to fructose and glucose and these monosaccharides are more slowly metabolized. For such a case, the large disappearance of sucrose with a corresponding small increase in cell mass will lead to calculated yields which are low. Also, when nutrients (e.g., phosphates) are rapidly taken up and stored within the cells, calculated growth yields may vary greatly.

C.2.4. Stoichiometric Models Based on Catabolic and Anabolic Processes. The approach of empirically relating cell growth to sugar consumption is simple, and rules of thumb are available concerning the efficiency of culture growth (e.g. $Y_{X/S} \approx 0.5$ g cell/g sugar consumed). However, this simple approach fails to incorporate any knowledge of cellular metabolism into the stoichiometric description. Most attempts to incorporate metabolic information into stoichiometric descriptions have considered metabolism in two parts: catabolism and anabolism. In anabolism, the biosyntheis of macromolecules requires both intermediates (e.g. amino acids and nucleotides) and energy in the form of ATP. These energy demands are met by catabolic processes. By accounting for all anabolic and catabolic requirements and using knowledge of biochemical pathways, it is possible to calculate how much glucose is required to support a given amount of growth. Further discussion of these approaches can be found in Penning de Vries et al. (1974), Atkinson (1977), Stouthamer (1979), Roels (1983), and Harold (1986). A general conclusion from these approaches has been that cells have an additional requirement for maintenance energy. Maintenance energy is required for such functions as protein turnover and to maintain concentration gradients across membranes.

Attempts to quantify maintenance energy requirements were made by Pirt (1965). For the case of glucose being consumed as both the carbon and the energy source, and when product synthesis is negligible, then the total rate of consumption of this substrate (r_S) can be considered to result from three processes: consumption for incorporation into cell matter during growth (r_{SG}); consumption to provide the energy for growth processes (r_{SEG}); and consumption for maintenance energy (r_{SM}):

$$r_S = r_{SG} + r_{SEG} + r_{SM} \qquad [3.16]$$

For an exponentially growing culture, the total rate of sugar consumption can be related to the growth rate of the culture by equations [3.14] and [3.6] such that

$$r_S = -\frac{1}{Y_{x/s}}\mu X \qquad [3.17]$$

where $Y_{X/S}$ is the overall, or observed, cell yield.

The first two terms on the right-hand side of equation [3.16] are the carbon and energy requirements for cell growth. Thus, this sum represents the actual glucose required for growth processes and can be combined as

$$r_{SG} + r_{SEG} = -\frac{1}{Y_T}\mu X \qquad [3.18]$$

where Y_T is the true growth yield.

Finally, the rate of energy substrate consumed per gram of cell for maintenance functions is considered to be constant, and therefore

$$r_{SM} = -mX \qquad [3.19]$$

where m is the maintenance coefficient.

Combining equations [3.16] - [3.19] and rearranging leads to

$$\frac{1}{Y_{X/S}} = \frac{1}{Y_T} + \frac{m}{\mu} \qquad [3.20]$$

The usefulness of this approach is twofold: (1) it provides a means to experimentally estimate the energy required for processes not directly related to growth (e.g., maintenance); and (2) it has been observed that Y_T and m remain relatively constant over a wider range of conditions than $Y_{X/S}$, and thus this maintenance model is probably better able to describe growth and nutrient consumption over a wider range of conditions.

Efforts to exploit these theories to quantify the metabolism of plant cells in culture have been limited. However, there is a growing body of literature concerning the use of these techniques to describe whole plant growth (see Thornley 1976; Johnson 1990). For instance, Penning de Vries has attempted to theoretically relate the stoichiometry of plant growth to the carbon and energy requirements for the various metabolic processes (Penning de Vries et al. 1974) and to estimate the maintenance requirements in terms of protein turnover and active transport processes (Penning de Vries 1975). Experimental estimates of such parameters as true growth yields and maintenance coefficients for plant growth have also been reported, and there appears to be reasonably good agreement between theoretical approaches and experimental observations. Theoretical predictions (Penning de Vries 1975) and experimental observation (Amthor 1984) suggest that the maintenance coefficients for plants are significantly less than maintenance coefficients observed for bacterial cultures. This may be due to the lower rates of growth and protein turnover in plants (Thornley 1970). However, it is interesting to note that the maintenance requirements are significant - Amthor (1986) estimates that 0.4 - 9.3% of the carbon of the plant or organ must be respired each day to meet its maintenance energy requirement.

In summary, these more advanced methods for relating the stoichiometry of growth to the underlying metabolic processes have been developed for microbial systems and have been adapted to describe whole plant growth. Experimental studies have shown significant similarities between these two different biological systems. This similarity is probably due to the fact that the underlying biochemistry is similar for the two systems. The application of these techniques to interpret the growth of plant cells has not been actively pursued.

C.2.5. Stoichiometry for Process Monitoring and Control. As studies on biological production processes move from flasks to bioreactors and ultimately to commercial operations, there is greater emphasis placed on monitoring and control. By monitoring more parameters (e.g., respiration rates and respiratory quotients), it is possible to obtain a better understanding of cell metabolism. Also, correlations can be established between measured results and culture performance, and thus these measurements can be used as an indication of the condition of the culture. For instance, Hahlbrock et al. (1974) and Ryu et al. (1990) observed that a decrease in conductivity of the medium could be correlated to nitrate uptake and plant cell growth. Because the conductivity measurement is fast and can be automated, this measure is better suited for process monitoring and control than the measurement of dry cell concentrations. If the growth of a culture has been disrupted for whatever reason, this can be rapidly detected by conductivity measurements, and thus this measurement can be used to signal the need for some remedial (i.e., control) action. Stoichiometry is also commonly used for process monitoring and control.

When stoichiometry is used to monitor and control, relationships between reactants and products are not based on information on plant cell biochemistry, but rather on material balances. The basis for this approach is the recognition that although biological systems can convert matter and energy into different forms, a net increase in either matter or energy is precluded by the laws of thermodynamics. Thus, for a culture growing on glucose as the carbon and energy source and ammonia as the nitrogen source, cell metabolism can be described by the stoichiometric equation

$$\underset{\text{(Glucose)}}{\alpha C_6H_{12}O_6} + \beta NH_3 + \delta O_2 \rightarrow \underset{\text{(Cells)}}{\varepsilon C_wH_xN_yO_z} + \underset{\text{(Extracellular Product)}}{\sigma C_sH_tN_uO_v} + \tau CO_2 + \phi H_2O \quad [3.21]$$

In the above equation, the Greek symbols represent the stoichiometric coefficients that are required to ensure that the reaction is balanced, $C_wH_xN_yO_z$ represents the molecular formula of the cell, and $C_sH_tN_uO_v$ represents the molecular formula of extracellular product(s).

When material balancing is used, a balance is typically written in terms of the element carbon. For instance, reaction [3.21] shows that for every mole of glucose consumed, the six moles of carbon which are converted must appear in the carbon-containing products. Mathematically, this relationship can be expressed as

$$6\alpha = w\varepsilon + s\sigma + \tau \qquad [3.22]$$

When negligible amounts of extracellular product(s) are produced (i.e. σ = o), equation [3.22] reduces to

$$6\alpha = w\varepsilon + \tau \qquad [3.23]$$

Thus, the amount (or rate) of cell growth (ε) can be directly related to the amounts (or rates) of glucose consumed (α) and carbon dioxide produced (τ). Thus, by measuring off gas carbon dioxide concentrations (either through infrared analysis or mass spectrometry) and by rapidly assaying for sugar consumption, it is possible to solve equation [3.23] for ε and thus determine the growth rate of the culture during the course of cultivation. Dalton (1984, 1985a, b) used this approach with sweet basil cells and showed that the calculated growth agreed well with actual measurements. These indirect methods for determining growth are generally accepted in the fermentation industry.

In addition to performing a material balance around carbon, it is also possible to balance the elements H, O, and N (Cooney et al. 1977; Wang et al. 1977, 1979). Although these techniques have not been exploited in plant cell systems, there has been considerable investigation of these techniques in microbial cultures.

D. QUANTIFYING PRODUCT SYNTHESIS

D.1. Kinetics

When attempting to quantify product synthesis, there are two aspects which need to be considered: the timing of product synthesis relative to growth, and the actual mathematical model used to describe production. With respect to the timing, production is generally related to the growth of the culture.

Products are often classified as being either growth-associated (i.e., growth and production occur simultaneously) or non-growth-associated (i.e., growth and production do not occur simultaneously). As illustrated by the following examples, this operational classification of products greatly oversimplifies cellular metabolism. It has been shown that medium composition (especially the 2,4-D levels) can alter the pattern of secondary metabolite production relative to growth (Davies 1969; Sahai and Shuler 1984). Zenk et al. (1977) observed that indole alkaloid production by *Catharanthus roseus* was growth-associated when the culture was grown in flasks, while production was non-growth-associated when the cells were cultivated in a larger scale bioreactor. Morris et al. (1989) observed that after repeated subcultures, the growth rate of a *C. roseus* culture was increased while alkaloid production was altered from a growth-associated to a non-growth-associated pattern. Finally, characterizing products as growth-associated or non-growth-associated ignores the heterogeneity of the cell population. Stafford et al. (1985) suggested that only a fraction of cells in a culture may be involved in product synthesis. The effect of a heterogeneous population on product synthesis will be further illustrated in Chapter 8 with the synthesis of products by root cultures.

Despite the inaccuracy in classifying products as growth- or non-growth-associated, this classification is convenient–if not essential–for developing appropriate cultivation strategies. For instance, cell growth should be promoted if the product of interest is produced by actively growing cells, whereas a stationary phase should be extended if production occurs after growth has ceased. Further discussion of how the timing of biosynthesis affects bioreactor optimization can be found in Chapter 6 and in Payne et al. (1987).

Mathematically, for a batch operation, the net rate of product synthesis (r_P) can be related to changes in the product concentration by the material balance (equation [3.4b]). A kinetic description of this rate is given by:

$$r_P = \frac{dP}{dt} = q_P X \qquad [3.24]$$

where q_P is the rate of synthesis per gram of cell in units of milligrams of product per gram of cell per day, and X is the cell concentration in grams of cells per liter. As previously mentioned, product degradation is not explicitly accounted for in equation [3.24]. If the product is susceptible to degradation (either through biological or chemical mechanisms), an additional term can be included as shown in equation [3.5].

As mentioned earlier in the chapter, when considering the culture's production rate or productivity, two productivities can be considered. The scientist often attempts to improve productivity by increasing the rate of production per cell (i.e., increasing q_P). This can be accomplished by obtaining an improved cell line or by using more conducive environmental conditions. q_P is often correlated to nutrient concentrations by Michaelis-Menten type equations. However, when the goal is to efficiently produce a product, it is desirable to ensure that the productivity is not being reduced by a limitation in any nutrient. Therefore, it is desirable to ensure that all nutrients are available at concentrations above the critical level (e.g., for analogy, see Figure 3.3). In addition to substrate dependencies, q_P will also depend on temperature, pH, and phytohormone types and concentrations.

Another strategy for improving the volumetric productivity (r_P) would be to operate the bioreactor at higher cell concentrations. Further, a process engineer would be less concerned with either q_P or r_P and more concerned with the overall volumetric productivity - the amount of product produced per reactor volume per total cultivation time. This is especially important for non-growth-associated products where

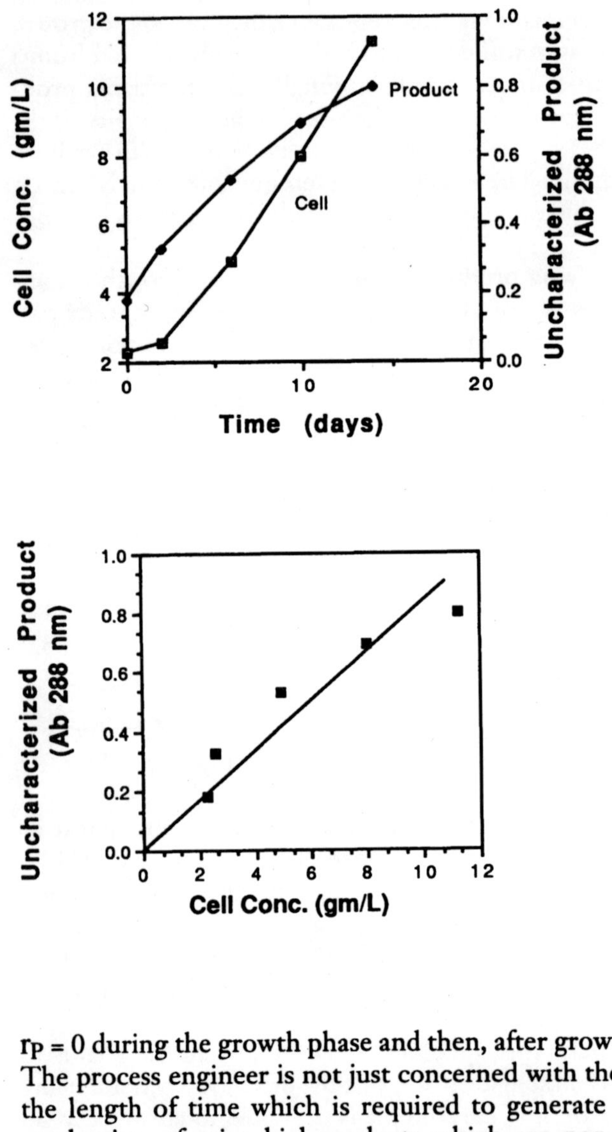

(a)

(b)

Figure 3.5. Growth-associated production of an uncharacterized UV-absorbing material by a carrot cell culture. (a) Growth and production profiles. (b) Differential plot of product synthesis versus growth. From Tate and Payne (unpublished).

$r_P = 0$ during the growth phase and then, after growth ceases, equation [3.24] is valid. The process engineer is not just concerned with the production phase, but also with the length of time which is required to generate the cell mass. Processes for the production of microbial products which are non-growth-associated are generally designed such that cell growth is very rapid initially, and then the cells are maintained in a highly productive state for an extended period.

D.1.1. Growth-Associated Production. As mentioned, products which are produced during active growth are operationally defined as growth-associated. Typically, growth-associated products are compounds which are directly involved in growth processes (e.g., membrane lipids or cell wall components). However, a few plant secondary metabolites such as rosmarinic acid (Razzaque and Ellis 1977) have been observed to be produced during active growth of the culture. (It should be noted that we are not considering the timing of biosynthesis in our definition of secondary metabolites.) To illustrate this pattern of product synthesis, Figure 3.5a shows the growth and production of an uncharacterized UV-absorbing material by a carrot cell culture. As can be seen from Figure 3.5a, during the period in which the culture was growing exponentially, this chromophore accumulated. An alternative approach to examining the timing of product synthesis is to use a differential plot of the product concentration as a function of the cell concentration (Payne et al. 1988). As shown in Figure 3.5b, this plot shows that as the cell concentration is increasing (i.e., the culture is growing), the amount of UV-absorbing material is also increasing. This plot also demonstates growth-associated product synthesis.

For growth-associated products, it is often believed that the faster the culture grows, the more rapid the product is synthesized. For this case, the specific production rate (q_p) would be related to the specific growth rate (μ) by

$$q_p = \alpha \mu \qquad [3.25]$$

where α is the proportionality relating growth and production. If the product was truly growth-associated and was retained within the cell, then α would represents the cellular content of that product.

Alternative mathematical descriptions are also possible. For instance, it has been observed that the fungal secondary metabolite penicillin is produced while the *Penicillium* culture grows. In this case, Righelato and Pirt (1967) observed that the specific productivity (q_p) was constant, independent of the growth rate.

D.1.2. Non-Growth-Associated Production. Many secondary metabolites are produced by cultures which are either not growing or are growing very slowly. As an example, Figure 3.6a shows the production of indole alkaloids by *C. roseus* culture. Rapid growth occurs until day 10, after which growth slows. Shortly before growth slows, on day 7, alkaloid production is observed. The non-growth-associated character of alkaloid production can be seen more clearly from Figure 3.6b, which shows that during cell growth, little alkaloid is produced. However, after growth has ceased, significant levels of production are observed.

As can be seen from Figure 3.6a, during the period of alkaloid biosynthesis, production occurs at a constant rate (i.e., the product concentration increases linearly with time). Thus, mathematical relationships describing production for non-growth-associated products are of the form

$$q_p = \beta \qquad [3.26]$$

where β is a constant in units of mg product/g cell-day. It should be noted that α in equation [3.25] and β in equation [3.26] are not true constants but rather depend on environmental conditions. Also, it should be recognized that the productivity often

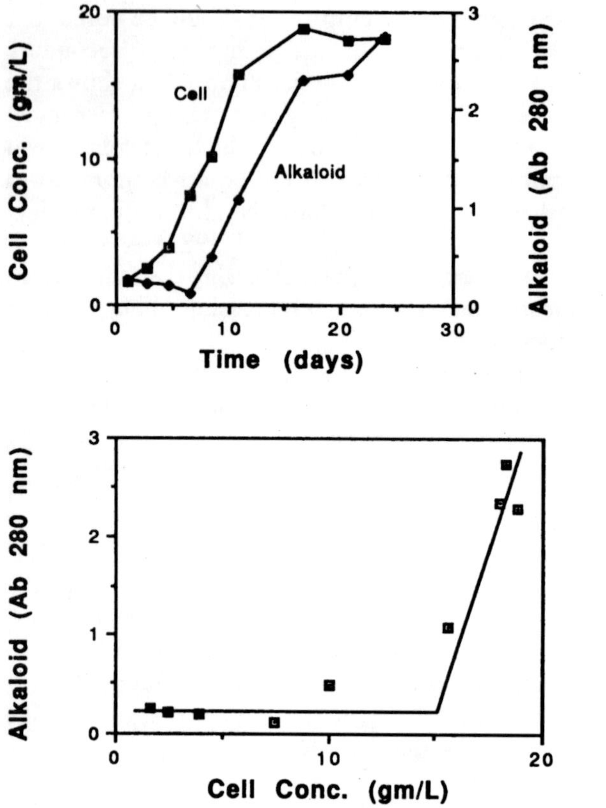

Figure 3.6. Non-growth-associated production of indole alkaloids by *C. roseus* cultures in a growth medium containing high initial glucose concentrations (70 g/L). (a) Growth and production profiles. (b) Differential plot of product synthesis versus growth. From Payne et al. (1988).

decays over extended periods. This productivity decay, which may be due to a loss in secondary metabolic enzymes or a general loss in metabolic activity, is not explicitly accounted for in either of the above two equations.

For non-growth-associated products, the goal for optimizing production is to extend the stationary phase responsible for biosynthesis. Because of the rather extreme non-growth-associated character of many plant products, several groups have used two completely separate media, the first supporting rapid growth, and the second medium supporting much less growth, but considerably greater secondary metabolite production (examples include shikonin, which is discussed in Chapter 6). Figure 3.7 shows a *C. roseus* culture after being transferred from a growth medium into a production medium. In the production medium, growth is slow and nonexponential while alkaloid is produced at a constant rate for an extended period.

D.2. Stoichiometry

Stoichiometry has been used in a few cases (mostly for microbial systems) to relate nutrient consumption and product synthesis. However, we believe stoichiometric

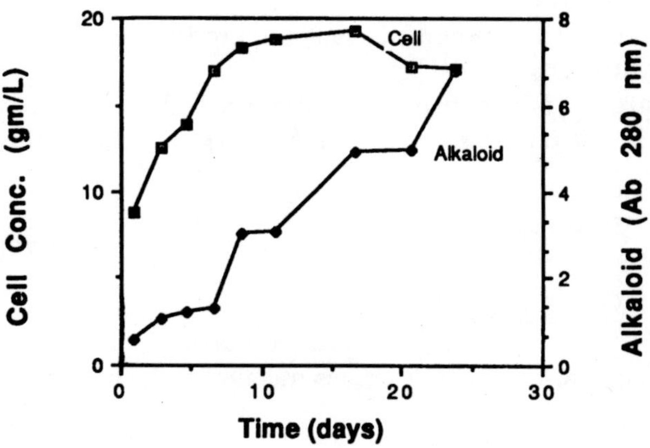

Figure 3.7. Cell and alkaloid levels for *C. roseus* cultures transferred to a production medium lacking phosphate, nitrogen, and the synthetic auxin 2,4-D. Data from Payne et al. (1988).

approaches may be useful for better characterizing the requirements for plant cell secondary metabolite synthesis. For growth-associated products, reaction [3.1] can be used to describe not only the formation of cells, but also the formation of the products. Thus, yields can be calculated based not only on the amount of growth obtained per gram of sugar consumed, but also on the amount of product formed per gram of sugar consumed.

For non-growth-associated products, the above stoichiometric procedure must be altered. As can be seen in Figure 3.6, for non- growth-associated products, cell growth and product synthesis occur at two distinctly different times. Cells at these two times are considerably different, and the overall stoichiometries describing their metabolism are also quite different. Thus, it may be convenient to consider two different stoichiometric equations, the first describing metabolism during the active growth phase and the second describing metabolism during the secondary metabolite-producing phase:

$$\text{Substrates} \rightarrow \text{Cells} \qquad [3.27a]$$

$$\text{Substrates} \rightarrow \text{Products} \qquad [3.27b]$$

When considering carbohydrate consumption, the rate of sugar consumption can be related to the relative amounts of consumption by these individual pathways and the amount of sugar consumed for maintenance energy requirements:

$$-r_S = \frac{1}{Y_T} r_X + \frac{1}{Y_P} r_P + mX \qquad [3.28]$$

where Y_P is the true product yield, or the amount of sugar required to provide the carbon and energy for product synthesis.

During periods when active growth is occurring (i.e., reaction [3.27a] describes metabolism), product synthesis is not occurring and therefore $r_P = 0$. Under this condition, equation [3.28] simplifies to

$$-r_S = \frac{1}{Y_T} r_X + mX \qquad [3.29]$$

which could also be derived from equation [3.16]. When production occurs for a non-growth-associated product (i.e., equation [3.27b] describes metabolism), then equation [3.28] simplifies to

$$-r_S = \frac{1}{Y_P} r_P + mX \qquad [3.30]$$

Theoretically, it should be possible to relate Y_P to the biosynthetic pathways for secondary metabolite production. This approach is limited by the sparse knowledge concerning both the biochemical pathways and the energetic requirements of the individual steps in the pathway. For microbial secondary metabolites, where sufficient pathway information is available to estimate Y_P, the amount of sugar required for biosynthesis is usually small relative to that required to meet the maintenance energy requirements (Cooney and Acevedo 1977; Cooney 1979; Heijnen et al. 1979; Payne and Wang 1989).

Because of the differences between microbes and plant cells, it is questionable whether quantitative similarities of stoichiometric parameters will be observed. However, we believe the general conclusion is valid: that to retain *de novo* biosynthetic activities - even for nongrowing cultures - significant energetic and biosynthetic needs must be met.

In addition to the carbon and energy requirements, which are typically met by the carbohydrates, elemental requirements may also be important if high levels of production are required. For instance, if high levels of sulfur-rich flavor precursors are desired, it is essential (although not necessarily sufficient) for the onion cultures to be supplied with adequate levels of sulfur-containing nutrients.

E. CONCLUSION

In this chapter, we developed kinetic and stoichiometric relationships to provide quantitative tools to assist in analyzing data and to facilitate decision-making. Despite finding examples to illustrate our points, the reader should be cautioned that often data are not so clearly described. For instance, few products can be classified as being strictly growth- or non-growth-associated. However, despite these failings, kinetic and stoichiometric quantities are useful and sometimes essential for the design and optimization of biological processes.

References

J. S. Amthor. 1984. The Role of Maintenance Respiration in Plant Growth. *Plant Cell Environ.* 7, 561-569.

J. S. Amthor. 1986. Evolution and Applicability of a Whole Plant Respiration Model. *J. Theor. Biol.* *122*, 473-490.

D. E. Atkinson. 1977. "Cellular Energy Metabolism and Its Regulation," Academic Press, New York, pp. 31-37.

C. L. Cooney. 1979. Conversion Yields in Penicillin Production: The Theory vs. Practice. *Process Biochem. 14*, 31-33.

C. L. Cooney, H. Y. Wang, and D. I. C. Wang. 1977. Computer-Aided Material Balancing for Prediction of Fermentation Parameters. *Biotechnol. Bioeng. 19*, 55-67.

C. L. Cooney and F. Acevedo. 1977. Theoretical Conversion Yields for Penicillin Synthesis. *Biotechnol. Bioeng. 19*, 1449-1462.

C. C. Dalton. 1984. The Effect of Sugar Supply Rate on Photosynthetic Development of *Ocimum basilicum* (Sweet Basil) Cells in Continuous Culture. *J. Exp. Bot. 35*, 505-516.

C. C. Dalton. 1985a. The On-Line Estimation of Yield and Photosynthetic Efficiency of Photoheterotrophic Plant Cells (*Ocimum basilicum*). *J. Exp. Bot. 36*, 1008-1016.

C. C. Dalton. 1985b. Application of Gas Analysis to Continuous Culture. In "Primary and Secondary Metabolism of Plant Cell Cultures," K.-H. Neumann, W. Barz, and E. Reinhard (eds.), Springer-Verlag, New York, pp. 58-65.

M. E. Davies. 1969. Polyphenol Synthesis in Cell Suspension Cultures of Paul's Scarlet Rose. *Planta 104*, 50-65.

D. K. Dougall, S. LaBrake, and G. H. Whitten. 1983. The Effects of Limiting Nutrients, Dilution Rate, Culture pH, and Temperature on the Yield Constant and Anthocyanin Accumulation of Carrot Cells Grown in Semicontinuous Chemostat Cultures. *Biotechnol. Bioeng. 25*, 569-579.

K. Hahlbrock, J. Ebel and A. Oaks. Determination of Specific Growth Stages of Plant Cell Suspension Cultures by Monitoring Conductivity Changes in the Medium. *Planta 118*, 75-84.

F. M. Harold. 1986. "The Vital Force: A Study of Bioenergetics," W. H. Freeman and Company, New York, pp. 158-163.

T. Hashimoto and S. Azecki. 1988. Bioreactors for the Large Scale Culture of Plant Cells. In "Biotechnology in Agriculture and Forestry," Vol. 4, Y. P. S. Bajaj (ed.), Springer-Verlag, Berlin, pp. 104-122.

J. J. Heijnen, J. A. Roels, and A. H. Stouthamer. 1979. Application of Balancing Methods in Modeling the Penicillin Fermentation. *Biotechnol. Bioeng. 21*, 2175-2201.

I. R. Johnson. 1990. Plant Respiration in Relation to Growth, Maintenance, Ion Uptake and Nitrogen Assimilation. *Plant Cell Environ. 13*, 319-328.

A. Kato, Y. Shimizu, and S. Nagai. 1975. Effect of Initial KLa on the Growth of Tobacco Cells in Batch Culture. *J. Ferment. Technol. 53*, 744-751.

P. Morris, K. Rudge, R. Cresswell, and M. W. Fowler. 1989. Regulation of Product Synthesis in Cell Cultures of *Catharanthus roseus*.V. Long-Term Maintenance of Cells on a Production Medium. *Plant Cell Tissue Organ Culture* 17, 79-90.

D. T. Nash and M. E. Davies. 1972. Some Aspects of Growth and Metabolism of Paul's Scarlet Rose Cell Susepnsions. *J. Exp. Bot. 23*, 75-91.

A. Pareilleux and R. Vinas. 1983. Influence of the Aeration Rate on the Growth Yield in Suspension Cultures of *Catharanthus roseus* (L.) G. Don. *J. Ferment. Technol. 61*, 429-433.

J. M. Park, W.-S. Hu, and E. J. Staba. 1989. Cultivation of *Artemisia annua* L Plantlets in a Bioreactor Containing a Single Carbon and a Single Nitrogen Source. *Biotechnol. Bioeng. 34*, 1209-1213.

G. F. Payne and H. Y. Wang. 1989. The Effect of Feedback Regulation and In Situ Product Removal on the Conversion of Sugar to Cycloheximide by Streptomyces griseus. *Arch. Microbiol. 151*, 331-335.

G. F. Payne, M. L. Shuler, and P. Brodelius. 1987. Large Scale Plant Cell Culture. In "Large Scale Cell Culture Technology," B. K. Lydersen (ed.), Hanser Publishers, New York, pp. 193-229.

G. F. Payne, N. N. Payne, and M. L. Shuler. 1988. Bioreactor Considerations for Secondary Metabolite Production from Plant Cells: Indole Alkaloids from *Catharanthus roseus*. *Biotechnol. Bioeng. 31*, 905-912.

F. W. T. Penning de Vries. 1975. The Cost of Maintenance Processes in Plant Cells. *Ann. Bot. 39*, 77-92.

F. W. T. Penning de Vries, A. H. M. Brunsting, and H. H. van Loar. 1974. Products, Requirements and Efficiency of Biosynthesis: A Quantification Approach. *J. Theor. Biol. 45*, 339-337.

S. J. Pirt. 1965. The Maintenance Energy of Bacteria in Growing Cultures. *Proc. R. Soc. London, Ser. B, 163*, 224-231.

A. Razzaque and B. E. Ellis. 1977. Rosmarinic Acid Production in *Coleus* Cell Cultures. *Planta 137*, 287-291.

R. C. Righelato and S. J. Pirt. 1967. Effect of Growth Rate on the Synthesis of Penicillin by *Penicillium chrysogenum* in Batch and Chemostat Cultures. *Appl. Microbiol. 15*, 1284-1290.

J. A. Roels. 1983. "Energetics and Kinetics in Biotechnology," Elsevier Biomedical Press, New York, pp. 99-128.

D. D. Y. Ryu, S. O. Lee, and R. J. Romani. 1990. Determination of Growth Rate for Plant Cell Cultures: Comparative Studies. *Biotechnol. Bioeng. 35*, 305-311.

O. P. Sahai and M. L. Shuler. 1984. Environmental Parameters Influencing Phenolics Production by Batch Cultures of *Nicotiana tabacum*. *Biotechnol. Bioeng. 26*, 111-120.

A. Stafford, L. Smith, and M. W. Fowler. 1985. Regulation of Product Synthesis in Cell Cultures of *Catharanthus roseus* (L) G. Don. *Plant Cell Tissue Organ Culture 4*, 83-94.

A. H. Stouthamer. 1979. The Search for Correlation between Theoretical and Experimental Growth Yields. *Microb. Biochem. 21*, 1-47.

J. H. M. Thornley. 1970. Respiration, Growth and Maintenance in Plants. *Nature 227*, 304-305.

J. H. M. Thornley. 1976. "Mathematical Models in Plant Physiology," Academic Press, New York, pp. 123-136.

L. Toivonen, M. Ojala and V. Kauppinen. 1990. Indole Alkaloid Production by Hairy Root Cultures of *Catharanthus roseus*: Growth Kinetics and Fermentation. *Biotechnol. Lett. 12*, 519-524.

W. M. van Gulik, J. J. Meijer, H. J. G. ten Hoopen, K. Ch. A. M. Luyben, and K. R. Libbenga. 1989. Growth of a *Catharanthus roseus* Cell Suspension Culture in a Modified Chemostat under Glucose-Limiting Conditions. *Appl. Microbiol. Biotechnol. 30*, 270-275.

H. Y. Wang, C. L. Cooney, and D. I. C. Wang. 1977. Computer-Aided Baker's Yeast Fermentation. *Biotechnol. Bioeng. 19*, 69-86.

H. Y. Wang, C. L. Cooney, and D. I. C. Wang. 1979. Computer Control of Bakers's Yeast Production. *Biotechnol. Bioeng. 21*, 975-995.

P. J. Westgate and A. H. Emery. 1990. Approximation of Continuous Fermentation by Semicontinuous Cultures. *Biotechnol. Bioeng. 35*, 437-453.

G. Wilson. 1976. A Simple and Inexpensive Design of Chemostat Enabling Steady-state Growth of *Acer pseudoplatanus* L. Cells Under Phosphate-Limiting Conditions. *Ann. Bot. 40*, 919-932.

S. B. Wilson, P. J. King, and H. E. Street. 1971. Studies on the Growth in Culture of Plant Cells. *J. Exp. Bot. 22*, 177-207.

M. H. Zenk, H. El-Shagi, H. Arens, J. Stockigt, E. W. Weiler, and B. Deus. 1977. Formation of the Indole Alkaloids Serpentine and Ajmalicine in Cell Suspension Cultures of Catharanthus roseus. In "Plant Tissue Culture and Its Biotechnological Applications," W. Barz, E. Reinhard, and M. H. Zenk (eds.), Springer-Verlag, Berlin, pp. 27-44.

CHAPTER 4

MASS TRANSFER

A. INTRODUCTION

The transport of material through fluid and solid phases and across phase boundaries is referred to as mass transfer or mass transport. In bioprocesses at any level, material has to be transported to the site of reaction before the reaction can proceed. Our interest in mass transfer is prompted by the concern that if mass transfer is slow, it may limit the overall velocity or rate of the reaction. The systematic and quantitative study of the interaction between mass transfer and a chemical or biological reaction is a seminal contribution of chemical engineering and is a subject of much research and several treatises (e.g., Satterfield 1970; Aris 1975; Cussler 1984; Froment and Bischoff 1990). This chapter provides a short introduction to mass transfer for readers not wholly familiar with the topic.

A.1. Motivation for Study

At least two scales of operation are implicitly involved in the design of a plant cell process. In the preliminary stages, experiments are conducted at the bench scale in small shake flasks. The objectives are generally twofold: (i) to determine the tissue type to be used, i.e., the particular cell line, and also whether free cells, immobilized cells, or organs should be used, and (ii) to determine the optimum environment, or sequence of environments, to which the cells should be exposed for maximum productivity. In the pilot- and process-design stages, the objective is to reproduce the appropriate environment(s) in a larger scale vessel.

It is perhaps easily appreciated that mass transfer is an important consideration for the engineering design and scale-up of bioreactors, since the uniform imposition of a desired environment throughout a large vessel is not a trivial task. Indeed, much engineering effort has been devoted to solving similar problems in the chemical and petrochemical industries, where chemically reactive, rather than biologically active, catalysts are used. This expertise and technology can be and has been directly applied to biological processes.

It has, however, only recently been appreciated that an understanding of the role of mass transfer is important even in the preliminary, shake-flask stages of experimental design. A clear understanding and avoidance of mass transfer limitations in the preliminary stages will reduce the ambiguities and uncertainties involved in larger scale bioreactor design and operation. Therefore, there is a strong need for all biotechnologists to have a working knowledge of the fundamentals of mass transfer.

A simplified and intuitive treatment of mass transfer is attempted in this chapter, and references are provided for those interested in a more in-depth study. The viewpoint of the practitioner is kept foremost in mind for the main objective is usually to avoid mass transfer limitations, not to analyze them! Accordingly, order-of-magnitude and rule-of-thumb analyses will be presented here rather than rigorous theoretical treatments. The principles outlined in this chapter can be extended with additional levels of complexity to reactor and process design (e.g., Satterfield 1970, Froment and Bischoff 1990).

A.2. Mechanisms of Mass Transfer

Mass transfer occurs by two mechanisms - diffusion and convection. Diffusion is the movement of an individual component by random molecular motion in response to a concentration gradient. Convective mass transfer refers to movement of a component due to bulk motion of the fluid itself. Both mechanisms contribute to overall mass transfer, though one or the other may be more important in a given situation. For example, whereas diffusion is likely to be more important in the stagnant, liquid-filled pores of an immobilized cell particle, convective transport is more important in mixing the liquid contents of a well-agitated shake flask. Next, we follow up these qualitative descriptions of convective and diffusive mass transfer with more quantitative descriptions.

A.2.1. Characteristic Times A "characteristic time" or a "relaxation time" for a process indicates how fast that process is capable of occurring. It is an extremely useful concept since it is simple, allows rough comparisons of rates, and provides quick estimates without the need for lengthy calculations. We shall invoke the characteristic time analysis in several contexts throughout the book.

There are subtle variations in the way a characteristic time is defined, and, moreover, there is no universally applicable definition for all processes. A certain amount of intuition is also involved in formulating a useful definition. The most common approach is to pose the following question: "How long does it take for an event to proceed to completion?" or "When a system is perturbed from a steady state, how long does it take the system to reach some arbitrary fraction (say, 70-90%) of its steady-state value?" Note that a process with a short characteristic time is fast, whereas one with a long characteristic time is slow.

The characteristic time for convection, τ_{con}, is defined by posing the question: How long does it take for a fluid element, traveling at a constant velocity U, to traverse a distance L? The answer is quite simply $\tau_{con} = L/U$.

As elaborated in Appendix 1, the characteristic time for diffusion, τ_{diff}, may be defined as L^2/D, where L is the diffusion length and D is the diffusivity of the molecule. In this case, τ_{diff} represents the time taken for ca. 70% of the steady-state concentration to be reached after a perturbation.

To demonstrate the utility of τ values and to gain an intuitive feel for the relative magnitudes of convective and diffusive transfer, let us consider an illustrative example.

Example 1

Case A: Consider the stagnant zones of an ill-designed reactor. Compare the rates of convective and diffusive oxygen transport over a 1-cm distance if

(i) $D_{O_2} = 2 \times 10^{-5}$ cm^2/s and the liquid is perfectly stagnant.

(ii) The liquid creeps very slowly, with an effective linear velocity of 2×10^{-3} cm/s.

$$\tau_{con} = L/U = (1\text{ cm}) / (2 \times 10^{-3}\text{ cm/s}) = 500\text{ s, i.e., } 8.3\text{ min}$$

$$\tau_{diff} = L^2/D = (1\text{ cm})^2 / (2 \times 10^{-5}\text{ cm}^2/\text{s}) = 50{,}000\text{ s, i.e., ca. } 14\text{ h}$$

Convective transport is two orders of magnitude faster than diffusion even when the liquid velocity is so low. In any well-mixed bulk liquid, where liquid velocities will be much higher than in this example, the contribution of diffusive transport to liquid mixing will essentially be negligible.

Case B: Consider the cell layers in a young developing stem. We would expect a nutrient such as oxygen to be transported radially inward across these cell layers primarily by diffusion. However, the cytoplasm is known to sometimes slowly swirl within the cell in a motion called "cytoplasmic streaming." Cytoplasmic streaming is obviously a possible mechanism for convective transport across plant cells. Assuming a streaming velocity of 2 x 10-3 cm/s (Armstrong 1979), compare convective and diffusive transport rates of oxygen ***across a single cell*** of 30-μm thickness.

$$\tau_{con} = L/U = \left(3 \times 10^{-3}\text{cm}\right) / \left(2 \times 10^{-3}\text{ cm/s}\right) = 1.5 \text{ s}$$

$$\tau_{diff} = L^2/D = \left(3 \times 10^{-3}\text{cm}\right)^2 / \left(2 \times 10^{-5}\text{ cm}^2/\text{s}\right) = 0.45 \text{ s}$$

In this case, the rates of diffusion and convection across a single cell are of the same order of magnitude, and diffusion is actually faster than convection.

Note that though the above two physical situations were different, the numerical values of the bulk velocity and diffusivity were the same in the two cases. Yet when relative rates were compared over a 1-cm distance, convective transport was two orders of magnitude faster, whereas over a distance of a few micrometers, the relative contributions were of the same order of magnitude. Clearly, the length scale over which transport processes occur is very important when the relative contributions of diffusion and convection are compared. Incidentally, Armstrong (1979) has shown that cytoplasmic streaming within individual cells does not contribute significantly to transport across the whole organ or tissue.

Though the intuitive and order-of-magnitude analyses presented so far are extremely useful, a more quantitative approach is necessary to tackle the problems that arise in biotechnological applications. These will often involve mass transfer not only in a single phase, but also across phase boundaries. Since the terminology used by engineers may often be confusing to the reader unfamiliar with mass transfer, formal quantitative definitions of mass transfer rates are presented below, with the units of each parameter expressed in terms of length (L), mass (M), and time (T).

A.2.2. Expression for Total Mass Transfer The total rate of mass transfer of component i is proportional to the concentration difference and to the area across which the transfer takes place. The constant of proportionality is defined as the mass transfer coefficient.

$$N_i = k A \ \Delta C_i \qquad [4.1]$$

where N_i is the rate of mass transfer (with units of M/T), k is the mass transfer coefficient (L/T), A is the area available for mass transfer (L^2), and ΔC_i is the concentration difference of component i (M/L^3).

The flux is defined as the rate of mass transfer per unit area:

$$j_i = \left(\frac{N_i}{A}\right) = k\,\Delta C_i \qquad [4.2]$$

where j_i is the flux of component i (M/L^2T).

The above description of the total flux combines the diffusive and convective contributions without separating them explicitly. The definition is a simple and convenient representation based on empirical arguments rather than on a mechanistic description of the transport process. The mass transfer coefficient is usually estimated from empirical correlations as will be discussed below.

A.2.3. Diffusion The diffusivity or diffusion coefficient of component i is defined by Fick's law:

$$j_i = D_i \frac{\partial C_i}{\partial x} \qquad [4.3]$$

where j_i is the unidimensional diffusive flux of component i (M/L^2T), D_i is the diffusivity or diffusion coefficient of component i (L^2/T), and $\frac{\partial C_i}{\partial x}$ is the concentration gradient along the x direction (M/L^4). (Note that the concentration gradient would be negative if the concentration decreased along the positive coordinate direction.) Fick's law is also assumed to hold for diffusion through the fluid-filled pores of a solid porous material:

$$j_i = D_{eff,i} \frac{\partial C_i}{\partial x} \qquad [4.4]$$

where $D_{eff,i}$ is called the effective diffusivity and incorporates the inherent complexities of diffusion through tortuous porous media. Examples of such situations include diffusion through immobilized cell particles or even through plant tissue itself. $D_{eff,i}$ must be obtained experimentally and is a function of the solute, the fluid, and the porous medium.

A.3. Outline of Problems

We shall consider four situations in which mass transfer is most relevant to plant cell cultures (see Figure 4.1).

1. Gas-liquid transport: This involves the transfer of volatile components such as oxygen from the gas phase to the liquid medium.

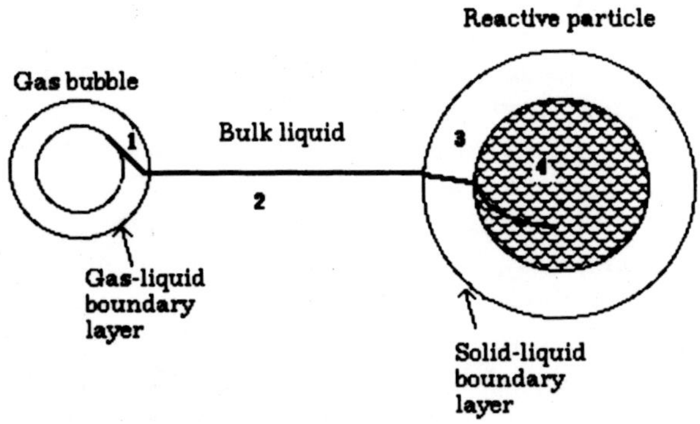

Figure 4.1 Outline of mass transfer problems encountered in plant cell culture. (1) Gas-liquid transport, (2) bulk liquid mixing, (3) external mass transfer, and (4) internal or intra-particle mass transfer.

2. Mixing of bulk liquid medium: Once a component is solubilized in the liquid medium, it must reach the cell before it can be consumed. For a finely dispersed cell suspension, adequate agitation has to be provided to ensure a well-mixed liquid phase, so that no concentration gradients of oxygen or other dissolved components exist in the liquid medium. The simplistic characteristic time analysis presented in case A of Example 1 may be extended to such a situation. The characteristic time for mixing, i.e., convective transfer, should be much higher than the characteristic time for reaction in the bulk liquid to ensure a well-mixed condition. A more complicated analysis called the residence time distribution analysis is usually used to gauge the degree of mixing in pilot- and industrial-scale reactors. The quantitative analysis of mixing is beyond the scope of this chapter, but many chemical engineering texts may be consulted for further information (e.g., Hill 1977; Levenspiel 1972).

For a very fine suspension culture, the above resistances are the only ones which need to be overcome to ensure adequate nutrient supply and product removal. However, if larger aggregates of cells, for example, natural aggregates, gel beads, leaves, or roots, are used, then other mass transfer limitations may become important.

3. External mass transfer: This refers to transport of dissolved components from the bulk liquid to the surface of the particle through a stagnant boundary layer surrounding the particle.

4. Internal mass transfer: If the reaction occurs within a stagnant environment such as a gel bead of immobilized cells, then diffusion is the sole mode of transport, and reaction and diffusion proceed together within the particle.

5. Mass transfer through porous reactive particles: An important situation occurs in organ cultures - consider, for example, a porous clump of tangled roots. Bulk flow and convective transport is possible inside the root clump and becomes the dominant mode of nutrient supply to the interior of the clump. However, external and internal mass transfer are still important at the level of each individual root.

B. GAS-LIQUID MASS TRANSFER

Since the major component to be transferred between the gas and liquid phases is oxygen, we will derive the mass transfer equations in terms of oxygen. However, these same relationships can be applied to describe gas-liquid transfer of other volatile components. Physically, the transfer of oxygen from the gas phase to the liquid phase is complex: oxygen must move through the bulk gas phase to the gas-liquid interface, dissolve into the liquid, move away from the interface, and finally be mixed throughout the liquid phase. It has been observed that the slow step in this process is the rate at which the oxygen moves from the interface into the liquid phase. To describe this process quantitatively, several models have been used. We will consider only the simplest and most intuitive of these models, film theory. According to film theory, the air bubble in Figure 4.2 is surrounded by a film of liquid which does not move. This stagnant film (or boundary layer) thus presents a barrier to the transfer of oxygen from the bubble into the liquid. If this hypothetical liquid film was truly stagnant, then oxygen transfer across this film could only occur by diffusion. To describe the diffusion of oxygen across the stagnant film Ficks' law for diffusion (equation [4.3]) is used. Since this film is thin, geometric considerations can be neglected and equation [4.3] can be integrated to give:

$$j_{O_2} = D_{O_2} \frac{\Delta C}{\Delta x} \tag{4.5}$$

where D_{O_2} is the diffusion coefficient of oxygen in the liquid, and j_{O_2} is the flux of oxygen from the gas into the liquid. The flux is the rate of oxygen transferred per unit of interfacial surface area [mmol/(cm^2 × s)]. ΔC is the concentration difference across the film of thickness Δx. According to film theory, the transfer coefficient k_L is defined as:

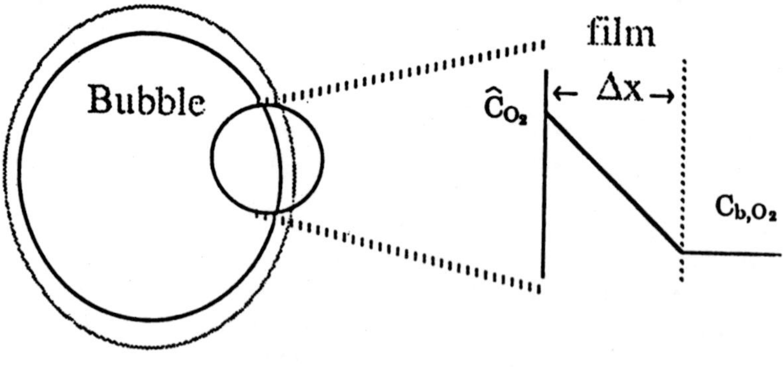

Figure 4.2 Schematic of a gas bubble, showing the oxygen concentration profiles in the gas-liquid film.

$$k_L = \frac{D_{O_2}}{\Delta x} \qquad [4.6]$$

Combining the above two equations, it can be seen that:

$$j_{O_2} = k_L \Delta C \qquad [4.7]$$

For most applications, it is desirable to describe oxygen transfer not in terms of the surface area but rather in terms of the liquid volume. To relate the above equation to the volumetric oxygen transfer rate (OTR), the flux must be multiplied by the interfacial area per unit volume of liquid:

$$OTR = j_{O_2}a = k_L a \Delta C \qquad [4.8]$$

where the interfacial area "a" accounts for the area of all the bubbles and has units of cm^2 per liter of liquid. Experimentally, it is very difficult to independently determine k_L and a, and therefore most workers use the combined term k_La to describe oxygen transfer in bioreactors. There is some discrepancy in the literature as to whether k_L or k_La is the oxygen mass transfer coefficient. In this text, we refer to the mass transfer coefficient as k_La. When the dissolved concentration is in units of mmol/L, then k_La has units of inverse time. In most plant cell applications, k_La is cited in units of h^{-1}.

ΔC is the driving force term or the "distance" from equilibrium and is given by:

$$\Delta C = \left(\widehat{C}_{O_2} - C_{b,O_2}\right) \qquad [4.9]$$

where C_{b,O_2} is the dissolved concentration of oxygen in the bulk liquid (e.g., far from the film) and $\widehat{C}_{O_2}$ is the liquid phase concentration which would exist if the liquid were in equilibrium with the gas.

Several methods are available for measuring k_La in bioreactors, and these methods will be briefly discussed [for further details, see a standard biochemical engineering text such as Bailey and Ollis (1986)]. In the absence of cells, k_La can be measured in one of two ways. In the sulfite oxidation method, oxygen that is transferred into the liquid is immediately reacted with sulfite according to the folowing reaction:

$$SO_3^{2-} + \tfrac{1}{2}O_2 \leftrightarrow SO_4^{2-} \qquad [4.10]$$

Since this reaction is fast, the dissolved oxygen (C_{b,O_2}) is maintained at essentially zero. Thus, the oxygen transfer rate (OTR) is given by:

$$OTR = k_La\widehat{C}_{O_2} \qquad [4.11]$$

The rate of oxygen transfer can then be related to the loss of sulfite which is measured over time.

A second method for measuring k_La in the absence of cells requires a dissolved oxygen probe. In this method, the saturation concentration of oxygen in the liquid ($\hat{C}_{O_2}$) is suddenly changed and the dissolved oxygen concentration is measured over time. The standard way to change $\hat{C}_{O_2}$ is to first sparge a reactor with nitrogen so that the dissolved oxygen concentration is reduced to a value approaching zero (at equilibrium with nitrogen, water would contain no oxygen so $\hat{C}_{O_2}$ = 0). The sparged gas is then changed (typically to air where the saturation concentration $\hat{C}_{O_2}$ is 0.25 mmol/L). Under this condition, the dissolved oxygen concentration will change according to the material balance:

$$\frac{dC_{b,O_2}}{dt} = k_La\,(\hat{C}_{O_2} - C_{b,O_2}) \qquad [4.12]$$

Integration of the above equation leads to:

$$\ln\left(\hat{C}_{O_2} - C_{b,O_2}\right) = -\ k_Lat + \text{Constant} \qquad [4.13]$$

and thus a semilogarithmic plot of ($\hat{C}_{O_2}$ - C_{b,O_2}) versus time yields a straight line of slope -k_La.

When cells are present, determination of k_La is complicated by the fact that the dissolved oxygen concentration depends not only on how oxygen is transferred from the gas to the liquid, but also on how rapidly the cells are consuming oxygen. If, as discussed in Chapter 5, the oxygen uptake rate (OUR) is given by

$$\text{OUR} = q_oX \qquad [4.14]$$

then the material balance is given by:

$$\frac{dC_{b,O_2}}{dt} = k_La\,(\hat{C}_{O_2} - C_{b,O_2}) - q_oX \qquad [4.15]$$

To measure k_La in the presence of cells, it is necessary to first determine the consumption rate of the cells and then to relate measured dissolved oxygen concentrations to the k_La using the above equation.

An alternative approach for measuring k_La can be taken if accurate measurements of the gas phase oxygen concentrations can be obtained (e.g., by mass spectrometry). Knowledge of the oxygen concentrations of the inlet and exit gases as well as the total gas flow rate can be used to directly determine the oxygen transfer rate, OTR. By combining this OTR with a measurement of the dissolved oxygen concentration, the k_La can be calculated directly.

By considering how k_La was derived, it is possible to see how this parameter can be affected by various operating conditions. Since the "a" term of k_La is dependent on the total gas-liquid interfacial area, then increases in this area will enhance k_La. Such increases in the interfacial area are achieved by sparging a larger volume of gas or by shearing the entering gas stream into more and smaller bubbles. These effects of aeration and agitation on k_La have been correlated many times (e.g., Bailey and Ollis 1986; Kargi and Moo-Young 1985) in equations of the form:

$$k_La = \alpha\,(\text{Agitation})^{\beta}(\text{Aeration rate})^{\gamma} \qquad [4.16]$$

where α, β, and γ are constants. The agitation can either be in terms of power input or impeller speed, and the aeration rate can be in terms of volumes of gas per minute per volume of liquid or a gas velocity.

k_La can also be affected by less readily altered parameters such as the medium viscosity and whether the medium promotes or inhibits gas bubble coalescence. Since these parameters can change during the course of cultivation, it is possible that k_La will not remain constant over time (e.g., see Kato et al. 1976).

C. EXTERNAL MASS TRANSFER

Once a component is solubilized in the liquid medium, mixing is used to minimize concentration gradients in the bulk medium (see Figure 4.1). The transport of dissolved components from the bulk medium to the surface of solid biocatalyst particles is hindered by an external mass transfer limitation. The aim of this section is to describe external mass transfer in quantitative terms. The analysis will lead to a simple criterion which can be used under a particular set of operating conditions to determine whether external limitations are significant or not. As shown in Figure 4.3, the effect of this limitation is to lower the surface concentration of a component i, $C_{s,i}$ to a value below the bulk concentration, $C_{b,i}$. Conversely, if external mass transfer limitations were negligible, then $C_{b,i} = C_{s,i}$.

C.1. Physical Picture

The film theory is a simple model which aids in an intuitive understanding of the external mass transfer limitation. The majority of the bulk medium is assumed to be well mixed and has no concentration gradients - the major limitation to transport is pictured to lie at the interface in the form of a stagnant film (also known as a boundary layer) through which transport occurs solely by diffusion. Intuitively, one expects that the thinner the boundary layer is, the better the rate of mass transfer. Furthermore, a high relative velocity at the solid-liquid interface leads to a thinner boundary layer, and hence to better mass transfer. Experimentally, one way to overcome external mass

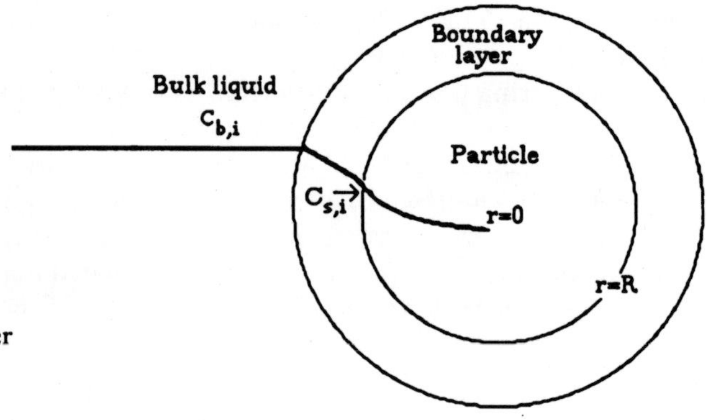

Figure 4.3 External mass transfer limitation through the stagnant solid-liquid boundary layer.

transfer limitations would be to increase the solid/fluid relative velocity, for example, increasing agitation to an extent that further increases in agitation do not yield a higher rate. However, one expects the mass transfer coefficient to also be affected by properties such as the fluid viscosity, the particle size, etc. To examine the quantitative relationships between such variables, a slightly more formal analysis is necessary.

C.2. Formulation of a Criterion

The purpose of this section is to formulate a simple criterion that will ensure that our ability to supply nutrient by external mass transfer exceeds the particle's nutrient demand. To quantify the rate of mass transfer, the concentration at the particle surface, $C_{s,i}$,has to be known (see equation [4.1]). $C_{s,i}$ is very difficult to measure, and in order to eliminate it as a variable, it is necessary to consider the mechanism by which transport occurs just inside the particle. Here we shall consider the case more commonly encountered in plant cell cultures, that is, diffusion and reaction take place within the particle. The case of external mass transfer followed by a reaction confined only to the surface is more commonly encountered in immobilized enzyme systems and is treated in detail by Bailey and Ollis (1986).

The principle of conservation of mass states that, at steady state, the flux of component i reaching the particle surface from the bulk medium must equal the flux of component i diffusing just inside the particle (see Figure 4.3).

The total mass flux of i from the bulk liquid to the solid surface, j_i, is given by:

$$j_i = k_s(C_{b,i} - C_{s,i}) \quad [4.17]$$

where k_s is the solid-liquid interfacial mass transfer coefficient, $C_{b,i}$ is the concentration of component i in the bulk medium, and $C_{s,i}$ is the concentration of component i at the particle surface.

The sole mode of transport inside the particle is diffusion. The steady-state diffusive flux of i into the particle evaluated at the solid surface follows by applying Fick's law at the interface:

$$j_i = D_{eff,i}\left(\frac{dC_i}{dr}\right)_{r=R} \qquad [4.18]$$

where $\left(\frac{dC_i}{dr}\right)_{r=R}$ is the concentration gradient just inside the particle's surface, C_i is the concentration of component i within the particle and is a function of the radial position within the particle, r, and R is the radius of the spherical particle.

At steady state, the magnitude of the two fluxes should be equal, so

$$k_s(C_{b,i} - C_{s,i}) = D_{eff,i}\left(\frac{dC_i}{dr}\right)_{r=R} \qquad [4.19]$$

Engineers often work with dimensionless variables. If we nondimensionalize the above equation by defining $C^* = C_i/C_{b,i}$ and $r^* = r/R$ as the dimensionless concentration and radius, respectively, then

$$\frac{C_{s,i}}{C_{b,i}} = 1 - \left(\frac{D_{eff,i}}{k_s R}\right)\left(\frac{dC^*}{dr^*}\right)_{surface} \qquad [4.20]$$

The dimensionless quantity $k_s\,(2R)\,/D_{eff,i}$, called the modified Sherwood number (Sh^*) or the Biot number, can be considered to be a measure of the ratio of total external mass transfer to diffusive mass transfer into the particle. Substituting Sh^* in equation [4.20] gives:

$$\frac{C_{s,i}}{C_{b,i}} = 1 - \left(\frac{2}{Sh^*}\right)\left(\frac{dC^*}{dr^*}\right)_{surface} \qquad [4.21]$$

The above equation essentially indicates the magnitude of the external mass transfer limitation. If $C_{b,i} = C_{s,i}$, then no limitation is present, for which the criterion is

$$\left(\frac{2}{Sh^*}\right)\left(\frac{dC^*}{dr^*}\right)_{surface} << 1 \qquad [4.22]$$

Two observations may be made:

1. If no concentration gradients exist within the particle, then the concentration gradient at the surface must, by definition, be extremely small, or approaching 0. This leads to a very useful rule of thumb: *If internal mass transfer limitations are not present, then external mass transfer limitations are also not present. However, if internal limitations are present, then external limitations may or may not be present depending on the value of the Sherwood number.*

2. The above analysis conveniently lumps those parameters which depend on the operating conditions into the dimensionless entity, that is, the Biot number or the modified Sherwood number, Sh*. The modified Sherwood number is related to the Sherwood number (Sh) by

$$Sh^* = (D_i/D_{eff,i}) \times Sh \quad [4.23]$$

with the effective diffusivity $D_{eff,i}$ being estimated or measured as discussed in Section D.2.3.

The Sherwood number physically represents the ratio of total mass transfer (i.e., convective plus diffusive contributions) to diffusive transport. Its lowest asymptotic value is therefore 1 under perfectly stagnant conditions where convective transport is negligible (or 2 for a sphere because the diameter, rather than the radius, is used as the length scale for the Sherwood number).

A rapid estimate of the Sherwood number allows the calculation of the modified Sherwood number and allows the practitioner to use the criterion of equation [4.22] to gauge the existence and severity of the external mass transfer limitation under a particular set of operating conditions. If external mass transfer is found to be a problem, then equation [4.21] provides an estimate of the numerical value of the Sherwood number that must be achieved to circumvent the limitation.

C.3. Sherwood Number Correlations

The Sherwood number would be expected to rely on fluid properties such as the viscosity and the density and also on the relative fluid velocity at the interface, the particle size, and the diffusivity. To simplify the interrelationships between several variables, engineers often lump variables together into dimensionless groups. Empirical correlations are made between these dimensionless groups. Such an approach enables a vast amount of experimental data to be summarized in a succinct form. Furthermore, it allows correlations to be extended to geometries and situations beyond the ones in which the original experiments were performed. Cussler (1984) has provided a highly readable and intuitive treatment of the use of dimensional numbers in mass transfer calculations.

The empirical correlations presented here have been developed in the context of traditional chemical engineering applications and are directly transferable to biological applications. Some readers may be unfamiliar with the dimensionless numbers and their physical significance but are nevertheless urged to use the formula as recipes and familiarize themselves with the calculations because the final information obtained can be extremely useful.

The important dimensionless groups encountered in correlations involving agitated particles (often called "forced convective mass transfer") are the Reynolds (Re) and Schmidt (Sc) numbers. The Sherwood number (Sh) is usually correlated empirically with the Reynolds and Schmidt numbers in the following form:

$$Sh = A + B\,(Re^{\alpha} \times Sc^{\beta}) \qquad [4.24]$$

where A, B, α, and β are arbitrary constants which are used to fit the correlation to experimental data.

The dimensionless numbers are defined as follows:

$$Sh = k_s\,(2R)/D_i \qquad [4.25]$$

$$Re = (d_p U \rho)/\mu = (d_p U)/\nu \qquad [4.26]$$

where Re is the Reynolds number (dimensionless), $d_p = 2R$ = particle diameter (L), U is the relative velocity of the fluid with respect to the particle (L/T), μ is the fluid viscosity (M/LT), ρ is the fluid density (M/L^3), and $\nu = \mu/\rho$ is called the kinematic viscosity (L^2/T).

$$Sc = \mu/\rho D_i = \nu/D_i \qquad [4.27]$$

where Sc is the Schmidt number (dimensionless), ν is the kinematic viscosity as defined above, and D_i is the diffusion coefficient of component i in the fluid.

The Reynolds number characterizes the flow, and the Schmidt number is a characteristic of diffusion in the fluid. Values for the Schmidt number are usually on the order of 10^3 for liquids, and values of the Reynolds number vary over a wide range. Correlations are usually good to about 10 to 20% accuracy provided they are used over the indicated ranges. Sherwood number correlations for commonly encountered situations are listed in Table 4.1. Further information may be found in, among other standard texts, Satterfield (1970), McCabe et al. (1985), and Cussler (1984).

In the following section, we will use an example to show how to gauge the importance of, and how to mitigate the role of, external mass transfer limitations.

D. INTERNAL MASS TRANSFER

D.1. Introduction

Once a component reaches the surface of a particle inside which cells are present, subsequent transport occurs by diffusion if fluid cannot flow through the particle. The interaction between diffusive mass transfer and reaction within a particle is referred to as internal or intraparticle mass transfer. Note that whereas external mass transfer is in series, reaction and diffusion occur in parallel. The analyses presented in this section are applicable to particles such as natural aggregates, immobilized cell preparations, or even differentiated structures such as leaves or roots. We shall often collectively refer to these particles as "reactive particles" or as "biocatalyst particles."

D.1.1. Description of Reaction/Diffusion Problem Consider a particle, containing cells which use up reactant i (such as dissolved oxygen) at a rate that is dependent

Table 4.1. Sherwood number correlations.

Mass transfer to single spheres

$Sh = (4 + 1.21\ Re^{2/3}\ Sc^{2/3})^{1/2}$	valid for Re < 10	[1]
$Sh = 2 + 0.6\ Re^{1/2}\ Sc^{1/3}$	valid for Re up to 1000	[2]

Mass transfer to a stirred tank of spheres

A conservative estimate of the Sherwood number may be obtained by using the particle's terminal velocity (which can be estimated from Stokes' law) to compute Re in equation [2]. The actual mass transfer coefficient may be from 1.5 to 8 times higher than the estimated value (Harriott 1962).

Mass transfer to packed beds

$$Sh = 1.17\ Re^{0.585}\ Sc^{1/3} \qquad [3]$$

The above equation is applicable to spheres as well as cylinders, and the appropriate diameters are used to compute Sh and Re.

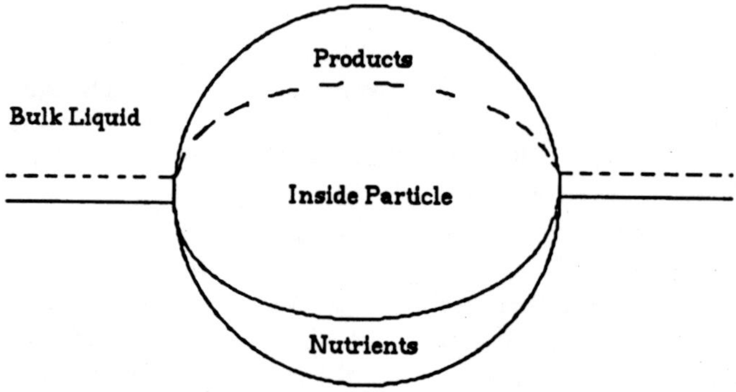

Figure 4.4 Intraparticle concentration gradients of nutrients and products. Note that external mass transfer limitations have been assumed to be insignificant.

on the local concentration C_i. If the rate at which cells are capable of consuming reactant i is faster than the rate of supply by diffusion, then concentration gradients will form within the particle (see Figure 4.4). Consequently, the local rate of reaction at different positions inside the particle will also vary if the rate is a function of the local concentration. The same argument applies in reverse to the buildup of products within the particle. The steepness of the gradient is an indication of the severity of the limitation, and the exact shape of the gradients will depend on factors such as the reaction kinetics, diffusion parameters, and geometry. If internal mass transfer limitations are severe, we may thus expect two consequences:

- Steep concentration gradients will form within the particle, and
- The total reaction rate may be reduced.

D.1.2. Objectives of the Practitioner The avoidance of mass transfer limitations in the shake-flask experiment provides the assurance that cells do indeed experience a uniform environment. The experimenter can exert complete control over this environment by manipulating bulk liquid conditions. Moreover, if internal mass transfer limitations are avoided, then the rate of reaction exhibited by the particle will truly reflect the cell's inherent reactive capabilities. Accordingly, in most situations, the main concerns of an experimenter testing the efficacy of an immobilized cell preparation for a given biotransformation or *de novo* synthesis of a secondary metabolite are:

1. Under the specific experimental conditions used, are mass transfer limitations important?
2. If so, how can mass transfer limitations be reduced? That is, how can the biocatalyst particles be made gradientless?
3. However, in certain situations, data may only be available from an experiment in which mass transfer limitations were severe. In some other situations, it may actually be desirable to conduct preliminary experiments with an immobilized cell preparation in which severe limitations exist. To interpret results meaningfully, one needs some sense of the concentration profiles which must have existed within the particle. The following question often becomes relevant: "How can concentration profiles within the particle be predicted *a priori* or *a posteriori*?"

Two analyses based on chemical engineering reaction/diffusion theory can be used to address the questions posed above. A simple but powerful tool developed by Weisz (1973) can be used to address questions 1 and 2. A more formal analysis is needed to address question 3. Both analyses will be discussed in turn.

D.2. Weisz's Analysis

Weisz's analysis is actually derived from Thiele's analysis described in Section D.4.2. It is extremely useful from the practitioner's point of view since it gives quick estimates of the presence and severity of internal limitations without recourse to lengthy calculations or the need for extensive additional experimental data. In fact, order-of-magnitude calculations may indicate whether or not to expect mass transfer limitations even before any experiments are performed.

Consider a steady-state reaction occurring in a particle of length scale L (to be defined below), immersed in medium containing reactant i at a bulk concentration $C_{b,i}$. If the effective diffusivity is $D_{eff,i}$ and the observed reaction rate per unit volume of catalyst is R_{obs}, then Weisz's modulus Φ_w, a dimensionless parameter, is defined by

$$\Phi_w = (R_{obs} L^2) / (D_{eff,i} C_{b,i}) \qquad [4.28]$$

To provide an intuitive explanation, Weisz's modulus may be considered to be a ratio of characteristic times for diffusion and reaction. As shown in Appendix 1, the characteristic time for diffusion is L^2/D_{eff}. The quantity $C_{b,i}/R_{obs}$ has the units of time and may be interpreted as the characteristic time for reaction. Since the characteristic time is the reciprocal of the rate,

$$\Phi_w = \frac{\tau_{diff}}{\tau_{reaction}} = \frac{\text{Characteristic reaction rate}}{\text{Characteristic diffusion rate}}$$

Weisz's general criterion, applicable for all geometries and kinetics is:

$\Phi_w << 1$, Internal limitations are not significant relative to the reaction rate.
Intraparticle concentration gradients are unlikely.
Reduction in reaction rate is unlikely.

$\Phi_w >> 1$, Internal limitations are significant relative to the reaction rate.
Intraparticle concentration gradients are likely.
Reduction in reaction rate is likely.

The criterion makes intuitive sense. When $\Phi_w << 1$, the characteristic diffusion rate is faster than the characteristic reaction rate, reactant supply can more than compensate for demand, and therefore mass transfer limitations are not going to be significant. When $\Phi_w >> 1$, the characteristic diffusion rate is slower than the characteristic reaction rate, reactant supply cannot match up to demand, and mass transfer limitations are therefore going to be significant.

For practical situations, Weisz suggests that the following criterion be used:

$\Phi_w < 0.3$, Internal limitations are insignificant.

$\Phi_w > 3$, Internal limitations are significant.

When Weisz's modulus is in an intermediate range (0.3 to 3), then a closer analysis and a more restricted criterion are necessary. Whether internal limitations are significant also begins to depend on factors such as the reaction kinetics and particle geometry. Before describing this narrower criterion, and before applying the analysis to experimental data, let us first look at each term of the modulus in detail. In doing so, we shall keep the practitioner's second question in mind: "How can gradients be reduced?"

Despite the simplicity of the above definition, the reader should be aware that there is considerable scope for confusion, and each parameter should be carefully and appropriately chosen.

D.2.1. The Length Scale The appropriate length scale for Weisz's reaction/diffusion analysis, L, is defined as the ratio (volume of the particle)/(surface area exposed to reactant). For the different geometries, L is defined as follows:

For a sphere, L= R/3, where R is the radius.

For a cylinder, L= R/2, where R is the radius.
For a flat plate with both sides exposed to the medium, L is the half-thickness of the plate.

Since Weisz's modulus has a strong dependence on L ($\Phi_w \alpha L^2$) , changing the particle size has a profound effect on mass transfer limitations. Increasing the particle size implies a larger Φ_w and, therefore, more severe limitations, that is, steeper concentration gradients and reduced reaction rates. Decreasing the particle size decreases Φ_w, and, therefore, lessens the importance of internal limitations and is a simple way to overcome them.

D.2.2. The Bulk Reactant Concentration Weisz's modulus depends inversely on the steady-state bulk reactant concentration ($\Phi_w \alpha 1/C_{b,i}$). Operating at high $C_{b,i}$ implies a lower Φ_w and, therefore, less severe mass transfer limitations. However, since components such as hormones are inhibitory to cells above certain concentrations, it may not always be possible to arbitrarily increase bulk concentrations in order to reduce gradients. The unmodified Weisz's modulus assumes that the reaction order is greater than or equal to zero. If a compound can be inhibitory, modifications to the analysis are required (see Section D.5).

Another important point to remember is that by using the bulk concentration $C_{b,i}$ in this analysis instead of the concentration at the particle's surface, $C_{s,i}$, we are implicitly assuming that external mass transfer limitations have been circumvented. If such is not the case, then the surface concentration may be estimated by using equation [4.21], and used in the modulus instead of $C_{b,i}$. An example of such a calculation appears in part B of Example 2 on page 96.

D.2.3. The Effective Diffusivity Diffusivities for most small molecules in water lie in the range of 10^{-5} to 10^{-6} cm^2/s. Effective diffusivities are typically less than the corresponding molecular diffusivities by a factor which depends on the support characteristics and the volume fraction of cells in the support. Accurate estimates of D_{eff} are difficult to obtain but may be necessary when concentration profiles within immobilized cell particles need to be accurately predicted. For use in Weisz's modulus, a conservative estimate is often sufficient. Table 4.2 summarizes values of D_{eff} for a few nutrients from a variety of sources. Empirical correlations for estimating the change of effective diffusivity with changing cell loading have been proposed by, among others, Axelsson and Persson (1988) and Scott et al. (1989).

Several studies have used the "transient method" for measuring the effective diffusivity (Hannoun and Stephanopoulos 1986; Tanaka et al. 1984; Chresand et al. 1988). Support particles of known dimensions containing inactivated cells are presaturated with the component whose D_{eff} is being measured. The particles are then transferred to a known volume of well-stirred medium initially free of the component, and the transient appearance in the medium is monitored. Alternatively, the transient diffusion of the component into initially component-free particles is

Table 4.2. Experimentally measured effective diffusivities for various nutrients.

Nutrient	Diffusivity in water (cm^2/s) (D_{water})[a]	System in which measured	(D_{eff} / D_{water}) x 100	Reference
Glucose	6.9 x 10^{-6}	Microbial aggregates	10 - 100	Matson and Characklis (1976)
		Yeast cake	43	Bringi and Dale (1990)
		Tumors	25 - 50	After Karel et al. (1985)
Sucrose	5.6 x 10^{-6}	Alginate/plant cells	20 - 50	Pu and Yang (1988)
		Alginate/plant cells	67	Furusaki et al. (1988)
Oxygen	2.5 x 10^{-5}	Microbial aggregates	20 - 100	Matson and Characklis (1976)
		Trickling-filter slimes	39	Chen and Bungay (1981)
		Agar/yeast	70	Sato and Toda (1983)
Ammonia	1.7 x 10^{-5}	Microbial aggregates	80	Matson and Characklis (1976)
Nitrate	1.6 x 10^{-5}	Microbial aggregates	90	Matson and Characklis (1976)

[a] Values obtained from "Perry's Chemical Engineering Handbook" (Perry and Green 1984). Values for ammonia and nitrate were obtained from Matson and Characklis (1976).

monitored. The equations of Crank (1975) are applied to the data and the effective diffusivities are calculated. The method is simple and requires no specialized apparatus. Some precautions need to be observed during the measurement:

1. The effective diffusivity must be measured with cells present inside the support since D_{eff} depends on the pore structure of the support. The significant barrier to diffusion is usually the cells themselves. In fact, effective diffusivity for sugar through empty alginate beads is almost identical to its value in water, showing that the gel by itself does not hinder transport (for example, see Tanaka et al. 1984; Chresand et al. 1988).
2. The component whose D_{eff} is being measured should not be consumed by reaction. Cells can be inactivated (but should not be lysed!), for example, by azide or glutaraldehyde, prior to the measurement if the component is rapidly depleted during the time scale of the experiment.
3. External mass transfer limitations must be overcome by adequate stirring; that is, the agitation should be increased until the efflux or influx rate no longer increases upon further agitation.

4. The theoretical model for data analysis ascribes the total efflux of the component to diffusion. A value of D_{eff} higher than the molecular diffusivity may indicate the possible additional contribution of convective transport through the porous catalyst as observed by Miura (1976) and as pointed out by Rodrigues and Quinta Ferreira (1988). When convective transport within the support is significant, an analysis such as that presented in Section E becomes appropriate.

D.2.4. The Observed Reaction Rate Weisz's modulus requires a knowledge of the observed *steady-state* reaction rate ***based on unit volume of biocatalyst support.*** Since Φ_w is proportional to R_{obs}, and since changing the number of cells per bead (the cell loading) changes R_{obs}, reducing the cell loading reduces Φ_w and therefore reduces internal mass transfer limitations.

For components whose concentrations in the bulk medium change slowly, for example, over a period of days, R_{obs} can be determined from a shake-flask experiment by measuring initial rate data. For a component that is quickly consumed, such as oxygen, severe transients may make the measurement of a steady-state reaction rate difficult in shake flasks. In such cases, more reliable data may be obtained from a flow-cell apparatus such as that described by Sato and Toda (1983) or that used by Prince et al. (1991) (Figure 4.5). The biocatalyst is allowed to reach a steady state at a particular bulk oxygen concentration under continuous-flow conditions. Oxygen supply is halted for a few minutes, and the initial rate of oxygen consumption is measured. A new steady-state is established at another bulk oxygen concentration and the procedure is repeated. Reaching a steady-state is important for reliable rate measurements. The figures given in Appendix 1 may be consulted for estimating the time required for a steady-state to be achieved.

Having considered each term of the Weisz's modulus in detail, we now examine the criterion for gauging the presence and severity of internal limitations.

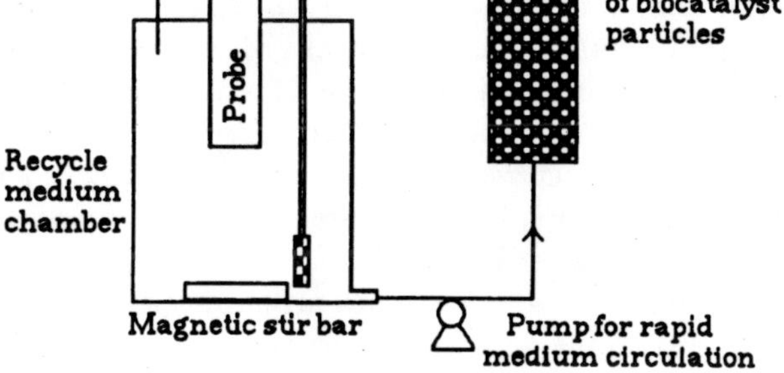

Figure 4.5 Apparatus for the measurement of steady-state oxygen uptake rates of biocatalyst particles such as immobilized cells or organ cultures. Aeration is achieved by sparging through a fritted glass tube. Rapid medium cycling ensures a well-mixed system. After a steady state is reached, aeration is stopped, gas bubbles are removed from the system, and the initial uptake rate is measured. Temperature control is important (jacketing not shown).

D.3. Weisz's Criterion for Gauging Limitations

Before quantitatively assessing the impact of mass transfer limitations on the biocatalyst particle performance, let us recapitulate the consequences of internal limitations:

- Concentration gradients form within the particle.
- The overall reaction rate may be reduced.

Concentration gradients within particles are very difficult to measure directly, and are hence difficult to quantify. Since the reaction rate is a much more easily measured parameter, mass transfer limitations are quantified by their effect on the reaction rate or, more precisely, on the effectiveness factor, η, defined as:

$$\eta = \frac{\text{Observed rate with mass transfer limitations}}{\text{Rate if mass transfer limitations were insignificant}}$$

If mass transfer limitations are not severe enough to depress reaction rates, then $\eta = 1$, and if mass transfer limitations are severe enough to do so, then $\eta < 1$. Further, with increasingly severe limitations, one would expect to observe decreasing effectiveness factors.

The presence or absence of concentration gradients is then indirectly surmised from the value of the effectiveness factor. Section D.3.2 outlines some subtleties in the relationship between concentration gradients and effectiveness factors.

D.3.1. Effectiveness Factor-Weisz's Modulus Charts Figure 4.6 depicts the dependence of η on Φ_W for different geometries and intrinsic kinetics obtained by numerical computations. Note that in keeping with Weisz's general criterion, η approaches 1 when Φ_W is less than 0.3 regardless of the geometry or kinetics and that when Φ_W is greater than 3, the curves for different geometries tend to merge. The asymptotic curves may be described by the equations:

$$\eta = 2/\Phi_W \quad \text{for zero-order kinetics}$$

$$\eta = 1/\Phi_W \quad \text{for first-order kinetics}$$

For values of Φ_W between 0.3 and 3, the chart must be consulted to gauge the effect of internal limitations on the overall rate. The following observations are relevant:

(i) The family of curves for saturation kinetics (i.e., Michaelis Menten kinetics) is bounded by the zero-order and first-order curves. If the dimensionless saturation constant, $\beta = C_i/K_m$, is less than 0.1, the curve becomes essentially first order, and if β is greater than 50, the curve becomes essentially zero order. Curves for intermediate values of β in spherical geometry can be found in Roberts and Satterfield (1965) or in Satterfield (1970).

(ii) For $\Phi_W < 0.3$, and $\Phi_W > 3$, the effect of geometry is not pronounced. For $\Phi_W \sim 1$, the η versus Φ_W curves depend on the geometry. Curves for cylindrical and

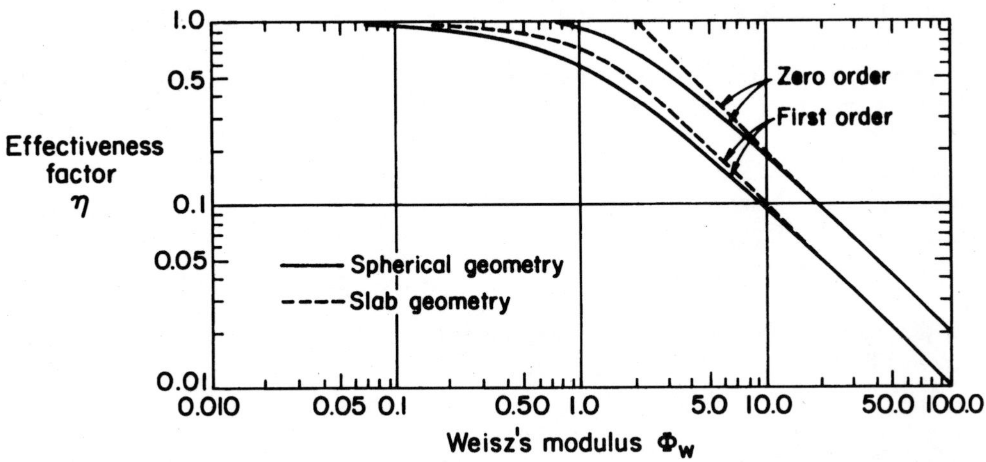

Figure 4.6 Effectiveness factor versus Weisz's modulus charts. Reproduced with permission from Bailey and Ollis (1986).

other irregular catalyst shapes lie between the flat-slab and spherical geometry curves.

(iii) For product-inhibited reactions, a more stringent criterion is applicable; see Froment and Bischoff (1990) or Satterfield (1970).

(iv) It is possible for η to exceed 1 for substrate-inhibited reactions and for exothermic reactions (e.g., see Aris 1975).

Having seen how internal limitations affect the effectiveness factor, let us see how internal limitations affect intraparticle concentration gradients.

D.3.2. Effectiveness Factors and Concentration Gradients We expect that internal mass transfer limitations will concomitantly affect both concentration profiles as well as reaction rates. Some subtleties in the relationship between the effectiveness factors and concentration gradients need to be clarified; in particular:

Does $\eta < 1$ always imply that concentration gradients exist within the particle?

Conversely, does $\eta = 1$ imply that concentration gradients do not exist within the particle?

For simplicity, we shall restrict ourselves to the asymptotic cases of zero- and first-order reaction kinetics.

Case A: First-Order Reaction

Since the local rate of reaction at any position within the particle is proportional to the local concentration ($V = KC_i$, K is the first-order rate constant), the concentration and reaction rate profiles will be similar, as shown in Figure 4.7. Any concentration gradient will cause a corresponding decrease in the reaction rate, and hence in the effectiveness factor.

Therefore, *for any first-order reaction, $\eta < 1$ implies the presence of concentration gradients, and $\eta = 1$ implies absence of concentration gradients.*

Case B: Zero-Order Reaction

Since the local rate is independent of the concentration, $V = V_{max}$ as long as any reactant is present. If no reactant is available, then, of course, $V = 0$. Concentration and reaction rate profiles are sketched in Figure 4.8 for increasingly severe internal limitations. For zero-order reactions,

A. When concentration gradients are not severe, $\eta = 1$.
B. When concentration gradients are severe, with $C_i = 0$ at the center point, η is still equal to 1. Incidentally, this point corresponds to $\Phi_w = 2$ for flat-plate geometry in Figure 4.6.
C. When concentration gradients become even more severe, a central region of the particle is deprived of reactant; that is, $C_i = 0$ throughout that region. In this case, η becomes less than 1, and, in fact, η is simply the volume fraction of the biocatalyst in which C_i is finite and greater than 0.

Thus, we see that *for a zero-order reaction, $\eta < 1$ implies severe gradients, and that a portion of the particle is deprived of reactant. Importantly, $\eta = 1$ does not imply that concentration gradients are not present* (Case B, Figure 4.8).

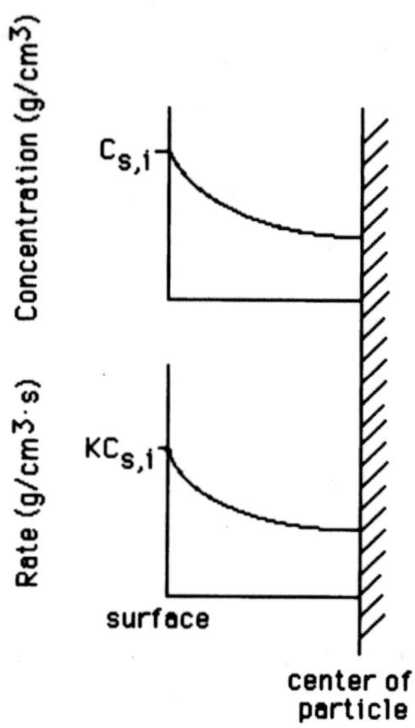

Figure 4.7 Intraparticle concentration and reaction rate profiles for a first-order reaction, $V = KC_i$.

In many cases, the practitioner would be interested in ensuring that the concentration at the center of the particle does not fall below some threshold value. Appendix 2 outlines the formal proof for a relationship between the Weisz's modulus and the dimensionless concentration at the center of the particle. For a sphere,

$$\Phi_w = 0.67\ [1 - (C_{center,i}/C_{s,i})] \tag{4.30}$$

The importance of the relationship between effectiveness factors and concentration gradients becomes apparent in the case where a single component participates in more than one reaction. For example, oxygen participates in many different cellular reactions, the most important of which is oxidative phosphorylation. Oxygen uptake is

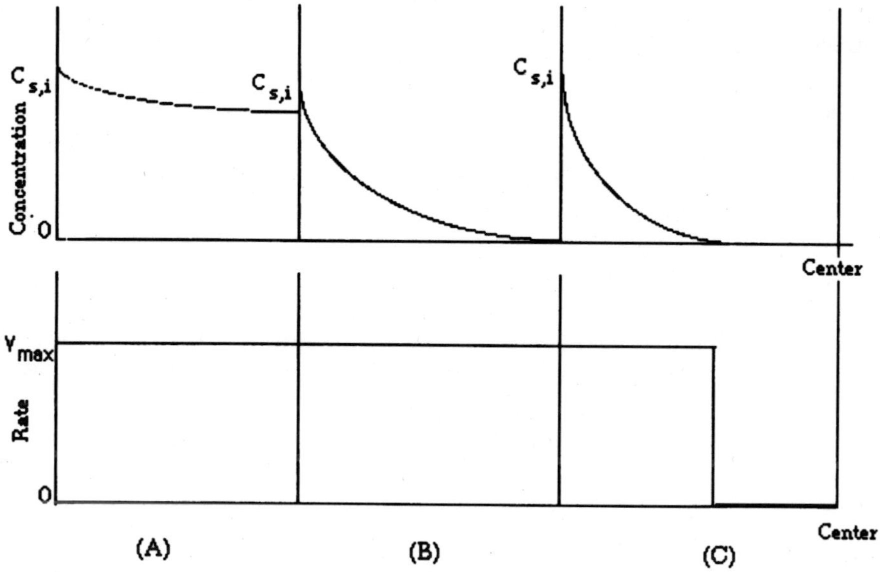

Figure 4.8 Intraparticle concentration (upper panel) and reaction rate (lower panel) profiles for a zero-order reaction, $V = V_{max}$. A, B, and C represent increasingly severe mass transfer limitations.

a high-affinity reaction (low K_m) which proceeds at V_{max} (i.e., it is zero order) over a wide range of oxygen concentrations. Now, oxygen may also be involved in another oxidase step in a secondary biosynthetic pathway, and the K_m for this oxidase step may be much higher than for uptake. For example, the requirement of an optimal oxygen concentration of 6 ppm (ca. 80% saturation) for shikonin biosynthesis (Fujita 1988) may indicate such an involvement.

The first inclination in designing an immobilized cell bead might be to ensure that oxygen supply is sufficient for respiration; that is, $\eta = 1$ for oxygen uptake. However, we have seen that $\eta = 1$ is not a sufficient condition for the absence of gradients in a zero-order reaction. The effectiveness factor for uptake will remain 1 as long as the concentration at the center is greater than the critical value, that is, $2K_m$. The effectiveness factor for secondary metabolite production will however be less than 1 because a significant portion of the particle will experience an oxygen concentration below that corresponding to the optimal (or threshold) value for secondary biosynthesis (6 ppm for shikonin).

Perhaps a better way to design the particle would be to specify that the concentration at the center of the particle, C_{center}, should always remain above the minimum oxygen concentration necessary for secondary metabolism. Equation [4.30] could then be used to determine whether this condition was being satisfied under a particular set of operating conditions. The effectiveness factors for both oxygen uptake and secondary metabolism would be 1.

Now that we have examined Weisz's analysis in detail, let us tackle two illustrative examples.

Example 2: Validity of Empirical Kinetic Measurements

The rate of oxygen uptake, like many other nutrient uptake reactions, follows saturation kinetics. The respiration rate q_o may be defined as:

$$q_o = \frac{q_{o,max}\, C_{O_2}}{K_m + C_{O_2}} \qquad [4.31]$$

where $q_{o,max}$ is the maximal respiration rate (milligrams of O_2 per gram fresh weight per hour); K_m is the saturation constant, defined as that concentration at which the respiration rate is half-maximal; and C_{O_2} is the dissolved oxygen concentration.

If $q_{o,max}$ and K_m represent the cells' intrinsic respiratory capabilities, they are referred to as intrinsic kinetic parameters. For example, if $q_{o,max}$ and K_m were empirically determined using a very fine cell suspension, it is likely that the observed parameters are intrinsic. However, if mass transfer limitations were imposed upon the system, for example, if the suspension culture consisted of large natural aggregates, would the observed and intrinsic parameters be different?

If mass transfer limitations were so severe that part of the aggregates were completely deprived of oxygen, then the observed oxygen uptake rate would be less than if no limitations were present. The dotted curve in Figure 4.9. shows the expected pattern under these circumstances. The effect of mass transfer limitations would be (i)

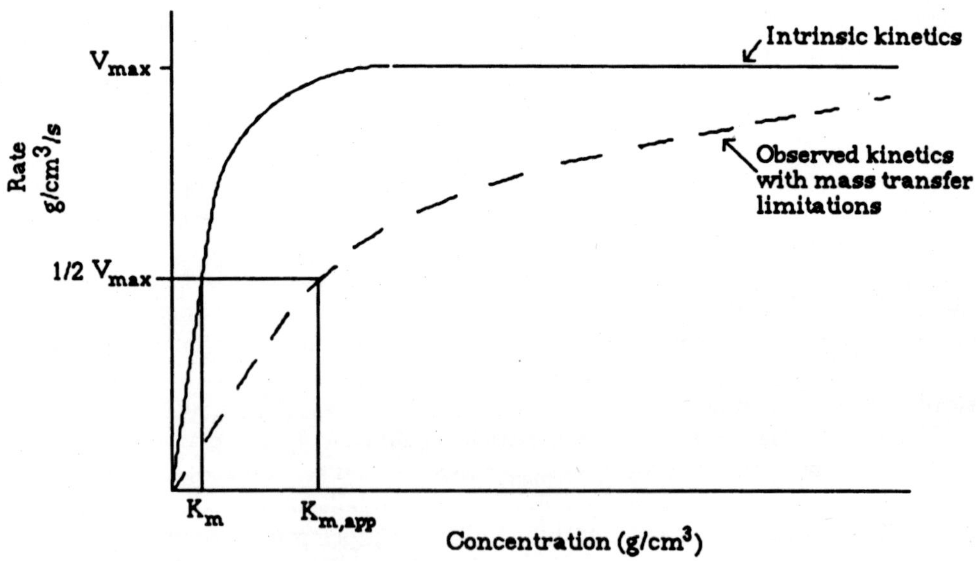

Figure 4.9 Effect of mass transfer limitations on observed kinetic parameters for Michaelis-Menten kinetics.

to decrease $q_{o,max}$ and (ii) to increase K_m compared to the intrinsic values (for more detail, see Hamilton et al. 1974). Therefore, before observed kinetics are assumed to be intrinsic, a check should be made to confirm the absence of mass transfer limitations. Failure to do so will cast doubt on the fundamental utility of the parameters. Let us work through a specific example.

Part A: An apparatus such as that shown in Figure 4.5 is used to measure intrinsic oxygen uptake parameters $q_{o,max}$ and $K_{m,app}$ (the "apparent" saturation constant) for a clumpy garlic suspension culture with a mean aggregate diameter of 1 mm. At an oxygen concentration of 8 mg/L, $q_{o,max}$ is measured to be 0.3 mg of O_2 per gram fresh weight of cells per hour. The oxygen concentration at which the observed uptake rate is half-maximal is 0.4 mg O_2/L. Assuming for now that external mass transfer limitations were insignificant, could intraparticle mass transfer limitations have masked the intrinsic kinetic parameters? Assume that the effective diffusivity of oxygen, D_{eff,O_2} = 1.8×10^{-5} cm²/s has the same value as the diffusivity in the bulk liquid. The specific gravity of cells may be assumed to be 1.

Solution: If internal mass transfer limitations were insignificant at the maximal and half-maximal observed uptake rates, then the observed parameters can be assumed to be intrinsic. We therefore compute Weisz's modulus at $q_{o,max}$ and at $\frac{1}{2}q_{o,max}$.

(i) Weisz's modulus at $q_{o,max}$:

The observed uptake rate, R_{obs}, is simply $q_{o,max}$ expressed in terms of unit volume of aggregate:

$$R_{obs} = 0.3 \frac{\text{mg } O_2}{\text{g fresh wt} \times \text{h}} = 8.33 \times 10^{-5} \frac{\text{mg } O_2}{\text{cm}^3 \text{ (aggregate)} \times \text{s}}$$

The appropriate length scale for a sphere is $L = R/3 = 0.016$ cm. The bulk oxygen concentration $C_{b,O_2} = 8$ mg/L $= 0.008$ mg/cm³. The effective diffusivity of oxygen $D_{eff,O_2} = 1.8 \times 10^{-5}$ cm²/s.

$$\Phi_W = \frac{R_{obs} L^2}{C_{b,2} D_{eff,O_2}} = \frac{8.33 \times 10^{-5} \times 2.77 \times 10^{-4}}{0.008 \times 1.8 \times 10^{-5}}$$

Φ_w at $q_{o,max} = 0.16$.

Since Weisz's modulus is less than 0.3, mass transfer limitations are unlikely to have been significant. From Figure 4.6, note that for a zero-order reaction in spherical geometry, η is equal to 1 at $\Phi_w = 0.16$. Since intraparticle limitations were not significant, the measured $q_{o,max}$ is closely reflective of the intrinsic value.

(ii) Weisz's modulus at $\frac{1}{2}q_{o,max}$:

$$R_{obs} = 4.16 \times 10^{-5} \frac{\text{mg } O_2}{\text{cm}^3 \text{ (aggregate)} \times \text{s}}$$

$$C_{b,O_2} = 0.0004 \text{ mg/ cm}^3$$

L and the effective diffusivity remain unchanged.

$$\Phi_w = \frac{4.16 \times 10^{-5} \times 2.77 \times 10^{-4}}{4 \times 10^{-4} \times 1.8 \times 10^{-5}} = 1.61$$

For this case, Φ_w is greater than 1. From Figure 4.6, the curve for zero-order spherical geometry shows that the effectiveness factor is about 0.8 at a Weisz's modulus of 1.6, indicating that internal mass transfer limitations were significant. Since part of the reduction of the observed rate can be attributed to the lowered effectiveness factor, and not to actual deviation from zero-order (saturated) kinetics, the measured $K_{m,app}$ is probably an overestimate of the intrinsic value.

Part B: From Section C.2, we know that external and internal mass transfer limitations are coupled and that external limitations become important with increasingly severe internal limitations. What is the error made by assuming that external limitations were insignificant at the maximal and half-maximal observed rates in Part A? The packed bed of Figure 4.5 was operated at a liquid recirculation velocity of 0.4 cm/s. Assume that the physical properties of the liquid medium are similar to those for water.

Solution: By assuming negligible external resistances, we assumed that $C_{s,O_2}/C_{b,O_2}$ was equal to 1. The deviation of this ratio from unity gives us the error made in the assumption.

From equation [4.21] we know that

$$\frac{C_{s,O_2}}{C_{b,O_2}} = 1 - \left(\frac{2}{Sh^*}\right)\left(\frac{dC^*}{dr^*}\right)^{?}_{surface}$$

Whereas the modified Sherwood number Sh^* can be estimated from empirical correlations, we first need an estimate of the dimensionless concentration gradient just inside the particle. Since we have no information about the concentration profile within the particle, we must express the concentration gradient at the particle's surface in terms of observable parameters.

Since the supply of oxygen to the particle takes place by diffusion via the surface of the particle, the mass conservation equation dictates that, at steady state, consumption must be balanced by the rate of diffusion through the liquid-particle interface, or

$$R_{obs}\left(\frac{4\pi R^3}{3}\right) = D_{eff,O_2}\left(4\pi R^2\right)\left(\frac{dC}{dr}\right)_{surface}$$

Nondimensionalizing the concentration gradient with the particle radius ($r^* = r/R$) and the bulk oxygen concentration $\left(C^* = C/\ C_{b,O_2}\right)$, and rearranging terms,

$$\frac{1}{3}\frac{R_{obs}\,R^2}{D_{eff,O_2}\,C_{b,O_2}} = \left(\frac{dC^*}{dr^*}\right)_{surface}$$

or, if we define the length scale $L = R/3$ for spherical geometry,

$$\left(\frac{dC^*}{dr^*}\right)_{surface} = 3\,\Phi_w.$$

Substituting in the criterion for external limitations,

$$\frac{C_{s,O_2}}{C_{b,O_2}} = 1 - \left(\frac{6}{Sh^*}\right)\Phi_w \text{ for spherical geometry.}$$

Next, we estimate the modified Sherwood number which is the same as the Sherwood number for this case because the diffusivity and the effective diffusivity are the same (see equation [4.23]). Using the empirical correlation from Table 4.1 for a packed bed,

$$Sh = 1.17\,(Re)^{0.585}\,(Sc)^{0.33}$$

The Reynolds number $Re = (d_pU\rho)/\mu$ where the particle diameter $d_p = 0.1$ cm; the liquid velocity $U = 0.4$ cm/s; the density $\rho = 1$ g/cm^3; and the viscosity $\mu = 0.0095$ g/(cm × s).

$$Re = (0.1 \times 0.4 \times 1)/(0.0095) = 4.2$$

The Schmidt number $Sc = \mu/\rho\,D_{O_2} = 528$

$$Sh^* = Sh = 1.17\,(4.2)^{0.585}\,(528)^{0.33} = 21.6$$

Gauging the extent of external limitations:

(i) At $q_{o,max}$ where $\Phi_W = 0.16$,

$$\frac{C_{s,O_2}}{C_{b,O_2}} = 1 - \left(\frac{6(0.16)}{21.6}\right) = 0.96$$

The surface concentration is less than the bulk oxygen concentration by only 4%, which suggests that external limitations were negligible under these operating conditions.

(ii) At $\frac{1}{2}q_{o,max}$ where $\Phi_W = 1.61$,

$$\frac{C_{s,O_2}}{C_{b,O_2}} = 1 - \left(\frac{6(0.61)}{21.6}\right) = 0.55$$

A significant external mass transfer limitation is evident in this case, since C_{s,O_2} is 45% less than the bulk oxygen concentration. Note that the Weisz's modulus calculation used the bulk concentration assuming that external limitations were insignificant. If the Weisz's modulus is recalculated on the basis of the surface concentration, we get a value of 2.9, at which, from Figure 4.6, η is about 0.5 for a zero-order reaction in spherical geometry. Therefore, the reduction in the observed rate from $q_{o,max}$ to $\frac{1}{2}q_{o,max}$ may have been due to a combination of external and internal mass transfer limitations, and not because the intrinsic value of K_m had been reached!

Example 3: Biocatalyst Particle Design for an Immobilized Cell Process

Kobayashi et al. (1989) studied the effects of biocatalyst particle size and cell loading on the production of berberine by immobilized cells of *Thalictrum minus.* Initially, they used gel beads of 5-mm diameter with 33% by volume cells. Under these conditions, they experimentally measured uptake rates of sugar, phosphate, ammonium, and nitrate for free and immobilized cells. The measured effectiveness factors were 1 (or even >1 for some cases!) for all four macronutrients. The interested reader should confirm that Weisz's criterion was satisfied for each nutrient and that the theoretical analysis would predict what was observed experimentally.

Part A: Oxygen supply to biocatalyst particles was, however, found to be insufficient, as evidenced by an experimentally observed effectiveness factor of 0.4. Calculate Weisz's modulus for oxygen to confirm that internal mass transfer limitations would indeed be expected. Does the predicted effectiveness factor agree with the experimental measurement? The experimental details were as follows:

Inoculum size: 1 g fresh weight of cells per 30 mL of medium
Bead size: 2.5-mm radius
Cell loading: 1 g fresh weight of cells per 3.33 cm^3 of bead volume

Effective diffusivity for oxygen: estimated at 1.62 x 10^{-5} cm^2/s
Bulk oxygen concentration at which rate was measured: 253 µmol/L
Observed uptake rate: 199 µmol/(L×h)

Solution: The observed uptake rate must first be expressed in terms of unit biocatalyst particle volume before Weisz's modulus can be calculated.

$$R_{obs} = \frac{199\ \mu mol}{3600 \times 1000\ mL \times s} \times \frac{30\ mL}{1\ g\ fresh\ wt} \times \frac{1\ g\ fresh\ wt}{3.3\ cm^3\ (bead)}$$

$$= 5.025 \times 10^{-4} \frac{\mu mol\ O_2}{cm^3 (bead) \times s}$$

$$\Phi_w = \frac{R_{obs} L^2}{C_{b,O_2}\, D_{eff,O_2}} = \frac{5.025 \times 10^{-4} \times (0.0833)^2}{0.253 \times 1.62 \times 10^{-5}} = 0.85$$

Since Weisz's modulus was not much less than 1, mass transfer limitations could not have been insignificant. From Figure 4.6, however, the expected effectiveness factor is about 0.95, which is much higher than the experimentally determined value of 0.4.

Part of the discrepancy may be attributed to inaccuracies in measuring and estimating the modulus parameters. Additionally, the possible role of external mass transfer limitations has not been accounted for. Its exact extent is difficult to estimate in this case since the oxygen uptake apparatus and stirring conditions have not been described in Kobayashi et al. (1989). As we have seen in the previous example (Part B), external limitations can significantly lower the expected effectiveness factor.

Another possible explanation for the discrepancy is based on physiological considerations and on the fact that oxygen uptake is predominantly a zero-order reaction. Even at an expected effectiveness factor of 0.95, concentration gradients of oxygen must have been present in the bead (see Figure 4.8). Furusaki et al. (1988) have speculated that cells may not be able to withstand lowered oxygen concentrations for prolonged periods. A small fraction of cells in the core of the particle may have lysed and released potentially toxic compounds. These compounds could have affected the metabolic vigor of neighboring cells, the consequence of which would be a lower effectiveness factor than would be predicted based solely on physicochemical principles.

Part B: To circumvent oxygen transport limitations, Kobayashi et al. (1989) changed the bead parameters such that the experimental effectiveness factor was 1. Confirm the experimental result with Weisz's analysis. What was the concentration in the center of the particle? The modified conditions were as follows:

Bead size: 0.1-cm radius
Cell loading: 1 g fresh weight of cells per 13.4 cm^3 of beads

Observed uptake rate: 507 μmol/(L × h)

Solution:

$$R_{obs} = 3.15 \times 10^{-4} \frac{\mu\text{mol } O_2}{\text{cm}^3 \text{ (bead)} \times \text{s}}$$

$$\Phi_w = \frac{R_{obs} L^2}{C_{b,O_2} D_{eff,O_2}} = \frac{3.15 \times 10^{-4} \times (0.033)^2}{0.253 \times 1.62 \times 10^{-5}} = 0.084$$

In this case, Weisz's modulus is much less than 1, and, from Figure 4.6, the expected effectiveness factor is 1, as observed. From equation [4.30], the concentration at the center may be estimated as 88% of that on the surface, suggesting that the particles were virtually gradientless. The take-home lesson from these examples is that Weisz's analysis provides a simple method of ascertaining whether or not mass transfer limitations are present in biocatalyst particles. However, it may be more difficult to quantitatively predict the deterioration of catalytic activity due to the presence of significant mass transfer limitations.

Next, we examine some aspects of predicting concentration profiles within particles and of predicting effectiveness factors and biocatalyst performance based on intrinsic kinetic data.

D.4. Predicting Concentration Profiles and Effectiveness Factors

If an experiment is knowingly or unknowingly conducted under conditions in which mass transfer limitations are severe, then some sense of the concentration profile within the catalyst may be desirable for proper interpretation of results. We shall encounter such situations in immobilized cell culture (Chapter 7) and in organ culture (Section E, this chapter). In other situations, one may wish to predict the performance of an immobilized cell preparation ***a priori***, that is, solely based on the knowledge of free-cell intrinsic kinetics (V_{max} and K_m of the fine suspension). This section is devoted to addressing the above situations.

The key step in predicting concentration profiles and effectiveness factors is the application of the law of conservation of mass. The mass conservation equation for reaction and diffusion in a single dimension in flat-plate coordinates can be written as (Bird et al. 1960):

Rate of accumulation = Rate of diffusion into particle – Rate of consumption

$$\frac{\partial C_i}{\partial t} = D_{eff,i} \frac{\partial^2 C_i}{\partial x^2} - V(C_i) \qquad [4.32]$$

where C_i is the concentration of component i inside the particle, $D_{eff,i}$ is the effective diffusivity, and $V(C_i)$ is the intrinsic rate of reaction as a function C_i.

Many simplifying assumptions are made in formulating and solving the above equation (see Karel et al. 1985):

- Convective transport does not occur within the particle.
- The particle is homogeneous.
- The diffusivity is independent of the concentration.
- The component i is not consumed by other reactions.

We next consider a further important simplification, namely, that the system is at steady state. Finally, we shall take a brief look at the unsteady-state problem in Section D.4.3.

D.4.1. Steady-State Solution: Concentration Profiles At steady state, the mass conservation equation reduces to

$$D_{eff,i} \frac{d^2C_i}{dx^2} - V(C_i) \qquad [4.33]$$

The boundary conditions are (see Figure 4.4):

(i) $C_i = C_{s,i}$, the surface concentration at the particle's surface.
(Note that when external mass transfer is insignificant, $C_{s,i} = C_{b,i}$.)

(ii) $\frac{dC_i}{dx} = 0$ at $x = 0$ due to symmetry at the center.

Case 1. Michaelis-Menten Kinetics
Since saturation kinetics are nonlinear, simple analytical solutions are not available, and a numerical computer solution is required to solve for $C_i(x)$, the concentration profile within the particle. This problem has been solved by, among others, Hulst et al. (1985) and Furusaki et al. (1988). Their results bear out the intuitive prediction that larger particles experience more severe concentration gradients, as do particles with higher cell loadings. The experimental verification of these predictions has not been reported for immobilized plant cell systems but has been reported for other immobilized biocatalyst systems (e.g., see De Beer and Van den Heuvel 1988).

Case 2. Asymptotic Cases: Zero- and First-Order Kinetics
If the value of C_i as compared to K_m is such that Michaelis-Menten kinetics can be largely approximated by zero-order ($C_i >> K_m$) or first-order ($C_i << K_m$) regimes, then the steady-state mass conservation equations can be solved to yield simple analytical expressions for the concentration profiles. Table 4.3 summarizes these expressions for various geometries.

D.4.2. Thiele Modulus We have seen how Weisz's analysis can be used to predict effectiveness factors when observed reaction parameters for immobilized cells are available. In another situation, intrinsic free-cell kinetic parameters (V_{max} and K_m) may

Table 4.3. Steady-state intraparticle concentration profiles for unidimensional transport for zero- and first-order reactions. C is the concentration inside the particle, and C_s is the surface concentration.

Zero-order reaction[a], $V = V_{max}$

Flat plate: $$C = C_s - \frac{1}{2}\left(\frac{V_{max}}{D_{eff}}\right)(b^2 - x^2)$$

Cylinder: $$C = C_s - \frac{1}{4}\left(\frac{V_{max}}{D_{eff}}\right)(R^2 - r^2)$$

Sphere: $$C = C_s - \frac{1}{6}\left(\frac{V_{max}}{D_{eff}}\right)(R^2 - r^2)$$

First-order reaction[b], $V = KC$

Flat slab: $$\frac{C}{C_s} = \frac{\cosh\{(K/D_{eff})^{1/2}\,x\}}{\cosh\{(K/D_{eff})^{1/2}\,b\}}$$

Cylinder: $$\frac{C}{C_s} = \frac{I_0\{(K/D_{eff})^{1/2}(r/R)\}}{I_0\{(K/D_{eff})^{1/2}\}}$$

Sphere: $$\frac{C}{C_s} = \frac{\sinh\{(K/D_{eff})^{1/2}\,r\}}{\sinh\{(K/D_{eff})^{1/2}\,R\}}$$

[a] Assumed that concentration is finite at the center; b is the half-thickness of the slab when both sides are exposed to the bulk liquid, and R is the radius of the cylinder or sphere.
[b] I_0 is a Bessel function of order 0. Values can be found in standard mathematical tables, e.g., Abramowitz and Stegun (1965).

be available. In the latter case, the Thiele modulus can be used to predict effectiveness factors for an immobilized cell particle *a priori.*

The Thiele modulus is obtained by nondimensionalizing the mass conservation equation and can be construed as a ratio of characteristic times for diffusion and reaction. As it was defined originally (Thiele 1939), a different modulus was used for different reaction orders and geometries, making its use very cumbersome. Generalized moduli have been defined (Aris 1969; Froment and Bischoff 1990) for more convenient use.

As we have seen in Section D.3.1, the parameter β, the dimensionless saturation constant ($C_{b,i}/K_m$) indicates whether the reaction tends toward either a zero-order asymptote ($\beta \approx 50$) or a first-order asymptote ($\beta \approx 0$) or lies in an intermediate kinetic range. We shall discuss two cases, namely, the asymptotic case and the intermediate kinetics case.

Part A: Asymptotic Kinetics: Zero- or First-Order Reaction

The generalized modulus for integer-order reactions is defined as follows (Aris 1965):

$$\phi = L\left[\left(\frac{n+1}{2}\right)\frac{K\,C_{b,i}^{n-1}}{D_{eff,i}}\right]^{1/2} \qquad [4.34]$$

where ϕ is the generalized Thiele modulus; L is the length scale of the particle, defined as the volume/surface ratio of the particle, as for the Weisz's modulus (see Section D.2.1); n is the order of reaction (this expression is valid for $n > -1$); $C_{b,i}$ is the bulk steady-state concentration of component i; $D_{eff,i}$ is the effective diffusivity of i in the biocatalyst particle; and K is the rate constant, defined as V_{max} for a zero-order reaction and V_{max}/K_m for a first-order reaction when Michaelis-Menten parameters are used in their asymptotic forms.

The effectiveness factor versus generalized Thiele modulus chart is shown in Figure 4.10. The following observations are relevant:

(i) The curve for zero-order kinetics is represented by two straight lines:

$\eta = 1$ for $\phi < 1$

$\eta = 1/\phi$ for $\phi > 1$

(ii) The first-order curve is a mean drawn through three closely lying curves for the different geometries. Analytical expressions can be found in Aris (1969) or Aris (1965).

(iii) Effectiveness factor values for Michaelis-Menten kinetics lie in between the corresponding asymptotic zero-order and first-order effectiveness factors.

(iv) The errors involved in estimating the modulus parameters are significant. Aris (1965) has suggested that it is better to view η versus ϕ curves as diffuse bands rather than as precise functions.

Part B: Intermediate Kinetics

If the operational conditions are such that the kinetics follow neither zero- nor first-order behavior ($\beta \approx 1$), then an empirical approach proposed by Kobayashi et al. (1976) may be useful for prediction of effectiveness factors.

D.4.3. Transient Reaction/Diffusion Processes If a system exhibits rapid transients such that a steady-state or even a pseudo-steady-state assumption is invalid, then Weisz's and Thiele's approaches are not strictly applicable, and a transient analysis is necessary. Such situations occur in batch plant cell culture when a biocatalyst particle rapidly depletes non-volatile nutrients from a finite volume of culture medium. For example, phosphates and even hormones have been observed to be significantly depleted from a shake flask within hours and are sequestered within the cells. Since the characteristic time for diffusion itself is on the order of hours, it is possible that the central portion of the particle may be exposed to a lower concentration of the component. Cells, depending on their position within the particle, may conceivably sequester different amounts of the component. Since phosphate and hormones play

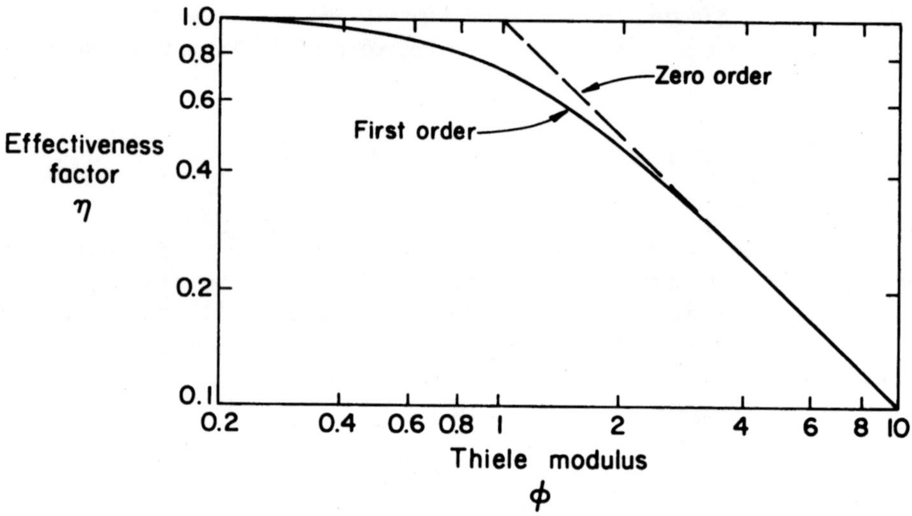

Figure 4.10 Effectiveness factor versus generalized Thiele modulus chart for zero- and first-order reactions. Thiele modulus is defined in equation [4.34]. Reproduced, with permission, from Aris (1965).

an important role in cellular metabolism and secondary metabolism, cells within a particle may exhibit spatial heterogeneity with respect to secondary metabolite production because of the spatial heterogeneity of phosphate or hormone distribution. We shall evoke such arguments during a discussion of the spatial heterogeneity in immobilized cell particles in Chapter 7.

Even though transient reaction/diffusion has not been quantitatively analyzed in the context of plant cell culture situations, a brief mathematical representation is provided for the interested reader. The mass conservation equations for the particle and the fluid phases are as follows:

Within the particle:

$$\frac{\partial C_i}{\partial t} = \frac{D_{eff,i}}{r^2}\left(\frac{\partial}{\partial r} r^2 \frac{\partial C_i}{\partial t}\right) - V(C_i) \qquad [4.35]$$

Boundary conditions: at $t = 0$, $C_i = 0$ (Initial condition)

$r = R$, $C_i = C_{b,i}(t)$ (No external mass transfer)

$r = 0$, $\dfrac{\partial C_i}{\partial r} = 0$ (Due to symmetry)

In the bulk liquid:

Component leaving liquid = Component entering particle by diffusion

$$V_{fluid} \frac{dC_{b,i}}{dt} = - D_{eff,i} (A) \left[\frac{\partial C_i}{\partial r}\right]_{r=R} \qquad [4.36)]$$

Initial condition: $t = 0$, $C_{b,i} = C_{bo}$

where A is the area of the particle exposed to the fluid, V_{fluid} is the volume of the fluid, and C_{bo} is the concentration of i at the start of the experiment.

The solution to the above equations is complex. Hatton et al. (1979, 1982) and Lewellen et al. (1982) have provided an analytical solution for transient reaction and diffusion in the presence of a first-order reaction. Their results and simulations confirm our intuitive expectation that steep concentration gradients of the reactant can indeed exist within the reacting volume, and in severe cases the central core of the particle may be deprived of the reactant.

D.5. Limitations of Reaction/Diffusion Analyses

Although the reaction/diffusion analyses presented in this section are extremely useful in analyzing and predicting biocatalyst performance, some limitations and restrictions should be kept in mind. Some of the restrictions pertain to the mathematical formalism employed. In principle, these restrictions can be circumvented by using more sophisticated mathematical models. Such models have not been developed, mostly because this level of sophistication has not been warranted as yet. Other restrictions arise from the simplification of a biological system to a simple and idealized physicochemical system [see Karel et al. (1990) for a more complete discussion].

D.5.1. The System Should Be at Steady-State Strictly speaking, Weisz's and Thiele's analyses are applicable only to steady-state situations. However, a true steady state is seldom reached in batch shake flasks or in batch reactors. The analyses can nevertheless be used for those reactions for which the pseudo-steady-state assumption is valid, that is, if the rates of reaction and diffusion are much faster than the rate of disappearance of reactant from the culture medium. The concentrations of many plant culture media components change appreciably only over a time frame of days whereas the characteristic times for reaction/diffusion phenomena are on the order of hours. Weisz's and Thiele's analyses can still be taken advantage of with batch plant cell culture systems. Possible exceptions include those non-volatile components that are very rapidly depleted from the medium, such as phosphate and hormones. The unsteady-state or transient analysis that was briefly alluded to in Section D.4.3 then becomes appropriate.

D.5.2. The Biocatalyst Particle Should Be Spatially Homogeneous The reaction/diffusion analyses are based on a simplified mass conservation equation in which the biocatalyst particle is considered to be spatially homogeneous with respect to the reaction rate and the effective diffusivity. If the biocatalyst particle under consideration is an organ such as a leaf or root, then the presence of a variety of cell types may contribute to spatial heterogeneity of the reaction rate and the effective diffusivity. When reaction/diffusion analyses are used in such systems, it is with the implicit simplifying assumption of spatial homogeneity, which is almost certainly incorrect in some cases.

For an immobilized cell preparation, spatial homogeneity implies that the cells should be uniformly well distributed within the particle. Such conditions are probably satisfied at the beginning of a reactor run or an experiment using freshly prepared immobilized cell beads. A common observation is that additional growth takes place around the periphery of the particle with prolonged culture, thus disrupting spatial homogeneity. Toda and Sato (1985) have developed a mathematical model that describes how spatial heterogeneities develop in response to concentration gradients of oxygen in an initially homogeneous immobilized yeast particle.

A related issue to which no convincing answer is available as yet is whether or not intrinsic kinetic constants are themselves dependent on the history of the cells. For example, if V_{max} and K_m for oxygen uptake have been determined for a cell population that has been subjected to adequate aeration, will these kinetic constants remain the same even when the cells have been subjected to lower oxygen concentrations for prolonged periods in the center of an immobilized cell particle?

D.5.3. Interacting Concentration Profiles A further set of implicit assumptions is involved in the reduction of an immensely complex biological system to a simplified, analytically tractable, physicochemical system.

(i) The uptake of each component of the medium is considered to be a simple chemical reaction which depends only on the concentration of its reactant. If the reaction is severely product inhibited, the curves shown in Figure 4.6 are not applicable, and Weisz's criterion has to be modified. The criterion for $\eta = 1$ becomes more stringent; that is, Φ_w has to be lower than for the corresponding non-product-inhibited reaction. Further details are provided in Roberts and Satterfield (1965) and Froment and Bischoff (1990).

(ii) The reactions are assumed to be mutually independent of each other. This assumption is restrictive since the kinetics of one reaction may depend in an indirect way on the concentration of a component not directly involved in the reaction. For example, consider the energy-requiring, active uptake of a component i from the medium by an immobilized cell preparation. It is not unreasonable to assume that the uptake rate will be dependent on the energy charge within the cells; that is, the kinetics will indirectly depend on the local oxygen concentration to which the immobilized cells are exposed. The presence of severe oxygen gradients within the bead would result in η for uptake of i being less than 1, even though an *a priori* prediction of η for component i may indicate otherwise.

E. CONVECTIVE MASS TRANSFER IN POROUS BIOCATALYSTS (ORGAN CULTURES)

E.1. Introduction

As outlined in Section A.2, mass transfer normally occurs by diffusion or by convection. In biological systems, convection, diffusion, and reaction can occur in various combinations, in series or in parallel. For example, in flow-impermeable catalyst particles such as gel beads, convective mass transfer brings nutrients to the boundary layer, followed by diffusion through to the bead surface, and then reaction and diffusion occur in parallel within the bead. For flow-permeable catalyst particles, a more complex situation may arise in which both convection and diffusion occur in parallel with reaction. This situation has been characterized for first-order kinetics (Nir and Pismen 1977), for zero-order kinetics in slab geometry (Rodrigues and Orfao 1984), and for creeping (very slow) flow around submerged objects (Stephanopoulos and Tsiveriotis 1989). Alternatively, it is possible for convective mass transfer to be much faster than diffusion, giving rise to the simpler situation of parallel convection and reaction within a porous catalyst.

Root or shoot clumps may be regarded as large porous biocatalysts within which it is possible to induce fluid flow and therefore convective mass transfer. When fluid flows freely around an object such as a root clump, a pressure difference results between the upstream side and the downstream side. Such a pressure difference across a porous material can result in internal convective flows depending on the "hydraulic permeability," κ, as given by Darcy's law (Bird et al. 1960):

$$v_s = \frac{\kappa dP}{\mu dx} \qquad [4.37]$$

where μ is the fluid viscosity, v_s is the superficial flow velocity (cm/s) - the velocity that would result if the entire cross-sectional area were available for flow - and dP/dx is the pressure gradient, which may be given by the pressure difference across a clump divided by the clump diameter.

How important is the convective contribution to total mass transfer into root or shoot clumps? Root or shoot clumps tend to be quite large (centimeters to tens of centimeters), and, on this scale, characteristic times for diffusion are long (L^2/D_{eff}, Section A.2). The Thiele modulus for zero-order oxygen uptake into a small 5-cm clump of root density 0.37 g fresh wt./mL and uptake rate 0.3 mg O_2/g fresh wt. per h is 11.6. From Figure 4.10, we see that less than 9% of the tissue would be expected to be oxygen sufficient based on diffusive mass transfer alone. As discussed further in Chapter 8, if organ cultures are to remain viable, grow, and form product, convective mass transfer must be sufficiently rapid.

On the scale of the clump as a whole, reaction appears to occur in parallel with convection; however, the situation becomes somewhat more complicated if we look at mass transfer and reaction on the scale of a rootlet within the clump. Rootlets are thick enough that internal and external (to the rootlet) mass transfer can be important, as they are for other immobilized cells (Example 2). As summarized in Figure 4.11, the entire mass transfer and reaction problem consists of (1) convective mass transfer to the clump surface, (2) convective mass transfer within the root clump, (3) diffusion across a stagnant boundary layer surrounding rootlets within the clump, and (4) diffusion and reaction within the rootlet itself.

Figure 4.11 also shows that, as fluid flows along a "streamline," oxygen is taken up such that the intraclump "bulk" oxygen concentration decreases steadily until the fluid emerges on the downstream side of the clump. Within a clump, the situation very much resembles a "packed bed" of cylindrical catalyst particles.

The objective in system design for organ cultures will normally be to maintain nutrient, especially oxygen, levels above some critical concentration. Given that (1) oxygen concentrations decrease along a streamline and (2) oxygen concentration within a rootlet is lowest within the root tip (Chapter 8), the design criterion will be to maintain oxygen levels above some critical concentration in a tip located somewhere on the downstream side of the clump.

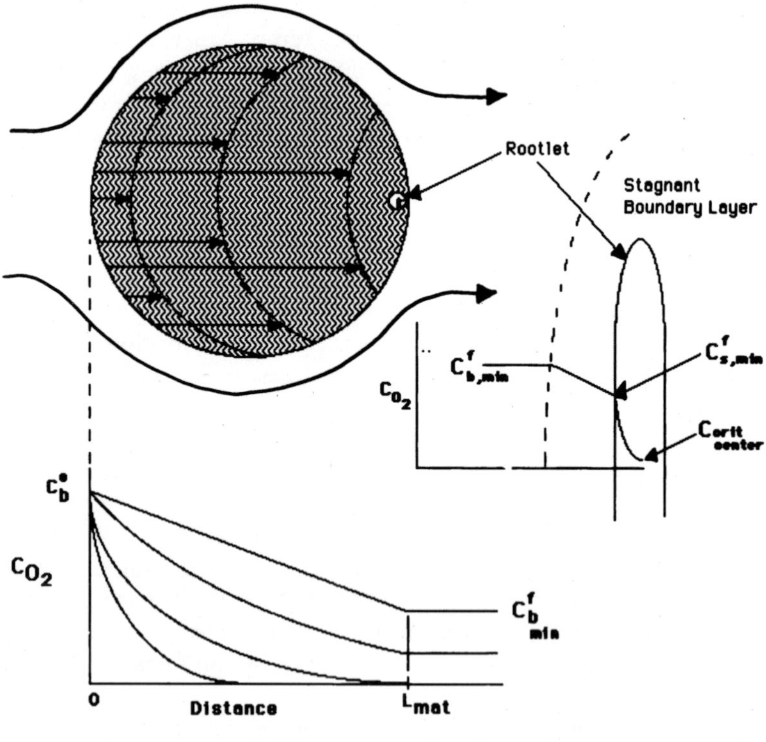

Figure 4.11 Convective and diffusive mass transfer and reaction in root clumps.

E.2. Design Analysis

It is useful to first determine the way in which the oxygen concentration decreases along a streamline. For large biocatalysts such as organ culture clumps, we may, for simplicity, neglect diffusive contributions to overall mass transfer into the clump. Consider a narrow tube of cross-sectional area A_p along the longest streamline within the clump. For a flow rate Q (cm^3/h) through the tube, we may consider a mass balance for oxygen entering at one end and being taken up before it passes out the other end.

$$\begin{pmatrix} O_2 \text{ entering} \\ \text{tube} \end{pmatrix} - \begin{pmatrix} O_2 \text{ leaving} \\ \text{at other end} \end{pmatrix} = (O_2 \text{ reacted in tube}) \qquad [4.38]$$

$$(C_b^o)Q - (C_b^f)Q = q_o \text{ (tube volume)}$$

where C_b^o = "bulk" oxygen concentration entering tube (mg O_2/cm^3); C_b^f = final "bulk" oxygen concentration leaving tube (mg O_2/cm^3); q_o= volumetric O_2 uptake at saturation [mg O_2/(mL × h)]; and tube volume = $A_p(d_{clump})$, where d_{clump} is the length of the tube or, approximately, the diameter of the clump.

We see that

$$(C_b^o - C_b^f)\frac{Q}{A_p} = d_{clump}(q_o) \qquad [4.39]$$

and since Q/A_p is the superficial velocity,

$$v_{s,min} = \frac{d_{clump}(q_o)}{(C_b^o - C_b^f)} \qquad [4.40]$$

We see from equation [4.40] that a minimum internal velocity may be defined which maintains the lowest oxygen concentration, C_b^f, above some minimum "safe" level, $C^f_{b,min}$.

What is this minimum "safe" concentration exiting a root clump? $C^f_{b,min}$ can be determined by working from the inside of the rootlet to the outside. As discussed in Chapters 5 and 8, a minimum oxygen concentration, C_{crit}, exists for a single cell to carry out aerobic metabolism. The lowest concentration occurs in the center of the root tip. Equations presented in Table 4.3 permit prediction of the concentration profile in a cylindrical catalyst such as a root:

$$C = C_s - \frac{1}{4}\left(\frac{V_{max}}{D_{eff}}\right)(R^2 - r^2) \qquad [4.41]$$

where C is the concentration at any radius r, R is the root radius, D_{eff} is the oxygen diffusivity in the tissue, and V_{max} [(mg/(mL × h)] is the volumetric uptake rate of the root tissue itself:

$$V_{max} = \frac{q_o}{(1 - \varepsilon)} \quad [4.42]$$

where ε is the "porosity" - the fraction of total clump volume which does not contain root tissue.

The minimum concentration on the surface of the root, $C^f_{s,min}$, to maintain the oxygen concentration in the center just at C_{crit} is therefore

$$C^f_{s,min} = C^{crit}_{center} + \frac{V_{max}R^2}{4D_{eff}} \quad [4.43]$$

Having determined the minimum concentration on the surface of a downstream rootlet, it is only necessary to calculate the corresponding minimum "bulk" concentration leaving the root clump, $C^f_{b,min}$. When the external mass transfer resistance through the "stagnant boundary layer" is significant, $C^f_{b,min}$ will be greater than $C^f_{s,min}$ such that there exists the necessary driving force for diffusion to the root surface. As discussed in Section C.2, C_b may be determined from C_s if the Sherwood number is known:

$$\frac{C_b}{C_s} = 1 + \frac{1}{Sh^*}\left(\frac{R}{C_s}\right)\left(\frac{dC}{dr}\right)_{surface} \quad [4.44]$$

Equation [4.44] is a simple modification of the expression derived for a sphere (equation [4.21]); each term has simply been multiplied by (C_s/C_b). Moreover, Sh* for a cylinder is defined based on the radius, rather than on the diameter as for the sphere. Sh* is the modified Sherwood number (Section C.2), in which D_{eff} in tissue is used instead of in liquid medium, and $\left(\frac{dC}{dr}\right)_{surface}$ is the concentration gradient just inside the surface of the rootlet, that is, the derivative of equation [4.41] evaluated at $r = R$,

$$\left(\frac{dC}{dr}\right)_{r=R} = \frac{V_{max}R}{2D_{eff}} \quad [4.45]$$

Substituting into equation [4.44] and with $C_s = C^f_s$,

$$C^f_b = C^f_s\left(1 + \frac{1}{Sh^*}\,\frac{V_{max}R^2}{2D_{eff}C^f_s}\right) \quad [4.46]$$

It can usually be assumed that the effective and molecular diffusivities are the same, so that Sh can be used in place of Sh*. The Sherwood number can be simply determined from empirical correlations (Table 4.1) with the Reynolds number (root radius, R, the superficial velocity, v_s, and the medium density and viscosity) and the Schmidt number (between 500 and 1000 for standard media):

$$Sh = 1.17\,Re^{0.585}Sc^{0.33} \quad [4.47]$$

The method for solving for the minimum internal flow velocity is as follows. Given the root radius R, the diffusivity of oxygen D, the oxygen uptake rate V_{max}, and some arbitrary safe value for the intrinsic critical oxygen concentration C_{crit}, equation [4.43] gives the minimum surface concentration $C^{f}_{s,min}$. An initial guess for v_s can be used in equation [4.47] to estimate the Sherwood number, which can be used in equation [4.46] to obtain the minimum "bulk" oxygen concentration leaving the root clump, $C^{f}_{b,min}$. $C^{f}_{b,min}$ can be used to estimate the minimum intraclump convective flow velocity from equation [4.40]. Finally, this value can be used iteratively to improve the estimate of Sh in equation [4.47] and to gain a more accurate estimate of the true minimum internal velocity, $v_{s,min}$. Velocities less than $v_{s,min}$ would be expected to result in root tip death over prolonged time periods (Chapter 8).

In the case of "packed bed" root reactors of the type described in Section H.5 of Chapter 8, v_s is under the direct control of the operator since short-circuiting of fluid around the root clump does not occur. Fluid is pumped through the root mass, and, therefore, the above analysis may be used directly for design purposes. For free-flowing liquid around suspended clumps, the above analysis is useful for determining the effect on $v_{s,min}$ of scaling up to larger root clumps. An example is outlined in Section H.4 of Chapter 8 for a five-fold increase in root clump diameter from 3 cm to 15 cm. For suspended root clumps, one would additionally like to be able to predict the required external velocity U (cm/s) which induces the required $v_{s,min}$.

E.3 Minimum Extraclump Velocities

Unfortunately, the complex fluid dynamics associated with flow around root clumps do not generally permit estimation of minimum extraclump velocities for a given minimum intraclump velocity. Stephanopoulos and Tsiveriotis (1989) succeeded in coupling internal to external flows only for slow flow rates (low Reynolds numbers - "creeping flow"). For organ culture applications, inertial effects associated with fluids moving at higher velocities must be considered (Prince et al. 1991). In such circumstances, it is possible only to qualitatively suggest the dependence of internal flows on external flow velocities. Qualitative descriptions may, however, permit discussion of the effect of scaling up to larger clump sizes on the magnitude of the extraclump velocity required to achieve $v_{s,min}$.

As mentioned above, the internal flow depends on the pressure distribution across the suspended clump. For a spherical clump suspended in a moving fluid (Figure 4.11), the "drag" force, F_d, divided by the cross-sectional area of the clump, A_p, can be correlated with the pressure difference across the clump (McCabe et al. 1985):

$$\Delta P \ \alpha \ \frac{F_d}{A_p} \ \alpha \ C_d U^2 \qquad [4.48]$$

where C_d is the drag coefficient as shown in Figure 4.12, and U is the extraclump flow velocity. The internal velocity v_s is proportional to the pressure gradient, which is given by

$$v_s \ \alpha \ \frac{\kappa \Delta P}{d_{clump}} \ \alpha \ \frac{C_d U^2}{d_{clump}} \qquad [4.49]$$

where d_{clump} is the diameter of the clump. Since we know that Re α (Ud_{clump}), and the dependence of C_d on Re is given (Figure 4.12), the dependence of U and d_{clump} on C_d may be inferred. In the "Stokes' law" region (low Reynolds number),

$$C_d \ \alpha \ \frac{1}{Re} \ \alpha \ \frac{1}{Ud_{clump}} \qquad [4.50]$$

Substituting into equation [4.49] and rearranging,

$$U \ \alpha \ v_s\left(\frac{d_{clump}{}^2}{\kappa}\right) \qquad [4.51]$$

The expression for $v_{s,min}$, equation [4.40], may be substituted into equation [4.51] to obtain an expression for the minimum U, U_{min}, in terms of measurable catalyst properties:

$$U_{min} \ \alpha \ \frac{q_o d_{clump}^3}{\kappa\left(C_b^o - C_b^f\right)} \qquad [4.52]$$

The minimum external velocity required to support good catalyst effectiveness is evidently very sensitive to biocatalyst size in the "Stokes' law" region.

In the "separated" region, C_d is independent of Re (and also therefore Ud_{clump}) and so, rearranging equation [4.50] for C_d = constant,

$$U \ \alpha \ \left(\frac{v_s d_{clump}}{\kappa}\right)^{1/2} \qquad [4.53]$$

Substituting again for v_s with $v_{s,min}$ (equation [4.40]),

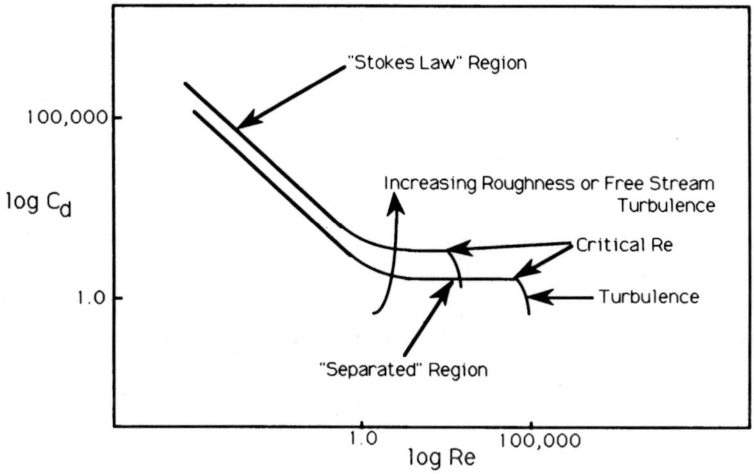

Figure 4.12 Schematic plot of typical behavior for the drag coefficient, Cd, in response to increasing Reynolds number. Increasing surface roughness or free stream turbulence can dramatically decrease the Reynolds number at which the transition to turbulence (ie. at the "critical Re") occurs. Beyond the critical Re, various behavior is observed however the drag coefficient can remain fairly constant.

$$U_{min} \propto \left(\frac{q_o}{\kappa (C_b^g - C_b^f)} \right)^{1/2} d_{clump} \qquad [4.54]$$

Beyond the "critical" Reynolds number, drag coefficients cannot be used to estimate pressure distributions over the spherical clump. Internal flow patterns become complex and may result in stagnant zones within the clump which will be difficult to supply with oxygen. The dependence of U on clump size in this flow regime is difficult to determine; however, it is likely be at least as great as it is in the separated region and may be still much more sensitive. Further evaluation of the difficulties of supplying oxygen to larger suspended clumps is given in Section H.4 of Chapter 8.

Appendix 1. Characteristic Time for Diffusion

The characteristic time for diffusion, τ_{diff}, may be defined by answering the following question:

> "If a particle (say, a flat plate of half-thickness b), initially free of component i, is submerged at time 0 into an infinitely large well-mixed volume of fluid in which the concentration of the component is $C_{s,i}$, then at what time will the concentration at the center of the particle $C_{center,i}$ reach a significant fraction, say, 70%, of $C_{s,i}$?"

Though the analytical solution for the concentration profile in the above problem is complex, the dependence of the concentration inside the particle, C_i, with time is as follows:

$$\frac{C_i}{C_{s,i}} \sim \exp\left[-\left(\frac{Dt}{b^2}\right)\right] \qquad [4.A1.1]$$

After a long time has elapsed (t very large), the exponential term decays to 1, and the concentration everywhere inside the particle is equal to the steady-state surface concentration in the absence of any chemical reaction. However, even when (Dt/b^2) is on the order of 1, that is, when t is on the order of b^2/D, a significant fraction of the steady-state value will be reached, as we shall see below. Therefore, the characteristic time for diffusion, τ_{diff}, is generally defined as L^2/D, where L is the length scale over which diffusion occurs.

Figure 4.13, modified after Crank (1975), shows a graphical representation of the transient, unidimensional concentration profiles inside the particle for three geometries. The length scales used in Figure 4.13 are the half-thickness for a plate (b) with both surfaces exposed to the fluid, and the radius (R) for a cylinder and a sphere. When $t = (b^2/D)$, that is, $(Dt/b^2) = 1$, note that ca. 90% of the steady-state surface concentration is reached at the center of a flat plate in Figure 4.13A. When $(Dt/R^2) = 1$, the steady-state value is already reached for cylinders and spheres (Figure 4.13B,C). Therefore, the definition of τ_{diff} as (R^2/D) is an overestimate of the characteristic time. The differences in the times taken to reach steady state arise due to differences in the volume-to-surface ratios for the different geometries.

Aris (1975) has proposed a transformed length scale which helps to minimize differences due to geometry and is now most commonly used in mass transfer analyses. The length scale L for any particle is defined as the ratio of the particle volume to the surface area exposed to the fluid:

For a flat plate with both sides exposed, L = the half-thickness, b.
For a plate with only one side exposed, L = the plate thickness.
For a cylinder, L = R/2.
For a sphere, L = R/3.

Finally, we redefine the characteristic time based on the transformed length scale and compute $(C_{center,i}/C_{s,i})$ after one characteristic time has elapsed for the different geometries:

(i) For a flat plate with both sides exposed, L = the half-thickness, as before, and after one characteristic time has elapsed, 90% of the steady-state surface concentration is reached at the center (Figure 4.13A).

(ii) For a cylinder, L = R/2. Therefore, the lapse of one characteristic time corresponds to $(Dt/R^2) = 0.25$ in Figure 4.13B, at which ca. 60% of the steady-state surface concentration is reached at the center of the cylinder.

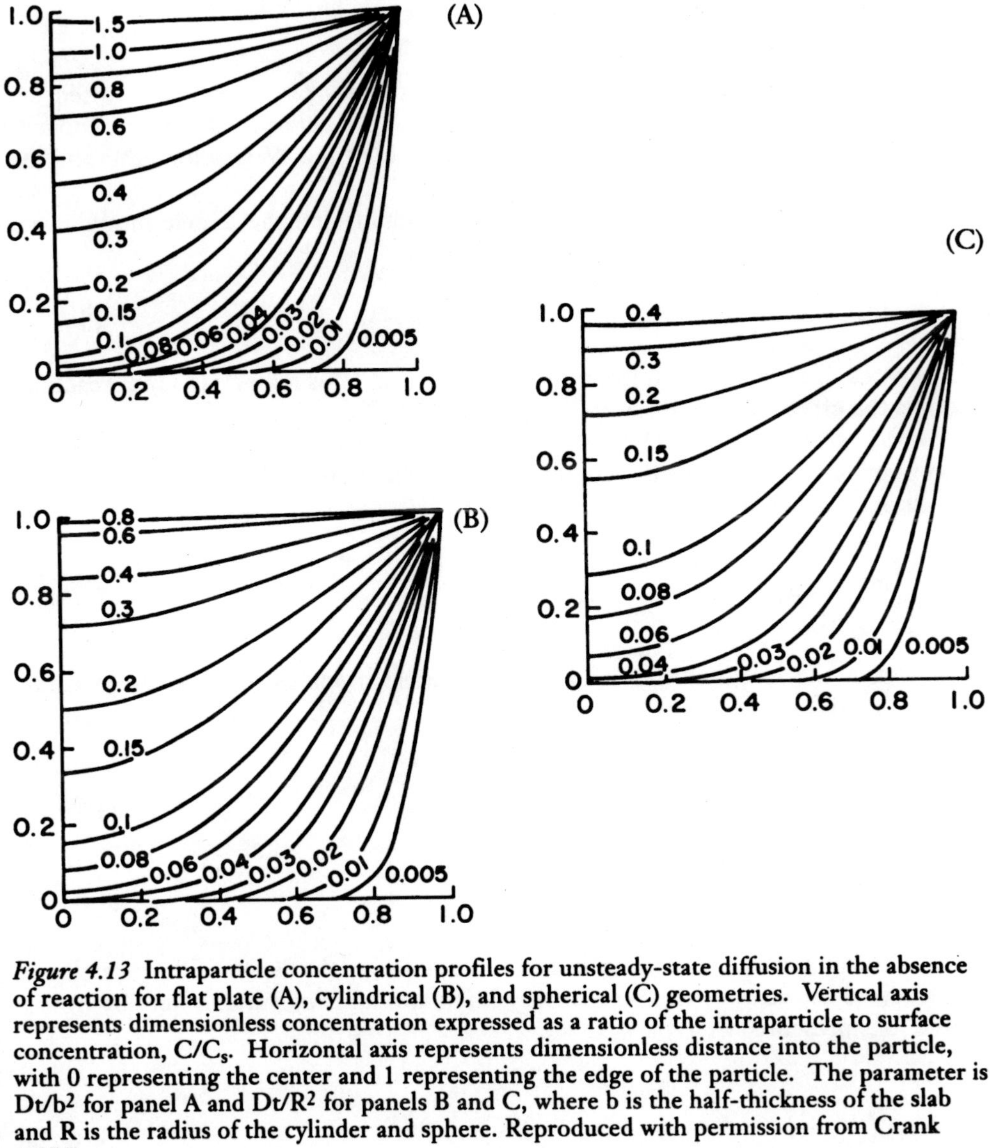

Figure 4.13 Intraparticle concentration profiles for unsteady-state diffusion in the absence of reaction for flat plate (A), cylindrical (B), and spherical (C) geometries. Vertical axis represents dimensionless concentration expressed as a ratio of the intraparticle to surface concentration, C/C_s. Horizontal axis represents dimensionless distance into the particle, with 0 representing the center and 1 representing the edge of the particle. The parameter is Dt/b^2 for panel A and Dt/R^2 for panels B and C, where b is the half-thickness of the slab and R is the radius of the cylinder and sphere. Reproduced with permission from Crank (1975).

(iii) For a sphere, the lapse of one characteristic time corresponds to $(Dt/R^2) = 0.11$ in Figure 4.13C, at which ca. 30% of the steady-state surface concentration is reached at the center of the sphere. After two characteristic times elapse, the concentration at the center reaches ca. 70% of its steady-state value.

Appendix 2. Relating Weisz's Modulus to the Concentration Profile

For a zero-order reaction, concentration gradients may be present even when the effectiveness factor is 1 (Section D.3.2). In order to predict or to specify the concentration at the center of particle, a relationship between the Weisz's modulus and the concentration at the center is required.

For an integer-order reaction, the Weisz's modulus and the Thiele modulus are related (see e.g., Roberts and Satterfield 1965):

$$\Phi_W = \eta\, \phi_o^2 \qquad [4.A2.1]$$

where Φ_W is the Weisz's modulus, ϕ_o is the Thiele modulus for a zero-order reaction, and η is the effectiveness factor.

We have seen in Section D.3.2 that for a zero-order reaction, $\eta = 1$, as long as the concentration at the center is finite (or $>2K_m$ for Michaelis-Menten kinetics). The above relation thus simplifies to

$$\Phi_W = \phi_o^2 \qquad [4.A2.2]$$

However, the length scales used for the Thiele modulus have not been transformed; therefore, the above equation has to be modified for each geometry as follows:

$$\Phi_W = \phi_o^2 \text{ for flat plates} \qquad [4.A2.3a]$$

$$4\Phi_W = \phi_o^2 \text{ for cylinders} \qquad [4.A2.3b]$$

$$9\Phi_W = \phi_o^2 \text{ for spheres} \qquad [4.A2.3c]$$

Now, the dimensionless concentration at the center of the center of the particle can be expressed in terms of the zero-order Thiele modulus by suitably modifying the expressions presented in Table 4.3:

$$C^*_{center} = 1 - \frac{\phi_0^2}{2} \text{ for a flat plate} \qquad [4.A2.4a]$$

$$C^*_{center} = 1 - \frac{\phi_0^2}{4} \text{ for a cylinder} \qquad [4.A2.4b]$$

$$C^*_{center} = 1 - \frac{\phi_0^2}{6} \text{ for a sphere} \qquad [4.A2.4c]$$

or

$$\phi_o^2 = 2\left(1 - C^*_{center}\right) \text{ for a flat plate} \quad [4.A2.5a]$$

$$\phi_o^2 = 4\left(1 - C^*_{center}\right) \text{ for a cylinder} \quad [4.A2.5b]$$

$$\phi_o^2 = 6\left(1 - C^*_{center}\right) \text{ for a sphere} \quad [4.A2.5c]$$

where C^*_{center} is defined as $C_{center}/C_{surface}$. Combining equations [4.A2.3] and [4.A2.5] yields the following desired relationships:

$$\Phi_w = 2\left(1 - C^*_{center}\right) \text{ for a flat plate} \quad [4.A2.6a]$$

$$\Phi_w = \left(1 - C^*_{center}\right) \text{ for a cylinder} \quad [4.A2.6b]$$

$$\Phi_w = 0.67\left(1 - C^*_{center}\right) \text{ for a sphere} \quad [4.A2.6c]$$

NOMENCLATURE

a	interfacial bubble area per unit volume of liquid
A	surface area of biocatalyst particle exposed to fluid
A_p	cross sectional area of particle
b	half thickness of a flat plate that has both surfaces exposed to fluid
C_{crit}	the critical oxygen concentration above which uptake is zero order
C_d	drag coefficient
C_i	concentration of component i
$C_{b,i}$	concentration of i in the bulk medium
$C_{center,i}$	concentration of i at the center of a biocatalyst particle
$C_{s,i}$	concentration of i at the biocatalyst particle surface
$\hat{C}_{O_2}$	concentration of oxygen in the liquid phase in equilibrium with gas
C^*	dimensionless concentration
C_b^o	bulk oxygen concentration in fluid stream entering porous catalyst
C_b^f	bulk oxygen concentration in fluid stream leaving porous catalyst
d_p	particle diameter
d_{clump}	diameter of porous particle or clump
D	diffusivity, with subscript i denoting component i
D_{eff}	effective diffusivity, with subscript i denoting component i
F_d	drag force
j_i	flux of component i
k	mass transfer coefficient; the subscript L is used for gas-liquid interfacial transport and S denotes solid-liquid transport
K	first-order rate constant
K_m	saturation constant; subscript "app" denotes apparent or observed value
n	reaction order
N_i	rate of mass transfer of component i
P	pressure
q_o	zero-order respiration rate; subscript "max" is sometimes also used to denote the maximal respiration rate

Q	volumetric flow rate
r	radial coordinate
r^*	dimensionless radial coordinate
R	particle radius
R_{obs}	observed reaction rate based on unit catalyst particle volume
Re	Reynolds number
Sc	Schmidt number
Sh	Sherwood number
Sh^*	modified Sherwood number or Biot number
U	external flow velocity around around catalyst particle
U_{min}	minimum external flow velocity to maintain internal nutrient sufficiency
v_s	superficial fluid velocity through porous biocatalyst clump
$v_{s,min}$	minimum internal flow velocity to maintain internal nutrient sufficiency
V	intrinsic reaction rate
V_{max}	maximal intrinsic reaction rate
ε	porosity
ϕ	Thiele modulus; a subscript is used to denote reaction order if the generalized definition is not used
Φ_w	Weisz's modulus
η	effectiveness factor
κ	hydraulic permeability
μ	viscosity
ν	kinematic viscosity
ρ	density
τ	characteristic time

References

M. Abramowitz and I. A. Stegun. 1965. "Handbook of Mathematical Functions," Dover, New York.

R. Aris. 1965. "Introduction to the Analysis of Chemical Reactors," Prentice-Hall, Englewood Cliffs, New Jersey.

R. Aris. 1969. "Elementary Chemical Reactor Analysis," Prentice-Hall, Englewood Cliffs, New Jersey.

R. Aris. 1975. "The Mathematical Theory of Diffusion and Reaction in Permeable Catalysts," Oxford University Press, London.

W. Armstrong. 1979. Aeration in Higher Plants. *Adv. Bot. Res.* 7, 225-332.

A. Axelsson and B. Persson. 1988. Determination of Effective Diffusion Coefficients in Calcium Alginate Gel Plates with Varying Yeast Cell Content. *Appl. Biochem. Biotechnol. 18*, 231-250.

J. E. Bailey and D. F. Ollis. 1986. "Biochemical Engineering Fundamentals," 2nd ed., McGraw-Hill, New York.

R. B. Bird, W. E. Stewart, and E. N. Lightfoot. 1960. "Transport Phenomena," John Wiley, New York.

V. Bringi and B. E. Dale. 1990. Experimental and Theoretical Evidence for Convective Nutrient Transport in an Immobilized Cell Support. *Biotechnol. Prog. 6*, 205-209.

Y. S. Chen and H. R. Bungay. 1981. Microelectrode Studies of Oxygen Transfer in Trickling Filter Slimes. *Biotechnol. Bioeng. 23*, 781-792.

T. J. Chresand, B. E. Dale, S. L. Hanson, and R. J. Gillies. 1988. A Stirred Bath Technique for Diffusivity Measurements in Cell Matrices. *Biotechnol. Bioeng. 32*, 1020-1036.

J. Crank. 1975. "The Mathematics of Diffusion," 2nd ed., Oxford University Press, London.

E. L. Cussler. 1984. "Diffusion: Mass Transfer in Fluid Systems," Cambridge University Press, Cambridge, U.K.

D. De Beer and J. C. Van den Heuvel. 1988. Gradients in Immobilized Biological Systems. *Anal. Chim. Acta 213*, 259-265.

G. F. Froment and K. B. Bischoff. 1990. "Chemical Reactor Analysis and Design," 2nd ed., John Wiley, New York.

Y. Fujita. 1988. Shikonin: Production by Plant *(Lithospermum erythrorhizon)* Cell Cultures. In "Biotechnology in Agriculture and Forestry," Vol. 4, "Medicinal and Aromatic Plants I," Y. P. S. Bajaj (ed.), Springer-Verlag, Berlin, pp. 225-236.

S. Furusaki, T. Nozawa, T. Isohara, and T. Furuya. 1988. Influence of Substrate Transport on the Activity of Immobilized *Papaver somniferum* Cells. *Appl. Microbiol. Biotechnol. 29*, 437-441.

B. K. Hamilton, C. R. Gardner, and C. K. Colton. 1974. Effect of Diffusional Limitations on Lineweaver-Burk Plots for Immobilized Enzymes. *AIChE J. 20*, 503-510.

B.J. Hannoun and G. Stephanopoulos. 1986. Diffusion Coefficients of Glucose and Ethanol in Cell-Free and Cell-Occupied Calcium Alginate Membranes. *Biotechnol. Bioeng. 28*, 829-835.

P. Harriott. 1962. Mass Transfer to Particles: Part 1. Suspended in Agitated Tanks. *AIChE J. 8*, 93-101.

T. A. Hatton, A. S. Chiang, P. T. Noble, and E. N. Lightfoot. 1979. Transient Diffusional Interactions between Solid Bodies and Isolated Fluids. *Chem. Eng. Sci. 34*, 1339-1344.

T. A. Hatton, P. C. Lewellen, and E. N. Lightfoot. 1982. Transient Diffusion and Reaction in Solids Immersed in Poorly-Stirred Fluids - I. Mathematical Developments. *Chem. Eng. Sci. 37*, 1315-1324.

C. G. Hill. 1977. "An Introduction to Chemical Engineering Kinetics & Reactor Design," John Wiley, New York.

A. C. Hulst, J. Tramper, P. Brodelius, L. J. C. Eijkenboon, and K. Ch. A. M. Luyben. 1985. Immobilized Plant Cells: Respiration and Oxygen Transfer. *J. Chem. Tech. Biotechnol. 35B*, 198-204.

S. F. Karel, S. B. Libicki, and C. R. Robertson. 1985. The Immobilization of Whole Cells: Engineering Principles. *Chem. Eng. Sci. 40*, 1321-1354.

S. F. Karel, P. M. Salmon, P. S. Stewart, and C. R. Robertson. 1990. Reaction and Diffusion in Immobilized Cells: Fact and Fantasy. In "Physiology of Immobilized Cells," J. A. M. de Bont, J. Visser, B. Mattiasson, and J. Tramper (eds.), Elsevier, Amsterdam, pp. 115-126.

F. Kargi and M. Moo-Young. 1985. Transport Phenomena in Bioprocesses. In "Comprehensive Biotechnology," M. Moo-Young (editor-in-chief), Vol. 2, "The Principles of Biotechnology: Engineering Considerations," C. L. Cooney and A. E. Humphrey (eds.), Pergamon Press, Oxford, U.K., pp. 5-56.

V. Kasche. 1983. Correlation of Experimental and Theoretical Data for Artificial and Natural Systems with Immobilized Biocatalysts. *Enzyme Microb. Technol. 5*, 2-13.

A. Kato, S. Kawazoe, M. Iijima, and Y. Shimizu. 1976. Continuous Culture of Tobacco Cells. *J. Ferment. Technol. 54*, 82-87.

T. Kobayashi, K. Ohmiya, and S. Shimizu. 1976. Approximate Expression of Effectiveness Factor for Immobilized Enzymes with Michaelis-Menten Kinetics. *J. Ferment. Technol. 54*, 260-263.

Y. Kobayashi, H. Fukui, and M. Tabata. 1989. Effect of Oxygen Supply on Berberine Production in Cell Suspension Cultures and Immobilized Cells of *Thalictrum minus*. *Plant Cell Rep. 8*, 225-258.

O. Levenspiel. 1972. "Chemical Reaction Engineering," 2nd ed., John Wiley, New York.

P. C. Lewellen, T. A. Hatton, and E. N. Lightfoot. 1982. Transient Diffusion and Reaction in Solids Immersed in Poorly-Stirred Fluids - II. Numerical Aspects. *Chem. Eng. Sci. 37*, 1325-1336.

J. V. Matson and W. G. Characklis. 1976. Diffusion into Microbial Aggregates. *Water Res. 10*, 877-885.

W. L. McCabe, J. C. Smith, and P. Harriott. 1985. "Unit Operations of Chemical Engineering," McGraw-Hill, New York.

Y. Miura. 1976. Transfer of Oxygen and Scale-Up in Submerged Aerobic Fermentation. *Adv. Biochem. Eng. 4*, 3-40.

A. Nir and L. M. Pismen. 1977. Simultaneous Intraparticle Forced Convection, Diffusion and Reaction in a Porous Catalyst. *Chem. Eng. Sci. 32*, 35-41.

R. H. Perry and D. Green. 1984. "Perry's Chemical Engineering Handbook," 6th ed., McGraw-Hill, New York.

C. L. Prince. 1991. Production of Plant Secondary Metabolites from Root Cultures: 1) Kinetics and Reactor Design and 2) Flavor Production in Allium species. Ph. D. Thesis, Cornell University, Ithaca, New York.

C. L. Prince, V. Bringi, and M. L. Shuler. 1991. Convective Mass Transfer in Large Porous Biocatalysts: Plant Organ Cultures. *Biotechnol. Prog.* 7, 195-199.

H. T. Pu and R. Y. K. Yang. 1988. Diffusion of Sucrose and Yohimbine in Calcium Alginate Gel Beads with or without Entrapped Plant Cells. *Biotechnol. Bioeng. 32*, 891-896.

J. M. Radovich. 1985. Mass Transfer Effects in Fermentations Using Immobilized Whole Cells. *Enzyme Microb. Technol.* 7, 2-10.

G. W. Roberts and C. N. Satterfield. 1965. Effectiveness Factors for Porous Catalysts: Langmuir-Hinshelwood Kinetic Expressions. *Ind. Eng. Chem. Fundament. 4*, 288-293.

A. E. Rodrigues and J. M. Orfao. 1984. Intraparticle Convection, Diffusion and Zero Order Reaction in Porous Catalysts. *Chem. Eng. Commun.* 27, 327-337.

A. E. Rodrigues and R. M. Quinta Ferreira. 1988. Convection, Diffusion and Reaction in a Large-Pore Catalyst Particle. *AIChE Symp. Ser. (#266) 84*, 80-87.

K. Sato and K. Toda. 1983. Oxygen Uptake Rate of Immobilized Growing *Candida lipolytica*. *J. Ferment. Technol. 61*, 239-245.

C. N. Satterfield. 1970. Mass Transfer in Heterogeneous Catalysis. MIT Press, Cambridge, Massacheussetts.

C. D. Scott, C. A. Woodward, and J. E. Thompson. 1989. Solute Diffusion in Biocatalyst Gel Beads Containing Biocatalyst and Other Additives. *Enzyme Microb. Technol. 11*, 258-263.

G. Stephanopoulos and K. Tsiveriotis. 1989. The Effect of Intraparticle Convection on Nutrient Transport in Porous Biological Pellets. *Chem. Eng. Sci. 44*, 2031-2039.

H. Tanaka, M. Matsumura, and I.A. Veliky. 1984. Diffusion Characteristics of Substrates in Ca-Alginate Beads. *Biotechnol. Bioeng. 26*, 53-58.

E. W. Thiele. 1939. Relation Between Catalytic Activity and Particle Size. *Ind. Eng. Chem. 31*, 916-920.

K. Toda and K. Sato. 1985. Simulation Study on Oxygen Uptake Rate of Immobilized Growing Organisms. *J. Ferment. Technol. 63*, 251-258.

P. B. Weisz. 1973. Diffusion and Chemical Transformation: An Interdisciplinary Excursion. *Science 179*, 433-440.

CHAPTER 5

SPARINGLY SOLUBLE VOLATILE COMPONENTS: OXYGEN, CARBON DIOXIDE, AND ETHYLENE

A. INTRODUCTION

In Chapter 2, the effect of the nutritional environment on the metabolism of the culture was discussed mainly in terms of the soluble nutrients (e.g., the carbohydrate and salt components of the medium). In this chapter, we will discuss the role of sparingly soluble volatile components on plant cells. However, to examine the role of volatile components, it is necessary to use the quantitative framework developed in Chapter 4 with respect to the transport processes between the gas and liquid phases. In the present chapter, we will focus predominantly on the role of oxygen since the importance of this nutrient for aerobic heterotrophic growth is well accepted. Also, we will consider other volatile components (i.e., carbon dioxide and ethylene) which have been proposed to be important in plant cell systems.

B. OXYGEN

There are two reasons why consideration of volatile components such as oxygen warrants treatment in a separate chapter: (1) oxygen is important to plant cell metabolism, and (2) meeting the metabolic demand of the culture is difficult. As discussed in Chapter 2, oxygen is the terminal electron acceptor in oxidative phosphorylation and, as a result, plays a critical role in the energy metabolism of the cells. In many plant cell and tissue systems, low levels of oxygen have been related to reduced cellular levels of ATP and a reduced energy charge (Saglio et al. 1983). In some cases, these reduced energy (or ATP) levels have been associated with altered metabolic or morphological changes (see Pradet and Raymond 1983). Faiz-ur-Rahman et al. (1974a, b) reported that ATP levels were reduced when aged slices of carrot were transferred from air to a nitrogen environment. Rasi-Caldogno and De Michelis (1978) reported that when gas phase oxygen levels were reduced from 20% to 3%, the energy charge of squash cotyledons was reduced and a fivefold reduction in protein synthesis was observed. Raymond et al. (1985) also reported a lower energy charge for seeds under anoxic conditions. In related work, this group reported reduced seed germination rates under low oxygen partial pressures (Al-Ani et al. 1985). Kessell and co-workers (1972, 1977) reported that carrot embryogenesis occurred at lower dissolved oxygen levels, while differentiation occurred via rhizogenesis at higher oxygen levels. This behavior was also examined in terms of the ATP levels within the tissue (Kessell et al. 1977). In addition to its role in the energy-generating systems of the cells or tissue, oxygen may also be involved in other metabolic events. Oxygenases, enzymes which utilize molecular oxygen as a substrate, are involved in several secondary metabolic pathways (Butt and Lamb 1981). Also, oxygen may play less defined roles in cell physiology. For instance, Shirsat and Nair (1981) reported that phenylalanine ammonia lyase (PAL) induction was prevented in potato tuber tissue when the oxygen level was reduced. These authors suggested that cinnamic acid accumulation under reduced oxygen concentrations was responsible for the reduced PAL activities.

Although oxygen is important in metabolism, this does not, by itself, justify treatment of volatile components in a separate chapter - the soluble nutrients are also important for cell survival. The second reason why volatile components warrant separate discussion is that supplying adequate amounts of oxygen to meet the metabolic demand is difficult due to the high metabolic rates relative to the solubility of oxygen. In fact, supplying adequate amounts of oxygen is a major concern in bioreactor scale-up. To illustrate the difficulty in meeting oxygen requirements, we compare oxygen with sugar metabolism. Table 5.1, shows that the consumption rates of these two nutrients are of the same order of magnitude. However, as shown in Table 5.1 the solubility of oxygen is orders of magnitude less than the concentration of sugar typically supplied in plant cell media. Thus, although sufficient sugar can be supplied to meet the requirements of a culture for a period of several days to a week, an aqueous solution saturated with oxygen contains only enough oxygen to provide for a few minutes of aerobic metabolism. Therefore, to prevent the culture from being depleted of oxygen, it is essential that oxygen be supplied continuously throughout the course of cultivation. This is accomplished in flasks by the transfer of oxygen from the head space into the liquid. In bioreactors, gas transfer in plant cell systems is generally accomplished by sparging air through the liquid phase. To ensure that oxygen transfer is sufficient to meet the culture's demand, the rates of these processes must be considered.

Table 5.1. Typical concentrations and consumption rates for plant cells. Data were obtained from carrot cell cultures. Adapted from Payne et al. (1990).

	Oxygen	Glucose
Concentration (mmol/L)	0.25 [a]	167 [b]
Consumption rate (mmol/g cell-h) [c]	0.2	0.17

[a] Solubility of oxygen in water at 25 °C.
[b] A typical initial sugar concentration of 30 g of glucose per L.
[c] Dry weight basis.

B.1. Biological Oxygen Requirements

The rate at which oxygen is consumed by a culture is dependent on the oxygen concentration which is available (i.e., the dissolved oxygen). This dissolved oxygen dependence can generally be described by saturation type kinetics of the form

$$q_o = \frac{(q_{o,max})\,C_{O_2}}{K_{m,app} + C_{O_2}} \qquad [5.1]$$

where q_o and $q_{o,max}$ are the respiration rate and maximum respiration rate [mmol/g cell-hr], C_{O_2} is the dissolved oxygen concentration (mmol/L), and $K_{m,app}$ is the dissolved oxygen concentration which leads to a respiration rate which is half the maximum value. Although this equation has been originally derived from enzyme kinetics, it should be noted that the $K_{m,app}$ in equation [5.1] is not the same as the intrinsic half-saturation constant for the cytochrome oxidase enzyme. Discrepancies in these values would result from mass transfer limitations, which are discussed in greater detail in Chapter 4 (page 96). For fine cell suspensions, $K_{m,app}$ approaches the intrinsic K_m for the cyctochrome oxidase enzyme, while $K_{m,app}$ increases as mass transfer limitations become more severe. It should be noted that since mass transfer characteristics can differ (e.g., mass transfer rates depend on the size of root balls or aggregates), then $K_{m,app}$ will depend on cultivation conditions (e.g., the level of shear) and may change over the course of cultivation (e.g., as root balls or'aggregates become larger).

Equation [5.1] is a traditional approach to describing oxygen transfer and uptake. The $K_{m,app}$ in equation [5.1] includes information on both the intrinsic oxygen affinity of the cells and the mass transfer operations for the diffusion of oxygen from the bulk liquid to cells in internal regions of aggregates, immobilized cell particles, and organs. This approach is contrasted by that taken in Chapter 4, in which individual transport and reaction steps are rigorously accounted for mathematically. Because equation [5.1] does not rigorously account for mass transfer, this equation is not generally valid. However, equation [5.1] is often preferred over more rigorous approaches because of its mathematical simplicity and because the value of $K_{m,app}$ can be intuitively interpreted.

As discussed in Chapter 3, the same form of equation [5.1] can be used to describe other key metabolic parameters such as the growth rate:

$$\mu = \frac{\mu_{max} C_{O_2}}{K'_{m,app} + C_{O_2}} \qquad [5.2]$$

where μ and μ_{max} represent the specific and maximum specific growth rates (hr^{-1}) and the prime indicates that the $K_{m,app}$ for growth and respiration may differ. figure 5.1 shows that the growth of *Catharanthus roseus* cells follows this saturation pattern with $K'_{m,app}$ approximately 3.4% oxygen in the gas phase. If the gas and liquid phases are in equilibrium, then this 3.4% corresponds to a dissolved oxygen concentration of 0.04 mmol/L. Pareilleux and Vinas (1983) reported a half-saturation value of 0.0088 mmol/L for C. *roseus* respiration. Differences in these values may reflect differences in mass transfer characteristics between the two experimental systems, differences in the cell lines, or differences in the response of growth and respiration to oxygen. Criddle et al. (1988), who examined the heat generation from tomato cells, also observed saturation behavior with respect to the oxygen concentration. Their results suggest that the half-saturation constant for heat evolution is similar to the above cited values.

Another concept which has been used extensively to characterize the oxygen re-

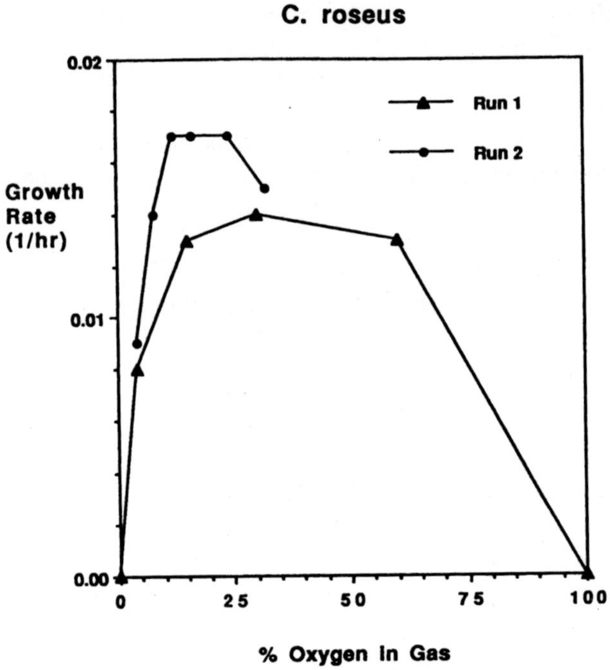

Figure 5.1. The specific growth rate of C. *roseus* as a function of the gas phase oxygen concentration. The experiment was conducted in flasks with gases of known oxygen concentration passed into the head space. This experimental system was developed such that the gas transfer capacities of the flask were large relative to the culture's oxygen consumption rate. Thus, the gas phase and liquid phases should be close to equilibrium. Data from Tate and Payne (1991).

quirements of cells is the critical dissolved oxygen concentration, $C_{O_2,crit}$ (or in the root culture literature, the critical oxygen presssure, COP). The critical oxygen level corresponds to the oxygen level above which no further effect is observed (e.g., the enzymes involved are saturated with oxygen). Typically, this critical oxygen level is between two and four times $K_{m,app}$. For instance, the COP for the growth of the *C. roseus* culture in figure 5.1 is approximately 15% oxygen in the gas (0.18 mmol/L if the gas and liquid were in equilibrium). With respect to roots and excised root tips, high levels of COP have been observed in submerged systems (Berry and Norris 1949; Bertani and Brambilla 1982; Saglio et al. 1984; Thompson et al. 1989). These high values are generally attributed to limitations associated with the transfer of oxygen from the liquid phase into the organs. Armstrong and Webb (1985) attempted to reduce these mass transfer limitations and measured COPs which were considerably lower (0.8 kPa vs. 30 kPa). Also, Saglio et al. (1984) reported that the COP for respiration was considerably lower than the COP for root growth.

In the above discussion, we have considered the oxygen requirements of the culture in terms of the affinity for oxygen. Practically, this is important because the critical dissolved oxygen value is used for designing appropriate bioreactor operating strategies. In almost all cases, it is desirable, if not imperative, to ensure that the metabolic activity of the culture is not suppressed by the oxygen limiting conditions. Thus, most bioreactors are designed to ensure that the culture never experiences dissolved oxygen levels below the critical value. A second important point, which is not as well characterized, is that the oxygen affinity of cyctochrome oxidase (in oxidative phosphorylation) may be greater than the affinity of other oxidases, and thus *the critical dissolved oxygen value for providing energy for growth and maintenance may be considerably different from the critical value for secondary metabolism.* This may explain why higher dissolved oxygen levels are sometimes required for secondary metabolite synthesis than are required for growth (see discussion of oxygen requirements for berberine synthesis in Chapter 6). finally, as mentioned in our discussion of $K_{m,app}$, the critical oxygen level will depend on the degree of aggregation, with larger aggregates having higher critical oxygen levels.

Table 5.2. Respiration rates ($q_{o,max}$) for *Catharanthus roseus* cells as measured in various laboratories.

$q_{o,max}$ (mmol/g - h)[a]	Reference
0.15	Pareilleux and Vinas (1983)
0.45	Bond et al. (1988)
0.15-0.30	Payne et al. (1988a)
0.36-0.38	van Gulik et al. (1989)

[a] Respiration rates are on a dry weight basis.

In addition to $K_{m,app}$ and COP, which describe the affinity of the cells or organs for oxygen, equation [5.1] contains a term describing the rate of oxygen consumption in the absence of oxygen limitations ($q_{o,max}$). Values for $q_{o,max}$ vary depending on the cell or tissue type and the metabolic activity of the tissue. Typical values for $q_{o,max}$ are listed in Table 5.2. These values are 10 to 100 times lower than bacterial respiration rates. Although $q_{o,max}$ is expected to change with changes in the metabolic activity of the culture, it has often been observed that these changes in $q_{o,max}$ are not large. Provided cells have had sufficient carbohydrates, we have observed that the respiration rate (per cell mass) is similar, regardless of whether the cells are rapidly growing or are in a nongrowth secondary metabolite-producing stage (e.g., respiration rates seldom differ by more than a factor of 2). As discussed in Chapter 3, high levels of oxygen are required during nongrowth stages to provide the high level of maintenance energy for the culture.

To quantify the volumetric oxygen demand of the culture, it is important not only to know the respiration rate per gram of cell (q_o), but also the concentration of cells present (X). Quantitatively, the oxygen uptake rate (OUR) is given by

$$OUR = q_o X \qquad [5.3]$$

where OUR is in units of mmol/L-hr.

In addition to its role in the kinetics of cell metabolism, oxygen can also affect the stoichiometry of nutrient conversion. Under conditions in which insufficient oxygen is available for NADH oxidation, many living systems respond by using carbonaceous intermediates (rather than oxygen) as electron acceptors. Under these conditions, reduced products such as ethanol and lactic acid accumulate. With respect to plant cells, current evidence suggests that ethanol is the major product of such fermentative metabolism. Ashihara et al. (1988) reported that alcohol dehydrogenase and pyruvate decarboxylase have higher activities than lactate dehydrogenase, which suggests that the activities for ethanol production exceed those for lactate production. Smith and ap Rees (1979a, 1979b) also reported that ethanol was the major product of anaerobic metabolism in excised roots. However, the extent of anaerobic metabolism appears to be limited. Smith and ap Rees (1979a) observed that the rate of ethanol formation was markedly reduced over the course of a few hours. Also as shown in figure 5.2, our

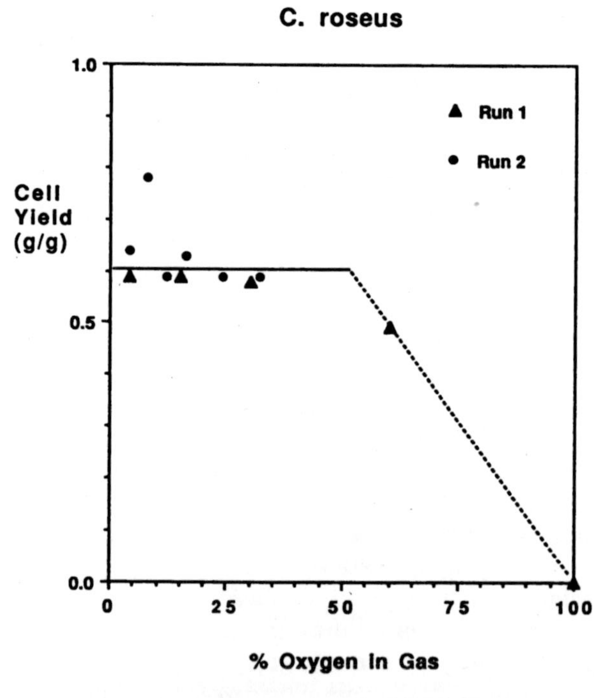

Figure 5.2. The growth yield (g dry cell weight produced/g glucose consumed) of a *C. roseus* culture grown under varying oxygen levels. Experimental system is described in the legend to figure 5.1. Data from Tate and Payne (1991).

results show that the growth yield for *C. roseus* cells was approximately constant even under conditions in which insufficient oxygen was available to support growth at μ_{max}. The constancy of this yield shows that both growth and sugar consumption were reduced proportionally. Thus, only small amounts of sugar appeared to have been diverted into the formation of anaerobic products such as ethanol. These results differ slightly from those of Pareilleux and Vinas (1983), who reported a small reduction in the growth yield for *C. roseus* cells growing under oxygen-limiting conditions.

B.2. Meeting the Biological Oxygen Demand

To meet the oxygen requirements of a culture, oxygen is transferred from a gas phase (e.g., air) to the liquid. This dissolved oxygen is then mixed throughout the liquid phase and brought into contact with the biological material (e.g., the cells, organs, or immobilized particle). Since mixing of dissolved components is fast, especially for smaller-sized bioreactors, then gas – liquid oxygen transfer is typically the slow step in bringing oxygen from the gas to the surface of finely suspended cells. For cell aggregates, immobilized cells, or organs, additional mass transfer steps (e.g., the diffusion of oxygen through an immobilized matrix to an entrapped cell) may be slow. Thus, to rigorously describe oxygen transfer to these latter systems, it is necessary to mechanistically describe the various mass transfer steps (see Chapter 4). Thus, the following discussion of gas - liquid oxygen transfer is generally sufficient to describe how the oxygen requirements of suspended cells or small aggregates are met, while this treatment is only sufficient to describe the first step in oxygen transfer to immobilized cells and organs.

As described in Chapter 4, gas - liquid oxygen transfer is described by the equation:

$$OTR = K_La\,(\hat{C}_{O_2} - C_{O_2}) \qquad [5.4]$$

where OTR is the volumetric oxygen transfer rate in mmol/L-h, K_La is the oxygen mass transfer coefficient in h^{-1} and $(\hat{C}_{O_2} - C_{O_2})$ is the driving force for oxygen transfer. $\hat{C}_{O_2}$ is the dissolved oxygen concentration which would be in equilibrium with the gas. For air (21% oxygen) at 1 atm and 25 °C, $\hat{C}_{O_2}$ is 0.25 mmol/L. C_{O_2} is the dissolved concentration in mmol/L.

Information on the oxygen transfer rate and the consumption rate are combined in a material balance:

$$\frac{dC_{O_2}}{dt} = K_La\,(\hat{C}_{O_2} - C_{O_2}) - q_o X \qquad [5.5]$$

where the term on the left-hand side accounts for the unsteady-state changes in the dissolved oxygen concentration. In many cases, this unsteady-state term is negligible in comparison to either of the terms on the right-hand side of equation [5.5]. If this unsteady-state term is unimportant, then this equation simplifies to

$$q_oX = K_La\,(\hat{C}_{O_2} - C_{O_2}) \qquad [5.6]$$

The above equation is intuitively expected – that is, the rate of oxygen uptake by the cells is balanced by the transfer rate.

Just as the $K_{m,app}$, the critical dissolved oxygen, and $q_{o,max}$ are used to characterize the oxygen requirements of the culture, the oxygen transfer rate and K_La are used to characterize the mass transfer capabilities of a cultivation system (either a flask or a bioreactor). Measurement of K_La and dependence of this value on levels of aeration and agitation are discussed in Chapter 4.

Before discussing oxygen transfer in flasks and bioreactors, there are three points that should be made. first, for most applications it is desirable to culture cells under conditions in which oxygen is not a limiting nutrient. That is, the dissolved oxygen should be equal to or exceed the critical dissolved oxygen. Under these conditions, the respiration rate is typically equal to $q_{o,max}$, and equation [5.6] can be rearranged to give:

$$K_La = \frac{q_{o,max}\,X}{(\hat{C}_{O_2} - C_{O_2,crit})} \qquad [5.7]$$

where $\hat{C}_{O_2,crit}$ is the critical dissolved oxygen concentration. This equation is useful since it can be used to calculate the required oxygen transfer coefficient of a culture system if the dissolved oxygen is to be maintained above the critical level. As will be seen, the K_La value is often used to characterize bioreactors.

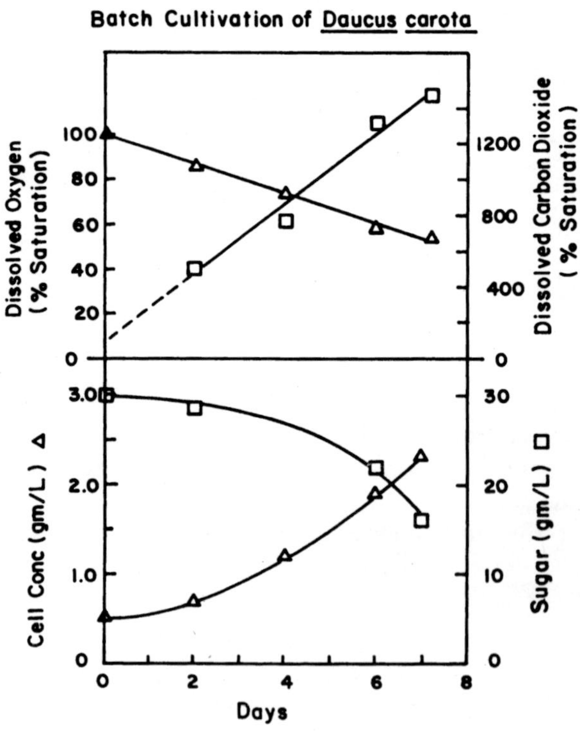

Figure 5.3. Concentration of dissolved gas species as a function of time during the batch cultivation of carrot cells in a bioreactor. Note that the dissolved concentrations in the upper plot are reported as a percent of their saturation value. For oxygen (triangles), 100% = 0.25 mmol/L. The carbon dioxide (squares) concentration refers only to the CO_2 species, and 100% = 0.01 mmol/L. The K_La for this reactor was measured to be 9 h^{-1}. Data from J. M. Smith (unpublished).

Second, since most plant cells are cultured under conditions in which the cell mass increases over time, then it can be seen that the left-hand side of equation [5.6] will be increasing with time. For the transfer rate (the right-hand side of equation [5.6]) to balance this increasing demand, then a reduction in the dissolved oxygen concentration generally occurs. This can be seen in figure 5.3, which shows that in a batch culture of *Daucus carota*, as the cell concentration increases over time, the dissolved oxygen level decreases. Also shown in figure 5.3 is the opposite trend with the dissolved carbon dioxide levels. Since carbon dioxide is produced (and not consumed by the culture), then the dissolved carbon dioxide concentration increases with increasing cell concentration. The point to be made here is that despite the fact that air was sparged to the bioreactor at a constant rate, the dissolved oxygen concentration decreased over time. This decrease resulted because the demand for this nutrient increased with increasing cell concentration. Thus, calculation of the desired K_La of a bioreactor from equation [5.7] is generally based on the maximum cell concentration. An example of such a calculation appears in Chapter 6.

A final point is that the K_La of a bioreactor is not necessarily constant. As expected, the K_La is dependent on the level of aeration and agitation. Also, the K_La depends on other less controllable factors such as the medium viscosity or whether the culture medium facilitates or suppresses air bubble coalescence. These factors can change from culture to culture and often also change during the course of cultivation due to changes in viscosity and/or release of surface-active compounds.

B.3. Oxygen Transfer in Flasks

Oxygen transfer in flasks can be limited by either the transfer of oxygen from the outside through the closure, or from the head space gas into the liquid. figure 5.4 shows that the closure on the flask can significantly restrict oxygen transfer to a culture. The impact of this reduced oxygen transfer can be seen in Table 5.3, which shows that the maximum cell concentration was reduced and the cell doubling time was increased for flasks with reduced oxygen transfer capabilities. Additional work to improve oxygen transfer into flasks is reported by Tunac (1989).

In conclusion, it can be seen that oxygen transfer in flasks can be limiting. Although the flask closure is especially important, the ratio of liquid to flask volume is also important. As the amount of liquid per flask is increased, the gas-liquid surface area per unit volume (the "a" in K_La) is reduced (Kobayashi et al. 1989; Snape et al. 1989).

B.4. Oxygen Transfer in Bioreactors.

Oxygen transfer in specific bioreactors will be considered in more detail in later chapters. However, it should be mentioned that several reports have correlated the performance of a culture (either in terms of cell growth or product synthesis) with the oxygen transfer rate or the K_La of the bioreactor. Typical or optimal K_La values have been reported to be 10-35 h^{-1} for cell growth (Kato et al. 1975; Yamakawa et al. 1983; Smart and Fowler 1984) and secondary metabolite production (Yamakawa et al. 1983; Tanaka 1987).

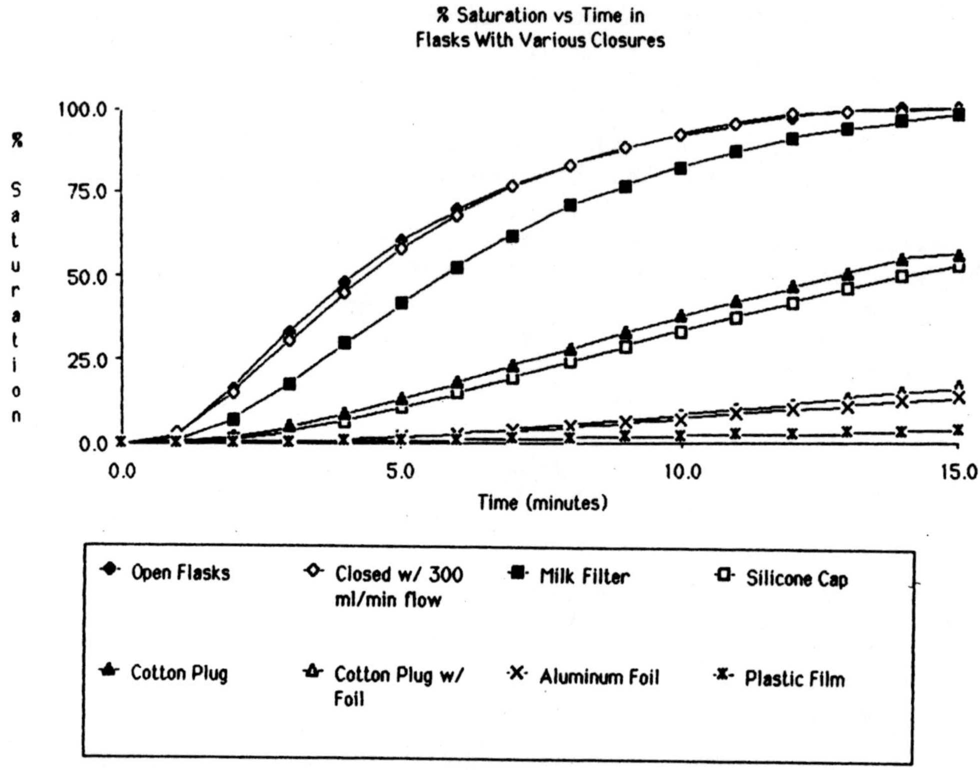

Figure 5.4. Oxygen transfer into flasks with different closures. Prior to this experiment, a dissolved oxygen probe was mounted in the side of a specially machined flask and the liquid was equilibrated with nitrogen which was flushed into the head space. After the dissolved oxygen concentration had fallen to zero, the nitrogen flush was stopped, the experiment was begun, and the dissolved oxygen concentration was followed over time. Data from Payne et al. (1988b).

C. CARBON DIOXIDE

The importance of carbon dioxide for plant cell and organ metabolism remains rather controversial, and few generalizations can be drawn from current literature. The concern about this issue arises because it has commonly been observed that plant cell growth is reduced under conditions of vigorous aeration. Although shear remains a possibility, it has also been suggested that at high levels of aeration, carbon dioxide may be stripped from the medium. If CO_2 is required by the culture, then excessive stripping may have an adverse effect on culture performance. In this section, we will provide a description of the metabolic effects which may involve carbon dioxide and then consider results from studies in which the importance of carbon dioxide has been more systematically examined.

Table 5.3. Carrot cell growth with varying oxygen transfer capacities. Experiments were conducted in 500-mL flasks containing 200 mL of medium. From Payne et al. (1988b).

Type of closure	Oxygen transfer rate (mmol/L-h)	Cell conc. at day 10 (g/L)	Doubling time (days)
Polypropylene	1.6	4.9	2.8
Silicone cap	34	6.4	2.2
Milk filter	122	6.1	2.0
Aerated flask (300 mL/min)	150	7.7	1.7

C.1. Possible Role for Carbon Dioxide in Plant Cell Metabolism

As a substrate for photosynthetic carbon assimilation, carbon dioxide levels have been shown to be important for autotrophically growing cells (Hanson and Edelman 1972). However, since most applications of plant cells will probably employ heterotrophically (rather than autotrophically) growing cultures, the role of carbon dioxide in photosynthesis will not be considered here.

Nonphotosynthetic carbon dioxide-fixing reactions also appear to be important in plant cells (for a review, see Basra and Malik 1985). For growth, the most important of these reactions appears to be the carboxylation of phosphoenolpyruvate (PEP) to oxaloacetate catalyzed by the enzyme phosphoenolpyruvate carboxylase (PEPCase). This reaction has been proposed to be important for maintaining intracellular pH (Bown 1985; Wright and Givan 1988), for detoxification of ammonia (see Givan 1979), and for intermediary metabolism. With respect to this latter role, figure 5.5 shows that the PEP carboxylase (PEPCase) reaction can supply TCA cycle intermediates. Significant amounts of these intermediates can be drained from the TCA cycle during periods of rapid ammonium assimilation and protein synthesis (Hunt and Fletcher 1976). Paul et al. (1978) reported that ammonia additions to isolated mesophyll cells resulted in an increased incorporation of radioactively labelled CO_2 into organic acids (e.g., malate) and amino acids (e.g., aspartic and glutamic). Nesius and Fletcher (1975) also observed incorporation into lipids and proteins in 5-day-old cultures of Paul's Scarlet Rose. Further supporting the importance of this reaction is the observation that during periods of rapid growth, high or increased levels of PEPCase are often observed (Nato and Mathieu 1978; Seeni and Gnanam 1982). In fact, Adams and co-workers (Adams and Rinne 1981; Adams et al. 1982) have estimated that a significant amount of the carbon which is catabolized through glycolysis in developing soybean seeds reacts through the PEPCase route as opposed to the more traditionally accepted route involving pyruvate kinase.

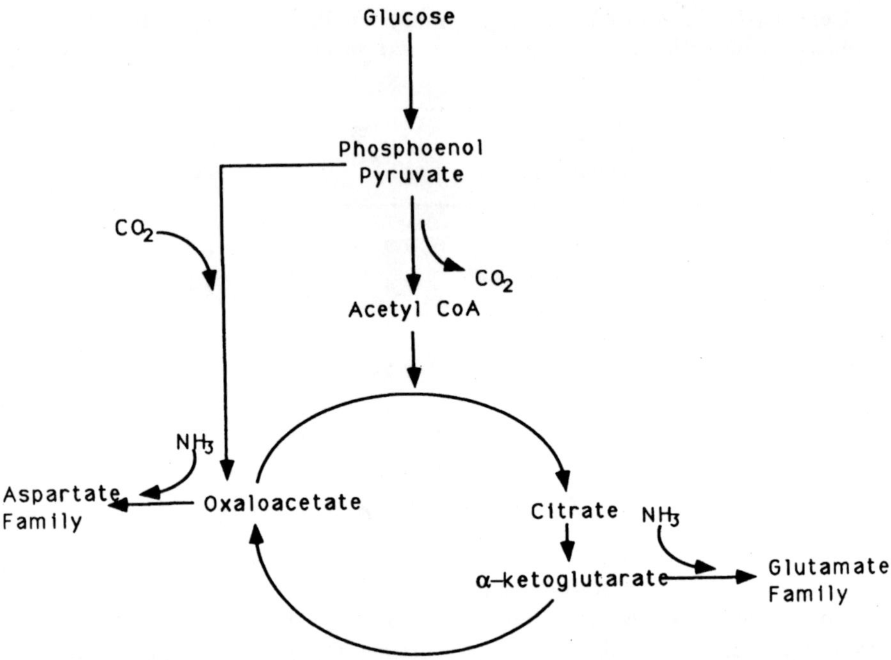

Figure 5.5. Schematic showing the role that PEPCase appears to play in replenishing TCA cycle intermediates.

It is also possible that carbon dioxide affects plant cell systems based on its interactions with ethylene synthesis and action (for a review, see Sisler and Wood 1988).

C.2. Systematic Study of Carbon Dioxide Effects on Plant Cells

Like oxygen, carbon dioxide can be exchanged between the gas and liquid phases. However, since carbon dioxide is a product of heterotrophic cell growth while oxygen is a substrate, then transport of these two species occurs in the opposite direction. That is, while oxygen is transferred from the gas to the liquid, carbon dioxide is typically stripped from the liquid into the gas. Mechanistically, the same gas-liquid mass transfer models derived in Chapter 4 apply for oxygen and carbon dioxide. However, since the diffusivities differ, then the K_La values for carbon dioxide are slightly smaller than those for oxygen transfer (Smith et al. 1990).

Conclusively establishing the role of carbon dioxide in plant cell metabolism is complicated by a couple of factors: (1) in aqueous solutions, carbon dioxide undergoes several reactions; and (2) it is unknown which carbon dioxide species is "biologically active." The reactions of inorganic carbon and their corresponding equilibrium relationships are shown in Table 5.4 (for further discussion see Helder 1988). At equilibrium with air at 25 °C, 0.01 mmol of CO_2 is dissolved into 1 L of water. This saturation concentration is low compared to that of oxygen (0.25 mmol of oxygen can be dissolved from air into 1 L of water at 25 °C) because of the much lower carbon dioxide content of air (air contains 0.03% carbon dioxide and 21% oxygen). As illustrated by the value of K_1 in Table 5.4, the equilibrium concentration of H_2CO_3 is very small relative to that of CO_2, and in most quantitative treatments reactions (a) and (b) are summed to give reaction (c) and the value of H_2CO_3 is not explicitly determined. Because of the involvement of the hydrogen ion in reactions (b) through (d) in Table 5.4, the equilibrium concentrations of the ionic species HCO_3^- and CO_3^{2-} are strongly dependent on the pH. For most plant cell applications where the pH is less than about 6.5, the CO_3^{2-} concentration is negligible. The HCO_3^- concentration

Table 5.4. Reactions of carbon dioxide in water.

	Reaction	Equilibrium constant
(a)	$CO_2 + H_2O \leftrightarrow H_2CO_3$	$K_1 = \frac{[H_2CO_3]}{[CO_2]} = 0.002$
(b)	$H_2CO_3 \leftrightarrow H^+ + HCO_3^-$	
(c)	$CO_2 + H_2O \leftrightarrow H^+ + HCO_3^-$	$K_2 = \frac{[H^+]\,[HCO_3^-]}{[CO_2]} = 4.45 \times 10^{-7}$
(d)	$HCO_3^- \leftrightarrow H^+ + CO_3^{2-}$	$K_3 = \frac{[H^+]\,[CO_3^{2-}]}{[HCO_3^-]} = 4.69 \times 10^{-11}$

Notes:
1. As with most equilibrium reactions involving water in an aqueous environment, the water term is not explicitly used in the equilibrium constants.
2 All concentrations in the above equilibrium equations are in units of mol/L.
3. Because of the low equlibrium concentrations of H_2CO_3, (i.e. K_1 is small), reactions (a) and (b) are often combined to give reaction (c).
4. A more thorough description of the reactions of carbon dioxide can be found in Helder (1988).

may be significant in plant cell applications as the equilibrium concentration of HCO_3^- is close to that of CO_2 at a pH of 6 while the equilibrium concentration of this ionic species exceeds that of CO_2 at higher pH values.

In addition to the complexity of having numerous reactions involving carbon dioxide, it is also unclear as to which chemical species is important biologically and how extracellular conditions can be related to physiologically important mechanisms. This complexity can be illustrated if we consider the carboxylation reaction catalyzed by PEPCase. If externally supplied (rather than endogenous) carbon dioxide is used in this reaction, then it is important to identify which species (CO_2 or HCO_3^-) is transported into the cell. Espie and Colman (1982) suggest that the neutral CO_2 species is transported. Once inside the cell, which species reacts with the PEPCase, CO_2 (Waygood et al. 1969) or HCO_3^- (Walker et al. 1988)? Further complicating this issue is the fact that multiple forms of the PEPCase enzyme may exist, each with different affinities (Ting and Osmond 1973), and these affinities may be altered by other metabolites (Walker et al. 1988) and by enzyme phosphorylation (Kluge et al. 1988).

In initial efforts to investigate the role of carbon dioxide on plant cell suspension growth, carbon dioxide-enriched gas was used in place of air. Several groups have observed that small additions of carbon dioxide to the gas (e.g., carbon dioxide levels of 2% in the gas) enhanced growth rates (Maurel and Pareilleux 1985, 1986; Hegarty et al. 1986). This stimulation of growth rate however is not universal as other reports have shown that carbon dioxide supplementation does not enhance growth rates (Ducos et al. 1988; Payne et al. 1990). Such contradictory results are not unexpected. If the culture responds to the dissolved concentration of carbon dioxide, then it is important to recognize that this dissolved concentration depends not only on the gas phase concentration, but also on the rate of carbon dioxide production by the culture and the rate at which carbon dioxide is transferred into or from solution by the sparged gas. Just as in the case of oxygen, figure 5.3 shows that as a culture grows the dissolved concentration of carbon dioxide does not remain constant (Hegarty et al. 1986; Ducos and Pareilleux 1986). Since it is only at the early stages of cultivation that the dissolved carbon dioxide concentration is low, it is likely that the adverse effects of an insufficient availability of carbon dioxde would be most important at that time (Vanderhoven and Zyrd 1978). Thus, it has been suggested that carbon dioxide may act as a conditioning factor to permit growth or reduce the lag phase for cultures growing at low cell concentrations (Gathercole et al. 1976; Ducos et al. 1988). In microbial systems, Bader (1986) suggested that carbonates are supplied in industrial media to provide a slow release of carbon dioxide. Presumably, these carbonates enhance growth in the early stages of the fermentation. Carbonates may also be important in animal cell systems (Geyer and Chang 1958).

It is also instructive to consider carbon dioxide effects in terms of the K_m for carbon dioxide-requiring reactions. Nesius and Fletcher (1973) suggested that due to the high affinity of PEPCase, Paul's Scarlet Rose cells were able to efficiently fix endogenously generated carbon dioxde. Waygood et al. (1969) reported that the K_m for for PEPCase was 0.0075 mmol/L. This value is significantly lower than the 0.1 to 2.8 mmol/L reported by Maurel and Pareilleux (1986) for *C. roseus* growth. Although such variation in K_m would suggest a mass transfer limitation, the fact that CO_2 is both produced and possibly consumed by the cells complicates any analysis. That is, a cell within a large clump which is actively producing CO_2 would experience a higher (not lower) localized concentration if serious mass transfer limitations existed. Such mass transfer limitations could mask any carbon dioxide requirements and may explain why different behavior has been noted for the effect of carbon dioxide on plant cell growth.

In addition to the effect of carbon dioxide on the growth kinetics, it is possible that carbon dioxide can affect the stoichiometry of growth. This is supported by observations of the incorporation of radioactively labeled carbon from carbon dioxide or carbonates. Also, such stoichiometric effects should be observed in calculations of the

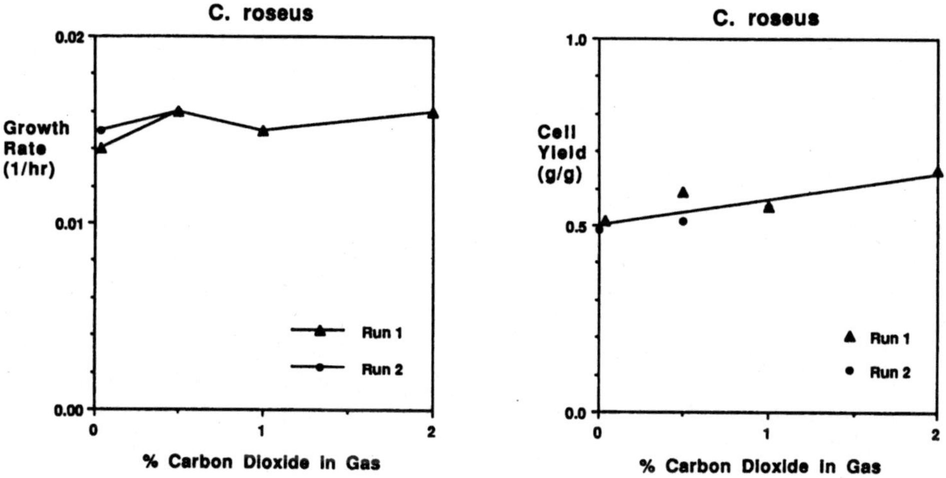

Figure 5.6. The specific growth rate and growth yield for *C. roseus* growing under varying levels of carbon dioxide in the gas. Experimental system is described in the legend to figure 5.1. Data from Tate and Payne (submitted).

growth yield. Figure 5.6 shows that when cells were grown at increasing levels of carbon dioxide, the cell yield increased. This is similar to observations of Maurel and Pareilleux (1985) and Ducos et al. (1988) . However, as shown in Figure 5.6 such increases in growth yield are typically small - and, in fact, Maurel and Pareilleux (1986) estimated that such carbon dioxide fixation can account for only about 2% of the cell carbon.

In summary, there are good reasons to suspect that carbon dioxide can affect the heterotrophic growth of plant cells. Although results are often contradictory with respect to the effect of carbon dioxide on the kinetics of growth, there appears to be general agreement in the literature that carbon dioxide can slightly affect the stoichiometry of growth. However, it should be stressed that the role of carbon dioxide appears to be considerably less significant than that of oxygen. Our studies suggest that cultivation conditions which permit moderate dissolved oxygen concentrations (e.g., 50 - 70% of saturation) are also favorable with respect to carbon dioxide.

D. ETHYLENE

Although ethylene is known to affect plant growth and development (see Lieberman 1979), few studies have examined possible applications involving ethylene. This may be due to the difficulties in working with such low concentrations of a volatile component. However, Cho et al. (1988) recently reported that ethephon, which slowly reacts to evolve ethylene, can be used to stimulate alkaloid production in suspension cultures.

It is important to recognize that ethylene can be important even in the absence of its purposeful introduction. Because plants and plant cells can produce ethylene, this phytohormone can accumulate to levels which can affect culture performance. For instance, Kumar et al. (1987) reported that ethylene buildup in flasks significantly affected morphogenesis.

E. CONCLUSION

In this chapter, we described how the sparingly soluble volatile compounds oxygen, carbon dioxide, and, to a lesser extent, ethylene affect plant cell metabolism and how the transfer of these components between the gas and liquid phases is characterized. Since insufficient or excessive levels of these components can profoundly affect plant cells, gas transfer considerations are important even in flask-scale experiments (e.g., see Table 5.3). As will be discussed in the next chapter, gas transfer considerations are essential for the proper design and scale-up of bioreactors.

References

C. A. Adams and R. W. Rinne. 1981. Interactions of Phosphoenolpyruvate Carboxylase and Pyruvic Kinase in Developing Soybean Seeds. Plant *Cell Physiol.* 22, 1011-1021.

C. A. Adams, T. A. Broman, and R. W. Rinne. 1982. Use of [3,4-^{14}C]Glucose to Assess *in vivo* Competition for Phosphoenolpyruvate between Phosphoenolpyruvate Carboxylase and Pyruvate Kinase in Developing Soybean Seeds. *Plant Cell Physiol. 23*, 959-965.

A. Al-Ani, F. Bruzau, P. Raymond, V. Saint-Ges, J. M. LeBlanc, and A. Pradet. 1985. Germination, Respiration, and Adenylate Energy Charge of Seeds at Various Oxygen Partial Pressures. *Plant Physiol. 79*, 885-890.

W. Armstrong and T. Webb. 1985. A Critical Oxygen Pressure for Root Extension in Rice. *J. Exp. Bot. 36*, 1573-1582.

H. Ashihara, T. Horikosi, X.-N. Li, K. Sagishima, and Y. Yamashita. 1988. Profiles of Enzymes Involved in Glycolysis in *Catharanthus roseus* Cells in Batch Suspension Culture. *J. Plant Physiol. 133*, 38-45.

F.G. Bader. 1986. Physiology and Fermentation Development. In "The Bacteria: Volume IX: Antibiotic-Producing *Streptomyces*," S. W. Queener and L. E. Day (eds.), Academic Press, New York, pp. 281-321.

A. S. Basra and C. P. Malik. 1985. Non-Photosynthetic fixation of Carbon Dioxide and Possible Biological Roles in Higher Plants. *Biol. Rev. 60*, 357-401.

L. J. Berry and W. E. Norris. 1949. Studies of Onion Root Respiration. I. Velocity of Oxygen Consumption in Different Segments of Root at Different Temperatures as a Function of Partial Pressure of Oxygen. *Biochim. Biophys. Acta 3*, 593-606.

A. Bertani and I. Brambilla. 1982. Effect of Decreasing Oxygen Concentration on Some Aspects of Protein and Amino-Acid Metabolism in Rice Roots. *Z. Pflanzenphysiol. 107*, 193-200.

P. A. Bond, M. W. Fowler and A. H. Scragg. 1988. Growth of *Catharanthus roseus* in Bioreactors: On-Line Analysis of Oxygen and Carbon Dioxide Levels in Inlet and Outlet Gas Streams. *Biotechnol. Lett. 10*, 713-718.

A. W. Bown. 1985. CO_2 and Intracellular pH. *Plant Cell Environ. 8*, 459-465.

V. S. Butt and C. J. Lamb. 1981. Oxygenases and the Metabolism of Plant Products. In "The Biochemistry of Plants," Vol. 7, E. E. Conn (ed.), Academic Press, New York, pp. 627-665.

G. H. Cho, D. I. Kim, H. Pedersen, and C. K. Chin. 1988. Ethephon Enhancement of Secondary Metabolite Synthesis in Plant Cell Cultures. *Biotechnol. Prog. 4*, 184-188.

R. S. Criddle, R. W. Breidenbach, E. A. Lewis, D. J. Eatough, and L. D. Hansen. 1988. Effects of Temperature and Oxygen Depletion on Metabolic Rates of Tomato and Carrot Cell Cultures and Cuttings Measured by Calorimetry. *Plant Cell Environ.* 11, 695-701.

J. P. Ducos and A. Pareilleux. 1986. Effect of Aeration Rate and Influence of pCO_2 in Large-Scale Cultures of *Catharanthus roseus* cells. *Appl. Microbiol. Biotechnol. 25*, 101-105.

J. P. Ducos, G. Feron, and A. Pareilleux. 1988. Growth and Activities of Enzymes of Primary Metabolism in Batch Cultures of *Catharanthus roseus* Cell Suspension under Different CO_2 Conditions. *Plant Cell Tissue Organ Culture 13*, 167-177.

G. S. Espie and B. Colman. 1982. Photosynthesis and Inorganic Carbon Transport in Isolated *Asparagus* Mesophyll Cells. *Plant Physiol. 70*, 649-654.

A. T. M. Faiz-ur-Rahman, A. J. Trewavas, and D. D. Davies. 1974a. The Pasteur Effect in Carrot Root Tissue. *Planta, 118*, 195-210.

A.T.M. Faiz-ur-Rahman, D.D. Davies, and A.J. Trewavas. 1974b. The Control of Glycolysis in Aged Slices of Carrot Root Tissue. *Planta 118*, 221-224.

R. W. Gathercole, K. J. Mansfield, and H. E. Street. 1976. Carbon Dioxide as an Essential Requirement for Cultured Sycamore Cells. *Physiol. Plant. 37*, 213-217.

R. P. Geyer and R. S. Chang. 1958. Bicarbonate as an Essential for Human Cells *In Vitro*. *Arch. Biochem. Biophys. 73*, 500-506.

C. V. Givan. 1979. Metabolic Detoxification of Ammonia in Tissues of Higher Plants. *Phytochemistry 18*, 375-382.

A. D. Hanson and J. Edelman. 1972. Photosynthesis by Carrot Tissue Cultures. *Planta 102*, 11-25.

P. K. Hegarty, N. J. Smart, A. H. Scragg, and M. W. Fowler. 1986. The Aeration of *Catharanthus roseus* L. G. Don Suspension Cultures in Airlift Bioreactors: The Inhibitory Effect at High Aeration Rates on Culture Growth. *J. Exp. Bot.* 37, 1911-1920.

R. J. Helder. 1988. A Quantitative Approach to the Inorganic Carbon System in Aqueous Media Used in Biological Research: Dilute Solutions Isolated from the Atmosphere. *Plant Cell Environ.* 11, 211-230.

L. Hunt and J. S. Fletcher. 1976. Estimated Drainage of Carbon from the Tricarboxylic Acid Cycle for Protein Synthesis in Suspension Cultures of Paul's Scarlet Rose Cells. *Plant Physiol. 57*, 304-307.

A. Kato, Y. Shimizu, and S. Nagai. 1975. Effect of Initial K_La on the Growth of Tobacco Cells in Batch Culture. *J. Ferment. Technol. 53*, 744-751.

R. H. J. Kessell and A. H. Carr. 1972. The Effect of Dissolved Oxygen Concentration on Growth and Differentiation of Carrot (*Daucus carota*) Tissue. *J. Exp. Bot. 23*, 996-1007.

R. H. J. Kessell, C. Goodwin, J. Philip, and M. W. Fowler. 1977. The Relationship between Dissolved Oxygen Concentration, ATP, and Embryogenesis in Carrot (*Daucus carota*) Tissue Cultures. *Plant Sci. Lett. 10*, 265-274.

M. Kluge, P. Maier, J. Brulfert, K. Faist, and E. Wollny. 1988. Regulation of Phosphoenolpyruvate Carboxylase in Crassulacean Acid Metabolism: *In Vitro* Phosphorylation of the Enzyme. *J. Plant Physiol. 133*, 252-256.

Y. Kobayashi, H. Fukui, and M. Tabata. 1989. Effect of Oxygen Supply on Berberine Production in Cell Suspension Cultures and Immobilized Cells of *Thalictrum minus*. *Plant Cell Rep. 8*, 255-258.

P. P. Kumar, D. M. Reid, and T. A. Thorpe. 1987. The Role of Ethylene and Carbon Dioxide in Differentiation of Shoot Buds in Excised Cotyledons of *Pinus radiata In Vitro*. *Physiol. Plant. 69*, 244-252.

M. Lieberman. 1979. Biosynthesis and Action of Ethylene. *Annu. Rev. Plant Physiol. 30*, 533-591.

B. Maurel and A. Pareilleux. 1985. Effect of Carbon Dioxide on the Growth of Heterotrophic Cell Suspensions of *Catharanthus roseus*. *Biotechnol. Lett.* 7, 313-318.

B. Maurel and A. Pareilleux. 1986. Carbon Dioxide fixation and Growth of Heterotrophic Cell Suspensions of Catharanthus roseus. J. Plant Physiol. 122, 347-355.

A. Nato and Y. Mathieu. 1978. Changes in Phosphoenolpyruvate Carboxylase and Ribulose-Biphosphate Carboxylase Activities during the Photoheterotrophic Growth of *Nicotiana tabacum* (CV Xanthi) Cell Suspensions. *Plant Sci. Lett. 13*, 49-56.

K. K. Nesius and J. S. Fletcher. 1973. Carbon Dioxide and pH Requirements of Non-Photosynthetic Tissue Culture Cells. *Physiol. Plant. 28*, 259-263.

K. K. Nesius and J. S. Fletcher. 1975. Contribution of Nonautotrophic Carbon Dioxide fixation to Protein Synthesis in Suspension Cultures of Paul's Scarlet Rose. *Plant Physiol. 55*, 643-645.

A. Pareilleux and R. Vinas. 1983. Influence of the Aeration Rate on the Growth Yield in Suspension Cultures of *Catharanthus roseus*. (L.) G. Don. *J. Ferment. Technol. 61*, 429-433.

J. S. Paul, K. L. Cornwell, and J. A. Bassham. 1978. Effects of Ammonia on Carbon Metabolism in Photosynthesizing Isolated Mesophyll Cells from *Papaver somniferum* L. *Planta 142*, 49-54.

G. F. Payne, N. N. Payne, and M. L. Shuler. 1988a. Bioreactor Considerations for Secondary Metabolite Production from Plant Cell Tissue Culture: Indole Alkaloids from *Catharanthus roseus*. *Biotechnol. Bioeng. 31*, 905-912.

G. F. Payne, J. L. Tate, and S. W. Davison. 1988b. Engineering Considerations for the Production of Plant Products from Tissue Culture. *SAAS Bull. Biochem. Biotechnol.* 1, 21-25.

G. F. Payne, S. W. Davison, and J. L. Tate. 1990. Experimental Constraints to Studying the Effects of Dissolved Oxygen and Dissolved Carbon Dioxide on Plant Cell Growth. *Dev. Ind. Microbiol. 31*, 293-301.

A Pradet and P. Raymond. 1983. Adenine Nucleotide Ratios and Adenylate Energy Charge in Energy Metabolism. *Annu. Rev. Plant Physiol. 34*, 199-224.

F. Rasi-Caldogno and M. I. De Michelis. 1978. Correlation between Oxygen Availability, Energy Charge, and Protein Synthesis in Squash Cotyledons Isolated from Germinating Seeds. *Plant Physiol. 61*, 85-88.

P. Raymond, A. Al-Ani, and A. Padet. 1985. ATP Production by Respiration and Fermentation, and Energy Charge during Aerobiosis and Anaerobiosis in Twelve Fatty and Starchy Germinating Seeds. *Plant Physiol.* 79, 879-884.

P. H. Saglio, P. Raymond, and A. Pradet. 1983. Oxygen Transport and Root Respiration of Maize Seedlings. *Plant Physiol.* 72, 1035-1039.

P. H. Saglio, M. Rancillac, F. Bruzan, and A. Pradet. 1984. Critical Oxygen Pressure for Growth and Respiration of Excised and Intact Roots. *Plant Physiol. 76*, 151-154.

S. Seeni and A. Gnanam. 1982. Carbon Assimilation in Photoheterotrophic Cells of Peanut (*Arachis hypogaea* L.) Grown in Still Medium. *Plant Physiol. 70*, 823-826.

S. G. Shirsat and P. M. Nair. 1981. Biochemical Mechanism for the Inhibition of Phenylalanine Ammonia-Lyase Induction in the Absence of Oxygen in Potato Tuber Tissue. *Phytochemistry 20*, 2315-2318.

E. C. Sisler and C. Wood. 1988. Interaction of Ethylene and CO_2. *Physiol. Plant.* 73, 430-444.

N. J. Smart and M. W. Fowler. 1984. Mass Cultivation of *Catharanthus roseus* Cells Using a Nonmechanically Agitated Bioreactor. *Appl. Biochem. Biotechnol. 9*, 209-216.

A. M. Smith and T. ap Rees. 1979a. Effects of Anaerobiosis on Carbohydrate Oxidation by Roots of *Pisum sativum*. *Phytochemistry 18*, 1453-1458.

A. M. Smith and T. ap Rees. 1979b. Pathways of Carbohydrate Fermentation in the Roots of Marsh Plants. *Planta 146*, 327-334.

J. M. Smith, S. W. Davison, and G. F. Payne. 1990. Development of a Strategy to Control the Dissolved Concentrations of Oxygen and Carbon Dioxide at Constant Shear in a Plant Cell Bioreactor. *Biotechnol. Bioeng. 35*, 1088-1101.

J. B. Snape, N. H. Thomas, and J.A. Callow. 1989. How Suspension Cultures of *Catharanthus roseus* Respond to Oxygen Limitation: Small-Scale Tests with Applications to Large-Scale Cultures. *Biotechnol. Bioeng. 34*, 1058-1062.

H. Tanaka. 1987. Large Scale Cultivation of Plant Cells at High Density. *Process Biochem.*, August, 106-113.

J.L. Tate, and G.F. Payne. 1991. Plant Cell Growth under Different Levels of Oxygen and Carbon Dioxide. *Plant Cell Rep. 10*, 22-25.

C. Thompson, B. J. Atwell, and H. Greenway. 1989. Response of Wheat Seedlings to Low O_2 Concentrations in Nutrient Solution. *J. Exp. Bot. 40*, 985-991.

I. P. Ting and C. B. Osmond. 1973. Multiple Forms of Plant Phosphoenolpyruvate Carboxylase Associated with Different Metabolic Pathways. *Plant Physiol. 51*, 448-453.

J. B. Tunac. 1989. High-Aeration Capacity Shake-Flask System. *J. Ferment. Bioeng. 68*, 157-159.

C. Vanderhoven and J.-P. Zyrd. 1978. Changes in Malate Content and in Enzymes Involved in Dark CO2 fixation during Growth of *Acer pseudoplatanus* Cells in Suspension Culture. *Physiol. Plant. 43*, 99-103.

W. M. van Gulik, J. J. Meijer, H. J. G. ten Hoopen, K. Ch. A. M. Luyben, and M. R. Libbenga. 1989. Growth of Catharanthus roseus Cell Suspension Culture in a Modified Chemostat under Glucose Limiting Conditions. *Appl. Microbiol. Biotechnol. 30*, 270-275.

G. H. Walker, M. S. B. Ku, and G. Edwards. 1988. The Effect of Phosphorylated Metabolites and Divalent Cations on the Phosphatase and Carboxylase Activity of Maise Leaf Phosphoenolpyruvate Carboxylase. *J. Plant Physiol. 133*, 144-151.

E. R. Waygood, R. Mache, and C. K. Tan. 1969. Carbon Dioxide, the Substrate for Phosphoenolpyruvate Carboxylase from Leaves of Maise. *Can. J. Bot.* 47, 1455-1458.

K. M. Wright and C. V. Givan. 1988. Regulation of Non-Autotrophic Carbon Dioxide Assimilation by Ammonia in Cultured Cells of *Acer pseudoplatanus* L. *Plant Sci. 58*, 151-158.

T. Yamakawa, S. Kato, K. Ishida, T. Kodama, and Y. Minoda. 1983. Production of Anthocyanins by *Vitis* Cells in Suspension Cultures. *Agric. Biol. Chem.* 47, 2185-2191.

CHAPTER 6

SUSPENSION CULTURE

A. INTRODUCTION

Suspension culture has been the primary route for studying plant cell secondary metabolism. It is also a form of tissue culture which has been taken to commercial production. Since this type of cultivation generally involves undifferentiated or partially differentiated cells, then suspension culture is inappropriate for the synthesis of products which are associated with some degree of tissue organization. For instance, as discussed in Chapter 2, several monomeric indole alkaloids can be produced by undifferentiated cells in suspension. However, a key precursor (vindoline) to dimeric alkaloid synthesis is only produced by tissue from the aerial portion of the periwinkle plant. Therfore, shoot (rather than suspension) culture may be required for dimeric alkaloid production. Organ culture will be discussed in subsequent chapters.

In addition to being appropriate only for the synthesis of products which are not associated with cellular organization, suspension cultivation is limited by other concerns. It has been observed that over extended subculture periods, the productivity of suspension cultures decreases. This reduction may result from genetic alterations, environmental conditions which limit the expression of genetic potential, or the selection of a rapidly growing cell type from the heterogeneous population of cells in the culture. Another concern is that because of the slow growth and low productivity of plant cells, suspension cultures must be maintained under aseptic conditions for several weeks. Although long-term operation is an inconvenience in the lab, it becomes an economic issue at the production scale. An additional concern is that culture performance may depend on the size distribution of the plant cell aggregates, and the aggregate size will be affected by the shear conditions experienced by the culture. As will be seen, our ability to control (and even to understand) shear is limited for suspension cultures. Thus, design and scale-up of suspension cultures may require considerable experimentation. A final concern with suspension culture is that the trend in using separate media for growth and secondary metabolite synthesis can be operationally troublesome on a large scale.

To develop suspension culture plant cell processes for secondary metabolite production requires optimization of the cell line, the cultivation media, and the bioreactor system. Cell line and media development have been illustrated in Chapter 2 in our discussion of indole alkaloid biosynthesis. Medium optimization in the context of other types of tissue is also developed in later chapters. In the present chapter, we will focus on the bioreactor and operating strategies for suspension culture.

Bioreactors have two advantages over flasks for culturing plant cells. The first advantage is that better control can be exerted on the system (i.e., pH and dissolved gas concentrations can be controlled). The second advantage is that since most bioreactors are scalable, they are better able to reproduce on a larger scale those conditions which were observed on a smaller scale to be most desirable for culture performance. Thus, if cultivation volumes are to exceed a few liters, bioreactors are generally desired. With respect to scale-up, maintaining constant environmental conditions at the various scales of operation is simple with respect to the soluble components (e.g., carbohydrates, salts, and growth regulators). However, reproduction of the physical environment (e.g., mixing and shear) and the dissolved concentration of sparingly soluble volatile components (e.g., oxygen and carbon dioxide) upon scale-up is practically impossible for suspension culture systems. Since compromises are generally required in choosing a bioreactor, it is important to understand how shear, mixing, and oxygen transfer affect the culture. This will be the topic of the first section of this chapter.

A second consideration in culturing cells in bioreactors is the operating mode. In the second section of this chapter, we will briefly compare the various operating modes (e.g., batch, fed batch, and continuous). Finally, we will attempt to illustrate our idealized discussion by examining a few cases of large-scale suspension cultivation.

B. IMPORTANT FACTORS FOR BIOREACTORS

Before discussing the individual bioreactor systems, we will provide some theoretical background necessary to understand the advantages of the various bioreactors. Much of this theory has been developed for either microbial or animal cell systems and in this section we will discuss the implications with respect to plant cells.

B.1. Gas-Liquid Mass Transfer

Because of the overriding concern for supplying sufficient oxygen for aerobic metabolism, most bioreactors are rated in terms of their oxygen mass transfer coefficient, K_La (for further discussion of oxygen transfer, refer to Chapters 4 and 5). In fact, scale-up of many bioprocesses is based on maintaining a constant K_La. Since microbial cultures have high respiration rates and are shear tolerant, the goal for culturing bacteria is often to maximize the oxygen transfer coefficient (K_La values of 100 h^{-1} are typical). This is achieved by using mechanical agitators at high rotational speeds and by sparging large volumes of gas (e.g., 1 volume of gas per minute per volume of liquid). Since plant cells have lower respiration rates, the oxygen transfer requirements are considerably less. For instance, if cells which respire at a rate of 0.2 mmol/g-h are to be grown to 10 g/L without allowing the dissolved oxygen concentration to fall below 20% of saturation, then equation [5.7] suggests that the K_La should be:

$$K_La = \frac{q_{o,max}\, X}{(\hat{C}_{O_2} - C_{O_2,crit})} \qquad [6.1a]$$

$$= \frac{(0.2\ \frac{mmol}{g-h})\,(10\ \frac{g}{L})}{[0.25\ \frac{mmol}{L} - 0.2\,(0.25\ \frac{mmol}{L})]} = 10\ h^{-1} \qquad [6.1b]$$

Plant cell bioreactors are typically operated at K_La values of 10-30 hr^{-1}. Operation at higher K_La values has often been observed to result in poorer plant cell growth (Tanaka 1981, 1987; Fowler 1982; Smart and Fowler 1984b) or product synthesis (Yamakawa et al. 1983). This poor performance at high K_La values may be due to either increased shear associated with the high-K_La conditions (Tanaka 1981, 1987) or to enhanced carbon dioxide stripping from the liquid (Maurel and Pareilleux 1985; Hegarty et al. 1986). Further discussion of carbon dioxide stripping can be found in Chapter 5.

It is important to recognize that conditions of constant K_La do not imply that the dissolved gas concentrations will be constant. As shown in Figure 5.3, when a constant K_La of 9 hr^{-1} was used in a carrot cell culture, the dissolved concentrations of both oxygen and carbon dioxide changed dramatically during the course of cultivation. It is important to recognize that the cells respond directly to the dissolved concentrations and that most bioprocesses are developed to ensure that the dissolved oxygen concentration does not fall below a critical value (see Chapter 5). When a bioreactor is equipped with a dissolved oxygen probe, it is common that the aeration or agitation rate (and therefore the K_La) is adjusted throughout the cultivation to maintain a more constant dissolved oxygen level. Thus, whereas dissolved oxygen is the intrinsically important parameter for the cell, the K_La is the parameter which is characteristic of the bioreactor system.

B.2. Shear

There have been many suggestions in the literature that plant cells are shear sensitive (Wagner and Vogelmann 1977; Tanaka 1981, 1987). At the present time, there is little fundamental knowledge on how shear affects the cells and how to quantify shear. In this section, we will briefly summarize what is known about shear.

It is difficult to define what is meant by shear. Formally, shear refers to forces exerted on the surface of a body in a direction parallel to the surface. This is contrasted to normal forces, which are exerted on a surface, but perpendicular to the surface. Mathematically, shear is described by the equation

$$\tau = \eta \gamma \qquad [6.2]$$

where τ is the shear stress (a force per unit surface area), γ is the shear rate (a change in velocity across a distance), and η is a viscosity (a coefficient which describes the resistance to flow).

To describe the shear damage to eucaryotic cells in mechanically agitated systems, several approaches have been taken. Initially, it was considered that shear damage resulted from "highly localized velocity gradients" (Midler and Finn 1966), and these localized gradients, or maximum shear rates, could be related to the speed with which the outer edges of the impeller blade traveled (Oldshue 1966). Impeller tip speed is given by the equation

$$\text{Tip Speed} = \pi N D_i \qquad [6.3]$$

where N is the impeller speed (rpm), and D_i is the impeller diameter. Midler and Finn (1966) observed that tip speed could be used to correlate protozoan disruption with various impeller sizes.

As animal cell culture became commercially more important, the need to understand and quantify shear damage was recognized. Sinskey et al. (1981) proposed that it may be better to use an integrated shear factor (ISF) to correlate shear damage to cells. The equation used in their study was

$$ISF = \pi N D_i / (D_t - D_i) \qquad [6.4]$$

where the numerator in the above equation is the tip speed (equation [6.3]) and the denominator is a measure of the distance between the impeller and the tank wall. D_t is the tank diameter. ISF thus represents a pseudo velocity gradient, or pseudo shear rate which exists between the impeller tip (where the fluid is presumably traveling at the tip speed) and the vessel wall (where the fluid is presumably not moving). This method was capable of correlating changes in animal cell growth for microcarrier cultures (Hu and Wang 1986; Croughan et al. 1987). Although the above correlations are useful for systems with similar impeller designs, they cannot be generally extrapolated to predict the damage resulting when alternate impellers are used.

In subsequent work, a more mechanistic approach was taken to describe shear damage to animal cells grown on microcarriers (Croughan et al. 1987; Cherry and Papoutsakis 1988). From fluid mechanical theories of turbulence, it is believed that energy imparted into the liquid by the agitation system is transferred from larger to smaller eddies and this energy is eventually dissipated. Since the added energy is dissipated by the smallest eddies, the power input (a measure of the added energy) can be related to the size of the smallest eddies. The greater the power added, the smaller is the size of the smallest eddy. When the cell or aggregate is small relative to the size of this smallest eddy, then it is carried around with the fluid in the eddy and is probably undamaged. However, when the cell or aggregate is larger than, or of the same size as the smallest eddies, then it can be caught between eddies such that the dissipation of the energy occurs at the surface of the cell. It is this energy dissipation at the surface of the cells or aggregates which is believed to lead to shear damage. These theories then suggest that shear damage can be related to power imparted into the liquid per unit liquid volume. If this approach is generally valid, then it may be applicable to various different types and sizes of mechanically agitated bioreactors.

Since the above theory suggests that shear damage occurs when the smallest eddy is smaller than or comparable to the size of the cell or aggregate, then the larger cells or aggregates would be more susceptible to shear damage. This may explain why microcarrier-bound animal cells (microcarrier size of 180 μm) are more susceptible to shear damage than suspended animal cells (cell sizes of 15 μm) (Aunins et al. 1986). With respect to plant cells, there are two important implications of this theory. First, shear forces may act on aggregates such that the aggregate size is reduced at increased turbulence. If productivity is a function of aggregate size, then changes in scale or operating conditions which alter the aggregate size may therefore affect productivities (Shuler et al. 1984). Such an aggregate size dependence was observed with respect to phenylalanine ammonia lyase (PAL) activities in cell cultures of tobacco (Kuboi and Yamada 1976) and parsley (Knobloch and Hahlbrock 1977). Mantell and Smith (1983) observed that steroidal sapogenin production in a 4 L reactor was greatly reduced compared to production in a 60-mL flask. These authors suggested that this reduced production resulted from the increased aggregation observed in the bioreactor system. The dependence of plant cell physiology on aggregate size is discussed in further detail in Chapter 7.

A second important implication of turbulent eddy theory for plant cell systems is that since plant cells enlarge as they age (e.g., Kato et al. 1972; Phillips and Henshaw 1977; Hashimoto and Azechi 1988; Snape et al. 1989), they will also become more shear sensitive. This may explain observations by Wagner and Vogelmann (1977) that although mechanical agitation did not adversely affect culture growth, it reduced production of secondary metabolites which are normally produced by stationary phase cultures. Schiel and Berlin (1987) observed that although shake-flask-grown *Catharanthus roseus* cells enlarged late in their growth cycle, cultures grown in a 25-L mechanically agitated bioreactor did not expand - possibly due to lysis of the more shear-sensitive cells. Bailey and Nicholson (1989) have recently attempted to correlate this behavior into a mathematical model.

In addition to hydrodynamic shear, recent reports suggest that cell damage may result from cell-cell and cell-impeller collisions (Cherry and Papoutsakis 1988; Prokop and Rosenberg 1989). However, the relative contributions of these mechanisms has not yet been established.

Shear can also result from gas sparging, even in the absence of mechanical agitation. Although estimates vary, most workers relate the shear rate in pneumatically agitated systems to the gas flow rate (Chisti and Moo-Young 1989).

A different approach to understanding shear damage to plant cells was proposed by Tanaka (1981, 1987). In this work, it was suggested that mechanisms which result in increased rates of dissolution of a solid into water can be related to mechanisms which lead to shear damage to plant cells. Thus, a hydrodynamic shear intensity index was developed based on the rate of dissolution of β-naphthol particles. A correlation was then established between this shear intensity index and growth suppression under conditions of high shear. Although the theoretical rigor may be questioned, this approach appears to be useful in providing a measurable parameter to characterize shear in a bioreactor, independent of the reactor type or geometry.

Finally, it should be mentioned that most studies have considered hydrodynamic effects of cells in terms of cell lysis. Interesting results from the animal cell literature have shown that sublytic levels of shear can have dramatic effects on cell behavior (Frangos et al. 1988; Diamond et al. 1989). The effects of sublytic levels of shear on plant cells have not been characterized.

B.3. Mixing

Mixing refers to the convective transport of matter (e.g., the transfer of solutes associated with bulk fluid motion). Since convective transport is fast relative to diffusion at the length scales associated with the bioreactor, the mixing of dissolved nutrients (e.g., salts and sugars) has generally not been a problem in suspension cultures. There are however three important mixing problems in suspension culture systems. The first problem occurs because of the large size of plant cells and especially cell aggregates, which leads to settling at the bottom of the bioreactor. These cells can then settle into dead zones, or unmixed regions of the bioreactor. These dead zones can become depleted of key nutrients (e.g., dissolved oxygen).

A second problem which can be broadly classified as a mixing problem is the attachment of cells onto surfaces above the level of the liquid. Since these cells are not bathed in the liquid medium, they can become deprived of nutrients. Although the problem of cell adhesion to surfaces above the liquid level is rarely discussed in the literature (Kato et al. 1975; Fowler 1982, 1984), it has been our experience that it is significant in plant cell suspensions in which gases are sparged.

The final mixing problem results because of the desire to culture cells at high concentrations. With moderate cell concentrations (packed cell volumes less than 50%), bringing cells in contact with the nutrient-containing liquid is often not a problem. However, incomplete mixing can be a serious problem when plant cells are cultured at high cell concentrations (Kato et al. 1976; Wagner and Vogelmann 1977; Tanaka 1981; Drapeau et al. 1986). Recently, there has been a trend toward using very high cell concentrations (e.g., packed cell volumes of 90%). Under these conditions, there may be significant problems encountered in contacting the liquid and the cells.

C. SCALE-UP

Problems of mixing, shear, and gas transfer form the basis for understanding why scale-up can be a difficult process. The overall objective of the process is to maximize sustainable volumetric productivity. This objective almost always imposes a very stringent limitation on design since operation is required to be at very high cell concentrations (50 to 90% packed cell volume) with cells at or near the stationary phase of growth. Such high cell concentrations make it difficult to attain complete mixing, good gas transfer, maintenance of aggregate size and minimal shear damage. Even if we can do this in a small-scale system (less than 10 L), can this be achieved on a larger scale?

The goal of scale-up is clearly to maintain the physical and chemical conditions as nearly the same in the large-scale vessel as in the smaller vessel. Maintenance of the physical environment in identically the same state in both large and small vessels is impossible with geometrically similar tanks (e.g., the same height-to-diameter ratios). Oldshue (1966) showed that shear, mixing times, and gas transfer cannot all be maintained constant with scale-up in the type of fermentors used in industrial microbiology.

Consider the case of high-density plant cell culture. Increases in shear even at modest levels can alter aggregate size and, as a consequence, the ratio of cell types. This change in aggregate size then might result in different levels of expression of biosynthetic pathways. Even higher levels of shear may cause lysis. Yet if shear is unchanged in the larger scale, then mixing times will increase, almost ensuring a nonuniform chemical environment. This non-uniformity will be particularly problematic with respect to the dissolved oxygen concentration.

An additional pitfall in scaling-up from laboratory data is cell adhesion. Since the surface-to-volume ratio decreases as the vessel size increases, the fraction of cells which adhere to vessel surfaces should be smaller in the large-scale vessel. Because cells which adhere in thick layers experience a less homogeneous chemical environment than those in suspension, a change in the fraction of adherent cells can result in changes in apparent biosynthetic capacity.

With an understanding of why scale-up can be difficult, we can better assess the potential advantages or disadvantages of various bioreactor designs for suspension cultures.

D. TYPES OF BIOREACTORS

The goal in this section is to provide a broad overview of the various bioreactors for suspension culture. We will focus more on fundamental concepts rather than on details of the design. A more detailed comparison - especially with respect to microbial fermentors - is provided by Bull et al. (1983), while discussion of various plant cell systems is given elsewhere (Fowler 1982, 1984; Payne et al. 1987; Kreis and Reinhard 1989).

The objective in culturing cells, whether in flasks or in bioreactors, is to provide conditions to the culture which allow it to perform in a reproducible and optimal manner. Various measures of culture performance exist, and the appropriateness of a measure depends on the desired results. For instance, if efficient cell growth is desired, then bioreactor performance is measured in terms of doubling times and maximum cell concentrations. If secondary metabolites are desired, then an appropriate performance measure would be the overall productivity, or the amount of product produced per total cultivation time. As mentioned previously, reproducing optimal conditions in bioreactors can be difficult with respect to the sparingly soluble volatile components and the physical factors of mixing and shear. In this section, we will consider two general types of bioreactors which differ according to how mixing is achieved - by mechanical or pneumatic agitation. The less conventional rotating drum bioreactor will be discussed in Section F.2.

D.1. Mechanically Agitated Bioreactors

The most common type of reactor for culturing cells (including microbial and animal cells) is the mechanically agitated vessel. This type of reactor employs impellers and mechanical energy for gas-liquid mass transfer and mixing. Various types of impellers

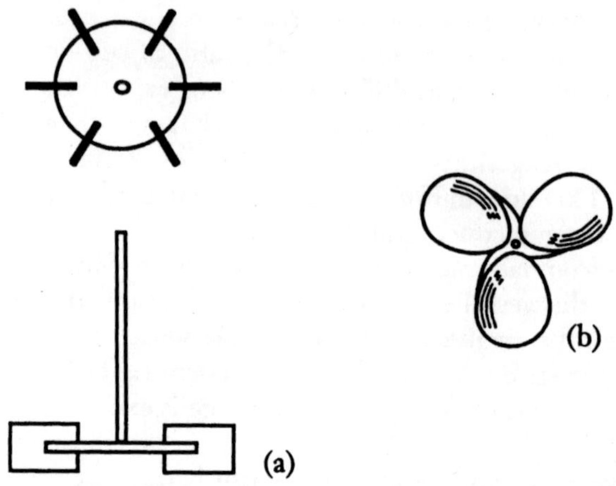

Figure 6.1. Various impellers used in mechanically agitated bioreactors. (a) Flat-blade turbine impeller. (b) Marine propellor.

have been used, and two traditional types are shown in Figure 6.1. The flat-blade turbine impeller (Figure 6.1a) and high agitation speeds are commonly used in bacterial cultivation because these conditions provide the high shear required to break the incoming air stream into many, small bubbles. Since plant cells cannot tolerate such high-shear conditions, and because mixing may be a more serious problem with plant cells (Wagner and Vogelmann 1977), alternate impeller types have been used.

An impeller better suited for low-shear mixing is the marine propellor (Figure 6.1b). As shown in Figure 6.2, the marine propeller provides axial mixing of the fluid. This is contrasted by the flat-blade turbine, which promotes radial fluid mixing. However, often neither the marine propellor nor the flat-blade impeller is suitable for large-scale systems, and recent work has focused on alternative low-shear impellers (e.g., paddle and helical types). These impellers have recently been shown to be effective for plant cell cultivation (Tanaka 1987; Kreis and Reinhard 1989).

In the typical mechanically agitated bioreactor, the rotational motion of the impeller can lead to the formation of a vortex. Vortex formation is eliminated by the use of baffles. Since baffles also provide a surface onto which the cells can attach, they

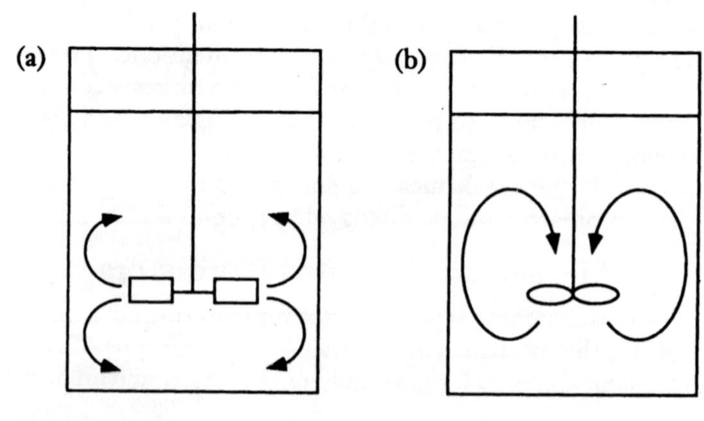

Figure 6.2. Mixing pattern in mechanically agitated bioreactors with (a) flat-blade turbine impeller and (b) marine propellor.

can aggravate problems associated with the adhesion of cells onto surfaces above the liquid level.

In microbial cultivation, agitation serves the dual purpose of mixing and oxygen transfer. Because of the shear sensitivity of plant cells, agitation speeds appropriate for plant cell cultivation are generally insufficient to break the incoming gas stream into small bubbles. Thus, obtaining sufficent oxygen transfer may require that the incoming gas stream be dispersed as fine bubbles by using an appropriate gas distributor. It is also possible to enrich the gas with oxygen if enhanced oxygen transfer is required.

D.2. Pneumatically Agitated Bioreactors

In addition to mechanical agitation, it has been suggested that the motion of the rising gas stream should be capable of providing the energy for fluid mixing. There are two types of pneumatically agitated bioreactors, the bubble column and the airlift. As shown in Figure 6.3, the pneumatically agitated reactors differ from the mechanically agitated systems not only by the absence of an impeller, but also the pneumatically agitated systems are tall and thin. Typically, the height-to-diameter ratio of pneumatically agitated reactors is 10 while this ratio is closer to 1 for mechanically agitated systems. A practical advantage of pneumaticaly agitated bioreactors is that there are no moving parts. Since plant cells must be cultured under aseptic conditions for several weeks without operational failure, the mechanical simplicity of the pneumatically agitated systems can become important (Smart and Fowler 1984a).

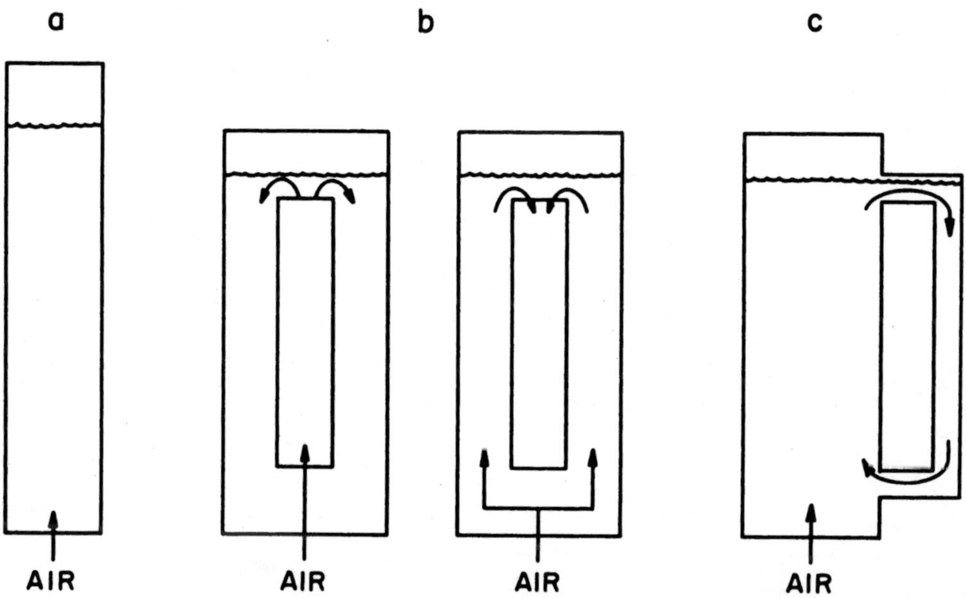

Figure 6.3. Pneumatically agitated bioreactors: (a) bubble column; (b) concentric (draft) tube airlift; and (c) external loop airlift. Liquid recirculation pattern in airlifts indicated by arrows.

The major difference between the bubble column and airlift designs is that the airlift provides for liquid circulation. In the riser section, where the gas is sparged, the fluid flows upward. After the gas disengages at the top of the column, the fluid then flows downward in the downcomer section. These sections may be separated using a baffle, a concentric cylinder, or an external loop. This liquid circulation in the airlift promotes better top-to-bottom mixing (Robinson 1986) and therefore has advantages in suspending cells and clumps. However, since it appears that little oxygen transfer occurs in the downcomer section, the airlift bioreactor is usually inferior to the bubble column with respect to an overall K_La (Bello et al. 1985). Also, if little or no oxygen is transferred in the downcomer section, then it is possible that actively respiring cells may deplete the oxygen and suboptimal oxygen levels may exist in this downcomer section. This mixing problem is only expected to occur in larger scale airlifts. It should be emphasized that the performance of an airlift is strongly dependent on the geometry of the system. Especially important is the ratio of the cross-sectional areas of the riser and downcomer sections (Kawase and Moo-Young 1986), as this ratio affects both mixing (Bello et al. 1984; Robinson 1986) and oxygen transfer (Bello et al. 1985).

D.3. Comparison of Bioreactors

Table 6.1 compares mechanically and pneumatically agitated bioreactors in the context of industrial microbiology. However, bioreactor performance is significantly affected by the conditions used (i.e., impeller type, agitation speed, and gas flow), and these conditions can be significantly constrained by the requirements of plant cells. Thus, for our discussion, Table 6.1 is used more as a point of reference since the generalizations therein must often be reevaluated in the context of plant cell culture.

Traditionally, it has been observed that gas-liquid transfer is better in mechanically

Table 6.1. A traditional comparison of various standard bioreactor systems.

	Mechanically agitated	Pneumatically agitated	
		Bubble column	Airlift
Oxygen transfer[a]	+++	++	+
Low shear[b]	+	++	+++
Mixing[c]	+++	+	++

[a] See Sigurdson and Robinson 1976; Bello et al. 1985; McManamey and Wase 1986.
[b] See Wase et al. 1985; Chisti and Moo-Young 1989.
[c] See Robinson 1986; Sukan and Vardar-Sukan 1987.

agitated bioreactors. However, when impeller speeds are reduced to prevent shear damage to the cells, the ability of the impeller to break the incoming gas stream into small bubbles is severely restricted. For instance, Hong et al. (1989) reported that when the rpm speed of a mechanically agitated bioreactor was set at a value just sufficient to suspend the cells, the observed K_La was smaller than the K_La observed in an airlift bioreactor. Thus, under conditions used for plant cell cultivation, oxygen transfer may depend less on the presence or type of impeller and more on the gas distributor used for the incoming gas. At the other extreme are problems of overaeration with pneumatically agitated reactors. Since agitation is coupled to aeration in pneumatic systems, it may be necessary to use higher gas flow rates to maintain the cells in suspension than would be required for oxygen transfer. Thus, it may be necessary to "overventilate" the suspension to achieve appropriate mixing. This potential problem is most important in the early stages of cultivation when the culture has the lowest oxygen demand and would be most sensitive to carbon dioxide stripping (see Chapter 5).

Typically, mechanically agitated bioreactors provide a higher shear environment than pneumatically agitated systems. However, with low impeller speeds and the use of alternative, low-shear impellers, this generalization from microbial fermentors may not be valid with plant cell suspensions.

Mixing can be a serious problem in plant cell suspensions - especially when high cell concentrations (packed cell volumes greater than 50%) and large volumes (greater than 100 L) are used. Tanaka (1981) suggested that for moderate cell concentrations (e.g., dry cell concentrations below 20 g/L), pneumatically agitated bioreactors are appropriate. Although the actual concentration limit may depend on the cell type, this generalization appears to be reasonable. When higher cell concentrations are used, mixing has been enhanced in pneumatic systems by adding agitators (Wagner and Vogelmann 1977). More recently, the trend has been toward the use of alternative impeller designs such as the paddle and helical screw (Tanaka 1981; Kreis and Reinhard 1989). With respect to the culture volume, initial work with tobacco cells used pneumatically agitated vessels up to 1,500 L (Kato et al. 1976). More recently large-scale processes for producing shikonin at 200 and 750 L (Yamada and Fujita 1983) and berberine at 4,000 L have employed mechanical agitation.

E. OPERATING MODE

Operating mode refers to how the nutrient and product streams are supplied or removed with respect to time. The choice of operating mode depends on the physiological characteristics of the culture. The most common way to determine the proper operating mode is to relate the timing of product synthesis to growth. Growth and non-growth-associated production are discussed in Chapter 3, and Table 6.2 shows how the timing of production is matched with the operating mode. If growth and product synthesis occur simultaneously (i.e., growth-associated production), then an operating mode should be chosen which supports growth over an extended period.

Table 6.2. Relationship between production pattern and appropriateness of bioreactor operating mode.

Operating mode	Growth-associated product	Non-growth-associated product
Batch	+	+
Fed batch	+	+
Repeated fed batch	+	+
Two-stage batch	-	+
Continuous		
Chemostat	+	-
Two-stage chemostat	-	+
Perfusion with cell retention	-[a]	+[b]

[a] As noted in the text, the use of perfusion for berberine production represents an unusual case (production is growth-associated and berberine is not excreted into the medium). In addition to perfusion, operation included nutrient feeding (i.e., fed batch), and occasional cell harvesting (i.e., repeated fed batch) for berberine production.
[b] Perfusion with cell retention is particularly suitable for extracellular products.

If product synthesis follows a period of rapid growth (i.e., non-growth-associated production), then an appropriate operating mode is capable of maintaining the culture in a non- or slowly growing state while retaining its productive capacity. For readers new to the field of secondary metabolism, it is easy to gain the erroneous impression that any condition which suppresses growth will be capable of enhancing production of non-growth-associated products. This, however, is not true, and thus classification of products as growth- or non-growth-associated has to be viewed as the first step to developing an optimal operating mode.

Before discussing individual operating modes, it is instructive to recognize that considerable efforts have been devoted to optimizing processes for the industrial production of microbial secondary metabolites. Much of this knowledge, especially with respect to the operating mode, can be applied to the development of plant cell processes. In fact, the idea of characterizing production in terms of the timing of product synthesis relative to growth evolved from industrial microbiology. However, there are two important differences to be considered. First, microbial processes employ highly mutated strains which can produce gram per liter quantities in less than a week of total cultivation time. Since most microbial secondary metabolites are produced in a non-growth-associated fashion, growth occurs during the first day and product is synthesized during the remainder of the fermentation. The fermentation is terminated when the culture's productivity decays. The major effort for improving productivity in microbial systems is to extend the production period by preventing this decay in productivity - either by preventing the loss of key biosynthetic enzymes or by preventing a loss in metabolic activity of the culture. An interesting study by

Shetty et al. (1989) illustrates how these same issues may apply to plant cell bioprocesses. In this study, it was observed that elicitors could be used to induce phenylalanine ammonia lyase (PAL) activities and to enhance the production of phenolics. However, shortly after induction, the PAL levels decreased. By adding various components (calcium, polyamines, osmolytes, and nutritional "viability stabilizers") with the elicitor, these authors observed that the high PAL levels could be maintained over longer periods and phenolics production could be significantly increased.

However, the typical productivity limitations in plant cell systems are somewhat different. Plant cells grow much more slowly than microbes - probably as a result of their smaller surface-to-volume ratio (Kreis and Reinhard 1989). Once production commences, plant cells then produce very slowly - achieving a few hundred milligrams per liter over several weeks. This low productivity may reflect our limited ability to induce plant cell secondary metabolite production. As pointed out by Kreis and Reinhard (1989), an exception is the *Coleus* culture, which was shown to be able to produce 5.5 g/L of rosmarinic acid over a six-day production phase. Thus, the goal for improving productivity in plant cell systems is not so much to make productive cells productive for a longer time, but rather to make less productive cells more productive, or to fill the reactor with more cells.

A second major difference is that microbial secondary metabolites are generally found in the extracellular medium whereas plant cell products may or may not be secreted. If the product is retained within the cells, then cell harvesting will be necessary for product recovery. If the cells secrete, or can be modified to secrete, the product, then it is possible to reuse the cells. The reuse of cells has been one motivation for the development of perfusion systems and cell immobilization. It is important to recognize that the reuse of cells is only appropriate if the following three criteria are met. First, the cells must secrete the product - otherwise, cells must be harvested for product recovery and thus cannot be reused. Second, the product must be produced in a non-growth-associated manner. If product synthesis is growth-associated, then cells are only needed for inoculation purposes and once a cell is finished growing and therefore is no longer productive, it is pointless to reuse it. Finally, it is essential that the biosynthetic capacity remain high if cell reuse is to be practical. Although most microbial secondary metabolite-producing cultures meet the first two criteria, they often fail to meet the third. Thus, most industrial processes for microbial secondary products are operated to obtain the highest productivity from the culture for the longest time, and no effort is made to then reuse the cells. If plant cells meet these criteria, then the reuse of cells is attractive because of the substantial investment made in growing the culture. Cell reuse is potentially attractive when cells are used to mediate a single or small number of biotransformation reactions.

E.1. Batch

When all the nutrients the culture requires are supplied initially, then this operation is termed batch. (It should be noted that oxygen is generally excluded from these definitions. As discussed in Chapter 5, a truly batch process with respect to oxygen could only remain aerobic for a short period due to the low solubilities of this nutrient.) As shown in Table 6.2, a batch mode could be used for growth- and non-growth-associated products. When a batch mode is used for growth-associated production, a medium is used which permits rapid growth and a high final cell concentration. To enhance non-growth-associated product synthesis in a batch process, a medium is used which permits rapid initial growth followed by an extended phase in which secondary metabolite synthesis occurs. Such media contain at least one component which becomes growth-limiting during cultivation, and it is the depletion of this nutrient which is believed to result in the cessation of rapid growth (e.g., Phillips and Henshaw 1977). Although the mechanistic details are unclear, under growth-limiting conditions the enzymes required for microbial secondary metabolism are often induced (Martin and Demain 1980). However, this induction may depend on which nutrient is growth-limiting (Lilley et al. 1981). For instance Dougall et al. (1983) observed that anthocyanin production by carrot cells was dependent on which nutrient was growth-limiting. In microbial systems, the growth-limiting nutrient is usually the carbon, nitrogen, or phosphorus source. In plant cell systems, it is possible that the growth regulators (e.g. 2,4-D) may serve this role (Roustan et al. 1982).

E.2. Fed Batch

In fed batch operation, nutrients are supplied either periodically or continuously throughout the course of cultivation. A product stream is not withdrawn in fed batch operation. This operating mode is an extension of batch operation and can also be appropriate for producing both growth- and non-growth-associated products. Fed batch mode is justified when it would be detrimental to supply all the nutrients required for the entire culture period at the time of inoculation. As mentioned in Chapter 2, phosphate limitations have been observed to be beneficial for the biosynthesis of many secondary metabolites (Knobloch and Berlin 1983). However, Schiel et al. (1984) reported that too little phosphate can also be detrimental, and these authors reported increased cinnamoyl putrescine production using phosphate feeding.

A second reason for operating in a fed batch mode is to supply precursors for biosynthesis. Hiraoka et al. (1973) observed that tropic acid (a precursor for tropane alkaloids) was inhibitory to *Datura innoxia* growth and that tropane alkaloid production could be improved if this precursor was added after the growth phase.

With respect to plant cell systems, a final reason for using fed batch operation is that if elicitor treatment results in only a transient induction, then it may be necessary to add elicitor at various times during cultivation (see DiCosmo and Misawa 1985).

E.3. Repeated Fed Batch (Semicontinuous or Variable Volume Continuous)

Often during fed batch operation it is necessary to remove, or harvest, a substantial portion of the liquid and to replace this volume with fresh medium. There are three general reasons why this repeated fed batch mode would be required. First, the bioreactor may become filled when large volumes of liquid are fed. Thus, some of the liquid is removed and the operation is resumed. Second, a repeated fed batch mode may be required to ensure proper conditions for biosynthesis. For growth-associated product synthesis, it is necessary to harvest a substantial amount of cells and add fresh medium when a culture reaches a stationary phase. For non-growth-associated products, the metabolic activity of the culture may decay after extended periods in which growth does not occur. It may be possible to "reactivate" the culture by allowing it to undergo a rapid growth phase. This is accomplished by draining a large volume of the liquid and adding fresh medium. Thus, the cells which remain in the bioreactor act as an inoculum for the subsequent production period. The advantage here is that the overall productivity of the operation can be improved since the bioreactor can be reused without the need for cleaning and resterilization. This appears to be the rationale for using repeated fed batch operation for the biotransformation of β-methyldigitoxin to β-methyldigoxin (Kreis and Reinhard 1989).

A third reason for using fed batch with draw-off is that it may be essential to remove a significant portion of the cells to ensure that the oxygen demand of the culture does not outstrip the oxygen transfer capacity of the bioreactor. For instance, in the penicillin fermentation, feeding schemes are implemented which permit the cells to grow slowly over extended periods. Periodically, however, the cell concentration, and therefore the oxygen demand of the culture, increases to a point that the required dissolved oxygen cannot be maintained above the critical level. For penicillin production, it was observed that if the dissolved oxygen fell below 30% of air saturation, then penicillin productivities were reduced, and below 10% saturation these productivitiy reductions were irreversible (Vardar and Lilly 1982). It appears that the possible detrimental effects of oxygen limitation have not been sufficiently recognized for many plant cell systems. Efforts to use large-scale systems have often focused on achieving the highest cell concentration, irrespective of whether these cells can be maintained under aerobic conditions.

E.4. Two-Stage Batch Operation

As shown in Table 6.2, two-stage operations are appropriate for non-growth-associated products. For many plant products, it has been observed that conditions which result in optimal secondary metabolite production are often significantly different from conditions which permit cell growth. Thus, the use of two separate media has been advocated in many plant cell systems: the first medium supports rapid growth whereas the second medium supports less growth but good product synthesis. Although the use of two media allows growth and product synthesis to be optimized independently, it is important to recognize that medium exchange on a large scale presents significant operational problems, and therefore two media should be used only when conditions for growth and product synthesis are mutually exclusive (Morimoto et al. 1988). With plant cells, this may occur if a component (e.g., 2,4-D) essential for growth inhibits product synthesis and this component is not metabolized by the cells.

E.5. Continuous Cultivation - Chemostat

Chemostat refers to operation in which fresh medium is continuously fed to the bioreactor, the bioreactor is well mixed, and a product stream containing both cells and spent medium is continuously withdrawn. To ensure that the cells are not completely washed out of the bioreactor, it is essential that the cells are able to grow at a rate sufficient to match the dilution rate of the incoming flow. The dilution rate is the ratio of the inlet liquid flow rate (L/h) to the liquid volume of the bioreactor (L). Since theoretical engineering analysis shows chemostat operation to be an efficient mode for growth-associated production, chemostats were used by Kato et al. (1976) for growing tobacco cells. Practical difficulties however have limited the use of chemostat operation (Hashimoto and Azecki 1988). With respect to non-growth-associated products, chemostat operation is generally inappropriate for efficient product synthesis. However, chemostats can be used as an experimental tool (Wilson and Marron 1978) to examine culture physiology. The major drawback of using chemostats is the difficulty in obtaining long-term, aseptic operation.

A two-stage chemostat may be appropriate for the production of non-growth-associated products since the first stage can be used to efficiently grow the cells while the second stage can be used for product synthesis. Initial results with two-stage chemostat operation have been reported by Tal et al. (1983) and Sahai and Shuler (1984).

E.6. Continuous Cultivation - Perfusion

In perfusion operation, fresh medium is added to the vessel and effluent is withdrawn such that a significant fraction of the cells are recycled or retained within the reactor. At one extreme, when all the cells are recycled or retained within the reactor, then there is no requirement for cell growth and this system effectively becomes an immobilized suspension culture. At this extreme, the perfusion reactor is appropriate for non-growth-associated product synthesis. At the other extreme, when there is no cell retention, then growth is essential for steady-state operation, and this system reduces to a chemostat. Although work with plant cells is limited, perfusion operation is common in mammalian cell systems. Obviously, to operate in a perfusion mode, a means to retain or recycle the cells is essential. Filtration and sedimentation are likely mechanisms for separating cells either within or outside the bioreactor. Such cell separations will be briefly discussed when indole alkaloid production is considered.

It should be mentioned that Mitsui Petrochemical Industries' commercial-scale process for berberine production uses perfusion cultivation. This process is somewhat unusual in terms of traditional bioprocess designs and warrants further comment. The goal of using perfusion is to operate at very high cell concentrations (packed cell volumes of about 90%). To supply sufficient nutrients to this culture, nutrient feeding was required. However, to prevent the cells from being diluted, liquid was removed using a membrane. Since berberine is a growth-associated product, significant growth occurs and occasionally a portion of the cells must be removed from the bioreactor. The intracellular bereberine can then be recovered from these harvested cells. Clearly, a considerable amount of effort was devoted to matching the bioreactor operation to the requirements specific to this process.

F. EXAMPLES

F.1. Tobacco Cells

One of the earliest efforts to culture plant cells on a large scale was for the production of smoking material from tobacco cells. Because the cells had a "characteristic and unfavorable taste on smoking" (Hashimoto and Azecki 1988), production was not commercialized. In a mixed vessel bioreactor, doubling times of 1 day and dry cell concentrations of 20 g/L were reported (Kato et al. 1972). However, at these high cell concentrations, the cells occupied up to 65% of the packed cell volume, and the viscosity of the culture was observed to increase markedly (Kato et al. 1978). It should be mentioned that such large viscosity increases have not been observed with all plant cell systems (Wayne Curtis, personal communication). At these high viscosities, oxygen transfer can be reduced (Kato et al. 1975). At K_La values less than 10 h^{-1}, tobacco cell growth was reduced whereas K_La values between 10 and 25 h^{-1} supported growth to 15 g of dry cells per liter (Kato et al. 1975). In subsequent work, mechanically and pneumatically agitated vessels were studied and continuous operation was reported. However, semicontinuous and continuous operations were difficult due to problems in maintaining asepsis and in maintaining high growth rates for long periods (Hashimoto and Azecki 1988). Finally, scale-up of tobacco cell growth from 20 to 20,000 L was accomplished such that similar cell growth rates were achieved. As will be seen from subsequent examples, maintaining similar secondary metabolite biosynthesis upon scale-up has been more difficult.

F.2. Shikonin Production from *Lithospermum erythrorhizon*

The first commercial process for producing a plant natural product by cell culture was developed in Japan for the production of the naphthoquinone shikonin. The development of this process has been reviewed by Yamada and Fujita (1983) and Tabata and Fujita (1985). Shikonin is used as a remedy for various skin ailments and as a dye for silk and cosmetics. Development of cell culture production methods was stimulated by the value of shikonin and difficulties encountered in meeting the demand using field-grown material (Tabata and Fujita 1985).

Figure 6.4 shows that shikonin is synthesized from the shikimic acid and isoprenoid pathways (Heide and Tabata 1987b). Initial efforts to develop high-producing cell lines (Mizukami et al. 1978) were facilitated by the fact that high producers are easily identified by their bright color. Thus, through visual inspection it was possible to identify and screen for high-producing cultures.

In medium development studies, it was determined that cytokinins, copper (Fujita et al. 1981b), high levels of sucrose, and low levels of nitrogen (Mizukami et al. 1977) were beneficial for shikonin production. The synthetic auxin 2,4-D (Tabata et al. 1974), gibberellin (Yoshikawa et al. 1986), ammonium (Fujita et al. 1981a), and

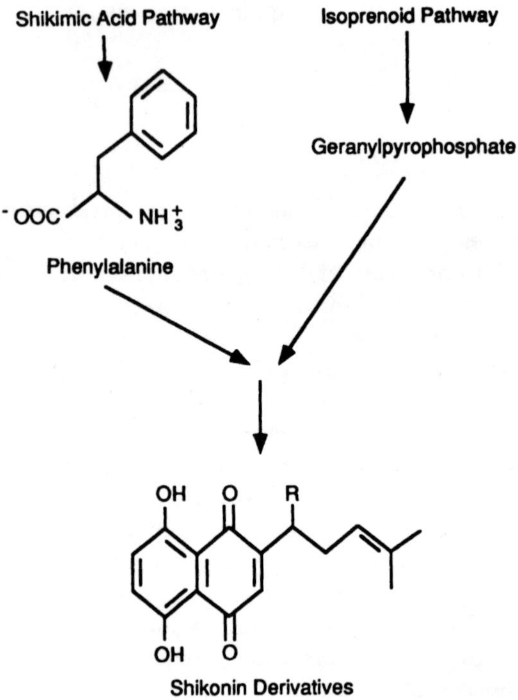

Figure 6.4. Abbreviated pathway of shikonin and its derivatives. Adapted from Heide and Tabata (1987b) and Tabata and Fujita (1985).

glutamine (Yazaki et al. 1987) were observed to suppress production. Interestingly, it was also observed that an acidic polysaccharide fraction stimulated production (Fukui et al. 1983), suggesting that shikonin may be produced by the cells in response to an elicitor (Tabata and Fujita 1985).

Since the medium (designated M-9) which supports high levels of shikonin production did not result in good growth (Fujita et al. 1981b), a two-stage process was developed. As shown in Figure 6.5, the cells are first grown in a growth medium (designated MG-5) for 9 days, then filtered and pumped to a second vessel in which the M-9 production medium is added. The cells are cultured for 14 additional days, after which they are recovered by filtration, and the shikonin is extracted from the cells (Tabata and Fujita 1985). Shikonin levels at that time were reported to be 4 g/L (Tabata and Fujita 1985).

With respect to the role of oxygen, shikonin production has been well studied compared to other plant cell systems. Tabata and Fujita (1985) reported the effect of oxygen on shikonin production. In flasks, growth and production were enhanced by increased K_La for values up to 18 h^{-1}. Tanaka (1987) reported that in a fermentor with a paddle type impeller, shikonin production was enhanced by increased K_La for values up to 10 h^{-1}. Also, Tanaka (1987) reported that scale-up of the shikonin process was successful when scale-up was based on maintaining constant K_La. In the two-stage batch process (Tabata and Fujita 1985), shikonin productivities were reported to be highest when the dissolved oxygen concentrations were 6.4 and 6 ppm in the indi-

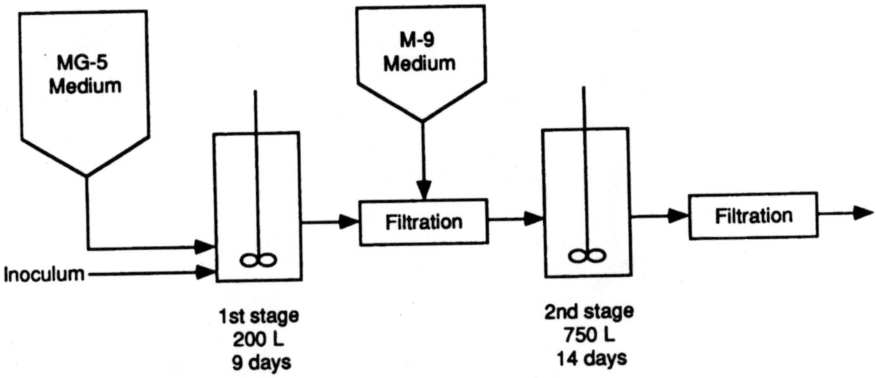

Figure 6.5. Schematic diagram of process for the commercial production of shikonin. Adapted from Tabata and Fujita (1985).

vidual stages. These oxygen concentrations are rather high, corresponding to 75-80% saturation with air.

In more recent work, Fujita and Tabata (1987) reported on some details on culturing *L. erythrorhizon* cells in bioreactors. When a paddle impeller was used, good cell growth was observed; however, shikonin production was reduced compared to that by shake-flask-cultured cells. Also, increases in the impeller speed further reduced shikonin concentrations. These authors attributed this reduction in production at the higher agitation to cell injury. Efforts to use an airlift type reactor were frustrated by the "bubbling-up" of the cells and the subsequent adherence of the cells to the tank wall. These authors then adopted a rotary drum type tank (Tanaka et al. 1983) which revolved slowly (Figure 6.6), thus washing the cells from the wall. This system has been scaled-up to 1,000 L without loss in shikonin production (Fujita and Tabata 1987). Scale-up from 5 to 1,000 L in either mechanically agitated or airlift systems was reported to result in reductions in shikonin production (Tanaka 1987).

Recent work has considered the use of *in situ* removal by extracting shikonin from the cell surface into nonpolar solvents (Heide and Tabata 1987a; Deno et al. 1987; Kim and Chang 1990).

F.3. Berberine Production

Berberine, a medicinal alkaloid, has also been considered for commercial production and has been scaled up to the 4,000-L scale by Mitsui Petrochemical Industries. This alkaloid is derived from two molecules of tyrosine as shown in Figure 6.7 (for a review, see Zenk et al. 1985). There are various producing cultures, and interestingly, berberine is stored within *Coptis japonica* cells, whereas this alkaloid is released into the medium with *Thalictrum minus* cells (Yamamoto et al. 1986).

As in the case of shikonin, efforts were made to obtain high-producing cell lines (Yamada and Sato 1981; Sato and Yamada 1984; Yamada and Morikawa 1985). High berberine producers were identified using fluorescence microscopy, and these cells

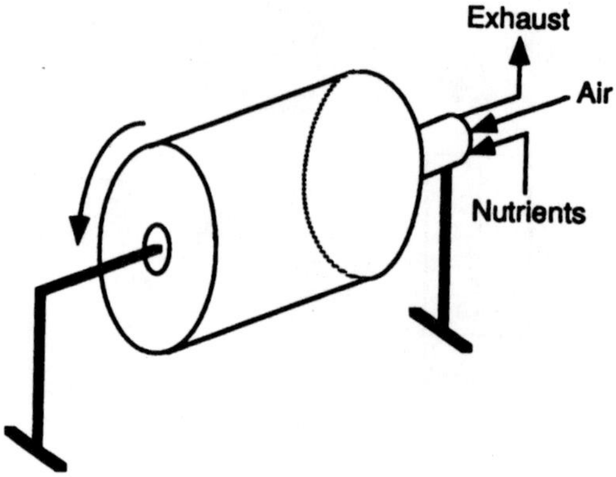

Figure 6.6. Rotary drum bioreactor for shikonin production. Adapted from Fujita and Tabata (1987).

were selected from small aggregates (Sato and Yamada 1984) and protoplasts (Yamada and Morikawa 1985).

Nakagawa et al. (1986) observed that although the synthetic auxin 2,4-D supported good growth of *T. minus* cells, berberine production was low. In contrast, when an auxin and a cytokinin were added, growth was inferior although berberine production was enhanced. Since optimal conditions for growth and product synthesis were different, these results suggested that a two-stage process would be appropriate. However, Fujita and Tabata (1987) reported that for *Coptis japonica* cultures, optimal conditions for growth and production were so similar that a single-stage process was appropriate.

Recently, Hara et al. (1988) reported that gibberellic acid stimulated berberine production in *C. japonica*. This increased production was associated with reduced starch formation, and the authors suggested that gibberellic acid may have "channeled" sucrose consumption from starch synthesis to berberine production.

With respect to nutritional levels, Sato and Yamada (1984) examined various basal media and various carbohydrates. They reported sucrose to be best for growth and berberine production for their *C. japonica* culture. Fujita and Tabata (1987) showed that high levels of nutrients inhibited berberine production in *C. japonica* cultures, and thus for their single-stage process they suggested that nutrient feeding would be appropriate. Using this fed batch mode, the authors reported obtaining a dry cell concentration of 70 g/L with a berberine content of 10%. As mentioned earlier in this chapter, the commercial-scale process involves perfusion culture with occasional cell harvesting. As with shikonin, it was observed that copper stimulated berberine production (Morimoto et al. 1988).

Brodelius and co-workers (Funk et al. 1987; Marques and Brodelius 1988; Gugler et al. 1988) reported that a polysaccharide preparation from yeast was able to elicit both berberine production as well as activities of a key biosynthetic enzyme, tyrosine decarboxylase, from *Thalictrum rugosum*.

In shake-flask studies, Yamada and Sato (1981) reported that increased aeration stimulated both *C. japonica* growth and berberine synthesis.

Breuling et al. (1985) reported the use of a 19-L airlift bioreactor for berberine production by ***Berberis wilsonae***. Although aeration between 0.1 and 0.5 volumes air per volume liquid per minute (dissolved oxygen levels between 20 and 50% saturation) had a small effect on growth, berberine production was increased at the higher dissolved oxygen levels. Kreis and Reinhard (1989) also discussed the dissolved oxygen requirements in mechanically agitated bioreactors.

Finally, since berberine is secreted by ***Thalictrum*** cells, Kobayashi et al. (1988) developed an immobilized cell bioreactor. Kobayashi et al. (1989) also used low cell

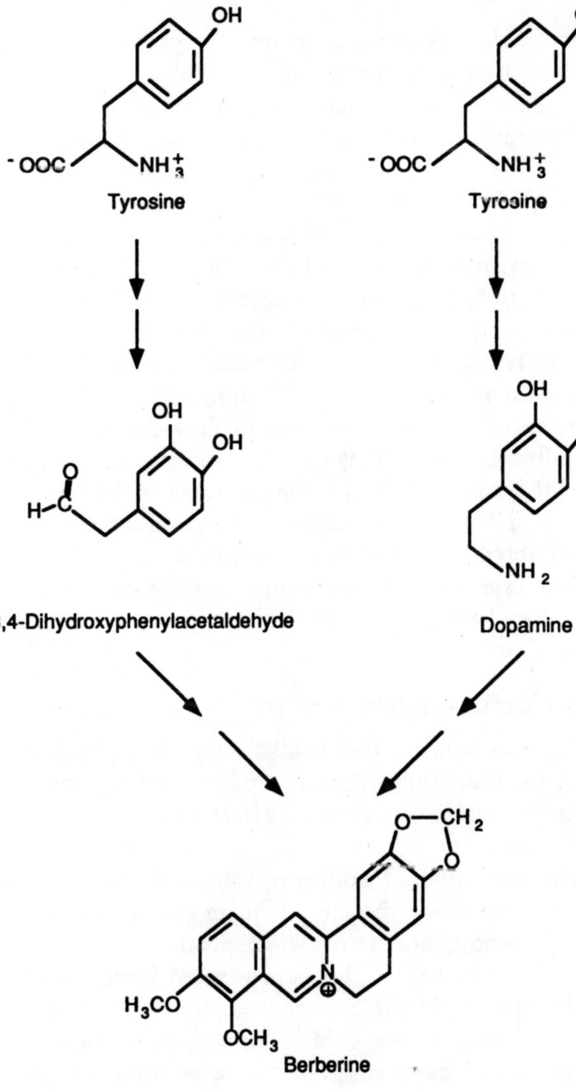

Figure 6.7. Abbreviated pathway for berberine biosynthesis. Adapted from Amann et al. (1986).

loading and small alginate beads to limit oxygen diffusional limitations for these immobilized cells. With respect to secretion, Kim et al. (1988) reported that although light enhanced growth and berberine synthesis in the later stages of the culture, berberine secretion was suppressed by light.

F.4. Rosmarinic Acid Production

The production of rosmarinic acid was investigated as a model for plant cell bioprocessing. Since antiphlogistic activity has been detected, "this seemingly unimportant compound may become an interesting candidate for production by cell cultures" (Berlin 1988).

Rosmarinic acid is produced from the distant precursors phenylalanine and tyrosine as shown in Figure 6.8. Although most of the rosmarinic acid was observed to be produced during periods of rapid growth, production only occurred in *Coleus* cultures when detectable levels of phenylalanine ammonia lyase (PAL) were present (Razzaque and Ellis 1977).

De-Eknamkul and Ellis (1984, 1985a,b) examined the dynamics of growth and rosmarinic acid production by cultures of *Anchusa officinalis*. Culture growth was observed to pass through lag, log, linear, and stationary phases. Phosphate and ammonia were observed to be depleted prior to the onset of the linear phase, and rosmarinic acid was produced during this linear phase. Also, these authors observed that conditions which enhanced growth generally improved rosmarinic acid production.

Initial studies with *Coleus blumei* in a single-stage airlift bioreactor showed that rosmarinic acid production was reduced compared to that in flask-grown cultures (Zenk et al. 1977). In later work, Ulbrich et al. (1985) reported serious mixing problems in a 32-L (liquid volume) airlift system. They thus opted to use a mechanically agitated vessel with a helical impeller. These authors reported the production of 5.5 g of rosmarinic acid per liter in a two-stage batch process. In the first stage, cells were grown for 8 days in a glucose fed batch system. After growth, cells were transferred to a second vessel in which they were diluted with a sucrose solution to initiate rosmarinic acid production.

F.5. Indole Alkaloids from *Catharanthus roseus*

In Chapter 2, it was noted that efforts to improve indole alkaloid production involve the selection of improved strains, medium optimization, and the development of appropriate bioreactor systems. The former two aspects have been discussed in Chapter 2 while the latter will be considered here.

Considerable work has focused on growing high concentrations of *C. roseus* cells in bioreactors. However, it must be stressed that although adequate growth is clearly a requirement for efficient alkaloid production, bioreactors which promote good growth are not necessarily optimal for product synthesis. This is illustrated by the work of Scragg et al. (1987), who observed that comparable growth was observed in flask and airlift systems; however, serpentine production was greatly suppressed in the airlift.

Although carbon dioxide stripping has been suggested as a possible source of

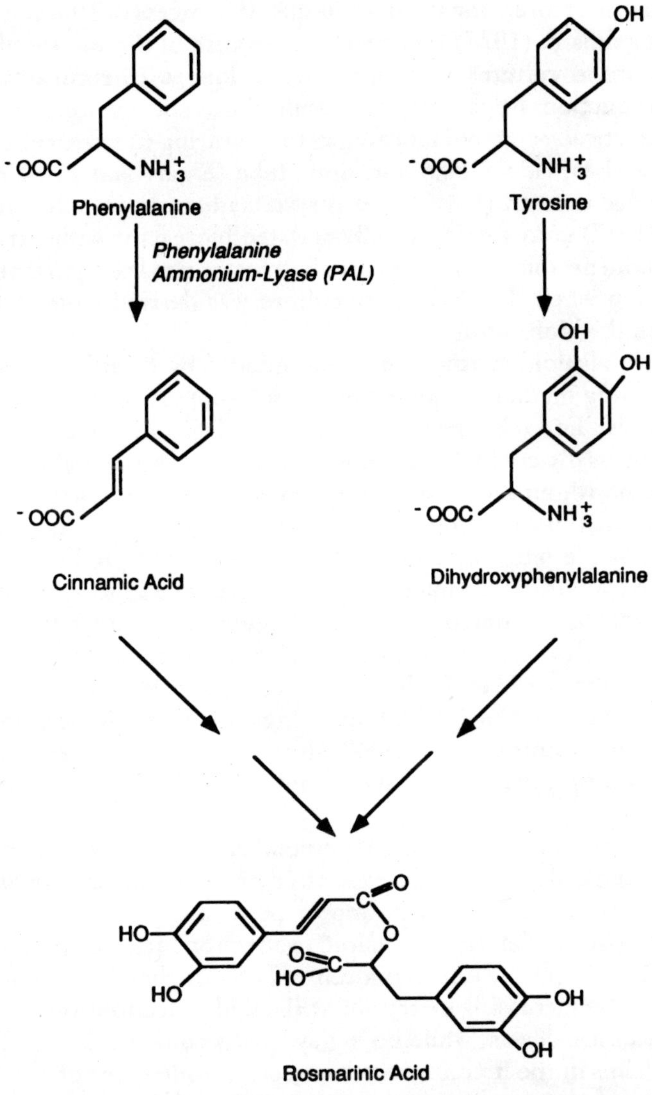

Figure 6.8. Abbreviated pathway for rosmarinic acid biosynthesis. Adapted from Ellis and Towers (1970).

variation between shake-flask and bioreactor growth, there appears to be no work describing the effect of oxygen and carbon dioxide concentrations on alkaloid production. Thus, it is not known if there is a link between changes in gas concentrations with scale-up and suppressed alkaloid production. It does seem likely that since *C. roseus* cells enlarge as they age (Snape et al. 1989), they may become more shear sensitive, and this increased shear sensitivity may explain observations of comparable growth but reduced secondary metabolite production. Although not conclusive, support for this view is provided by work of Fowler (1982), who observed that a shear-tolerant *C. roseus* cell line grew in a bioreactor at agitation speeds in excess of 300 rpm. In contrast, Wagner and Vogelmann (1977) reported that because of the increased shear sensitivity of older *C. roseus* cultures, agitation rates as low as 28 rpm were inappropriate for alkaloid production. Apparently as a result of this shear sensitivity, many groups have used pneumatically agitated bioreactors for culturing *C. roseus* cells. Both external loop (Smart and Fowler 1984a) and draft tube (Smart and Fowler 1984b) airlifts as well as bubble column (Payne 1988a) systems have been used. In contrast, Schiel and Berlin (1987) used a mechanically agitated bioreactor with flat-blade turbine impellers. Alkaloid production in this latter bioreactor was less than that observed in shake flasks – even when the shake-flask culture was derived from the same inoculum as was used in the bioreactor.

As previously noted, indole alkaloid synthesis is often induced by transferring a culture from a growth-promoting medium to a medium which supports less growth but enhanced product synthesis. To exchange media, an external filtration step was performed in the shikonin process. By coupling filtration to aeration, it was possible to filter the cells from the medium without the need for operations outside the bioreactor (Payne et al. 1988a). This internal filtration operation should reduce contamination risks and permit the use of a single bioreactor for both the growth and production phases. A similar system - the spin filter - couples agitation with filtration to achieve cell separations within a mechanically agitated bioreactor (Ammirato and Styer 1985; Styer 1985). Sedimentation has also been used for internal cell separation with *C. roseus* cultures (Pareilleux and Vinas 1984). It should be noted that Morris (1986) reported that when NAA is used in place of 2,4-D, then a single medium can be used for both growth and production. Smith et al. (1987) also studied indole alkaloid production in a single medium. Bioprocessing is significantly simplified if medium exchange is not required.

In addition to suspension cultures, various means of immobilization have been used with *C. roseus*. Since immobilized cell culture is discussed in detail in Chapter 7, only three points will be emphasized here. First, it is important to measure both the intracellular and extracellular space to determine alkaloid production rates and possibly even to determine if alkaloid production has occurred. With membrane-entrapped cells, we observed that the increasing extracellular alkaloid concentration was balanced by decreasing intracellular levels, while no net synthesis could be detected (Payne et al. 1988a). Many claims in the literature are based on measurement of only one of the alkaloid pools. Second, the response of the cells to immobilization appears to depend strongly on the method of immobilization (Brodelius and Nilsson 1980; Asada and Shuler 1989). Finally, since biosynthesis of the dimeric indole alkaloids appears to require tissue differentiation, immobilized undifferentiated cells may lack the ability to produce these dimeric compounds. To our knowledge, there have been no reports in which immobilized cells have been used for *de novo* synthesis of the dimeric alkaloids vincristine or vinblastine.

A newer approach which has been used to enhance both the production and excretion of alkaloids is *in situ* product removal. Brodelius et al. (1979) recirculated medium from an immobilized cell reactor through chloroform to extract lipophilic compounds (e.g., alkaloids). Solid sorbents have also been observed to enhance alkaloid production by suspended *C. roseus* cells (Payne et al. 1988b). Subsequent work by Asada and Shuler (1989) showed that ajmalicine production could be improved in a synergistic manner by combining alginate immobilization, elicitor treatment, and *in situ* adsorption. Ajmalicine levels observed in this work (90 mg/L) are among the highest extracellular levels observed for any individual indole alkaloid. However, these concentrations remain more than an order of magnitude lower than concentrations observed with shikonin- berberine- and rosmarinic acid-producing systems.

G. CONCLUSION

There are two major conclusions which are important for the design and scale-up of suspension culture bioreactors. First, to culture plant cells in a bioreactor, adequate attention must be paid to meeting the dissolved oxygen needs and preventing shear damage of the culture. With knowledge of the biological needs and the constraints of plant cells, it is possible to determine appropriate reactor designs (e.g., impeller types, impeller speeds, and gas sparging rates). Scale-up requires compromises with respect to shear, mixing, and oxygen transfer. For instance, to achieve the same level of mixing in conventional bioreactors, cells in the larger scale reactor are typically exposed to greater shear intensities. A second point for the development of bioprocesses is that the operating mode (e.g., batch, fed batch, or perfusion) should be chosen based on the culture's behavior. Typically, the operating mode is chosen based on whether growth and production occur simultaneously or sequentially.

Finally, when considering the commercialization of plant cell systems, it is instructive to consider why certain processes have been successful. From a technical standpoint, it appears to us that the shikonin process became a commercial reality because of the dedicated efforts devoted to solving the problems which were encountered. Obviously, screening for high-producing cell lines by simple visual inspection was helpful. However, extensive study was required to develop an optimized medium and to design appropriate bioreactor systems. It is impossible to estimate the rewards that would ensue if similar efforts were devoted to other plant cell systems.

References

M. Amann, G. Wanner, and M. H. Zenk. 1986. Intracellular Compartmentalization of Two Enzymes of Berberine Biosynthesis in Plant Cell Cultures. *Planta 167*, 310-320.

P. V. Ammirato and D. J. Styer. 1985. Strategies for Large-Scale Manipulation of Somatic Embryos in Suspension Culture. In "Biotechnology in Plant Science: Relevance to Agriculture in the Eighties," M. Zaitlin, P. Day, and A. Hollaender (eds.), Academic Press, New York, pp. 161-178.

M. Asada and M. L. Shuler. 1989. Stimulation of Ajmalicine Production and Ecxretion from *Catharanthus roseus*: Effects of Adsorption *in situ*, Elicitors and Alginate Immobilization. *Appl. Microbiol. Biotechnol. 30*, 475-481.

J. G. Aunins, M. S. Croughan, D. I. C. Wang, and J. M. Goldstein. 1986. Engineering Developments in Homogeneous Culture of Animal Cells: Oxygenation of Reactors and Scaleup. *Biotechnol. Bioeng. Symp. 17*. 699-723.

C. M. Bailey and H. Nicholson. 1989. A New Structured Model for Plant Cell Culture. *Biotechnol. Bioeng. 34*, 1331-1336.

R. A. Bello, C. W. Robinson, and M. Moo-Young. 1984. Liquid Circulation and Mixing Characteristics of Airlift Contactors. *Can. J. Chem. Eng. 62*, 573-577.

R. A. Bello, C. W. Robinson, and M. Moo-Young. 1985. Gas Holdup and Overall Volumetric Oxygen Transfer Coefficients in Airlift Contactors. *Biotechnol. Bioeng.* 27, 369-381.

J. Berlin. 1988. Formation of Secondary Metabolites in Cultured Plant Cells and Its Impact on Pharmacy. In "Biotechnology in Agriculture and Forestry," Vol. 4, Y. P. S. Bajaj (ed.) Springer-Verlag, Berlin, pp. 37-59.

M. Breuling, A. W. Alfermann, and E. Reinhard. 1985. Cultivation of Cell Cultures of *Berberis wilsonae* in 20-L Airlift Bioreactors. *Plant Cell Rep. 4*, 220-223.

P. Brodelius and K. Nilsson. 1980. Entrapment of Plant Cells in Different Matrices. *FEBS Lett. 122*, 312-316.

P. Brodelius, B. Deus, K. Mosbach, and M. H. Zenk. 1979. Immobilized Plant Cells for the Production and Transformation of Natural Products. *FEBS Lett. 103*, 93-97.

D. N. Bull, R. W. Thoma, and T. E. Stinnett. 1983. Bioreactors for Submerged Culture. *Adv. Biotechnol. Proc. 1*, 1-30.

R. S. Cherry and E. T. Papoutsakis. 1988. Physical Mechanisms of Cell Damage in Microcarrier Cell Culture Bioreactors. *Biotechnol. Bioeng. 32*, 1001-1014.

Y. Chisti and M. Moo-Young. 1989. On the Calculation of Shear Rate and Apparent Visosity in Airlift and Bubble Column Bioreactors. *Biotechnol. Bioeng. 34*, 1391-1392.

M. S. Croughan, J.-F. Hamel, and D. I. C. Wang. 1987. Hydrodynamic Effects on Animal Cells Grown in Microcarrier Cultures. *Biotechnol. Bioeng. 29*, 130-141.

W. De-Eknamkul and B. E. Ellis. 1984. Rosmarinic Acid Production and Growth Characteristics of *Anchusa officinalis* Cell Suspension Cultures. *Planta Med. 51*, 346-350.

W. De-Eknamkul and B. E. Ellis. 1985a. Effects of Macronutrients on Growth and Rosmarinic Acid Formation in Cell Suspension Cultures of *Anchusa officinalis. Plant Cell Rep. 4*, 46-49.

W. De-Eknamkul and B. E. Ellis. 1985b. Effects of Auxins and Cytokinins on Growth and Rosmarinic Acid Formation in Cell Suspension Cultures of *Anchusa officinalis. Plant Cell Rep. 4*, 50-53.

H. Deno, C. Suga, T. Morimoto, and Y. Fujita. 1987. Production of Shikonin Derivatives by Cell Suspension Cultures of *Lithospermum erythrorhizon* VI. Production of Shikonin Derivatives by a Two-Layer Culture Containing an Organic Solvent. *Plant Cell Rep. 6*, 197-199.

S. L. Diamond, S. G. Eskin, and L. V. McIntire. 1989. Fluid Flow Stimulates Tissue Plasminogen Activator Secretion by Cultured Human Endothelial Cells. *Science 243*, 1483-1485.

F. DiCosmo and M. Misawa. 1985. Eliciting Secondary Metabolism in Plant Cell Cultures. *Trends Biotechnol. 3*, 318-322.

D. K. Dougall, S. LaBrake, and G. H. Whitten. 1983. The Effects of Limiting Nutrients, Dilution Rate, Culture pH, and Temperature on the Yield Constant and Anthocyanin Accumulation of Carrot Cells Grown in Semicontinuous Chemostat Cultures. *Biotechnol. Bioeng. 25*, 569-579.

D. Drapeau, H. W. Blanch, and C. R. Wilke. 1986. Growth Kinetics of *Dioscorea deltoidea* and *Catharanthus roseus* in Batch Culture. *Biotechnol. Bioeng. 28*, 1555-1563.

B. E. Ellis and G. H. N. Towers. 1970. Biogenesis of Rosmarinic Acid in *Mentha. Biochem. J. 118*, 291-297.

M. W. Fowler. 1982. The Large Scale Cultivation of Plant Cells. *Prog. Ind. Microbiol. 16*, 207-229.

M. W. Fowler. 1984. Large-Scale Cultures of Cells in Suspension. In "Cell Culture and Somatic Cell Genetics of Plants," Vol. 1, I. K. Vasel (ed.), Academic Press, New York, pp. 167-174.

J. A. Frangos, L. V. McIntire, and S. G. Eskin. 1988. Shear Stress Induced Stimulation of Mammalian Cell Metabolism. *Biotechnol. Bioeng. 32*, 1053-1060.

Y. Fujita and M. Tabata. 1987. Secondary Metabolites from Plant Cells - Pharmaceutical Applications and Progress in Commercial Production. In "Plant Tissue and Cell Culture," C. E. Green, D. A. Somers, W. P. Hackett, and D. D. Biesboer (eds.), Alan R. Liss Inc., New York, pp. 169-185.

Y. Fujita, Y. Hara, T. Ogino, and C. Suga. 1981a. Production of Shikonin Derivatives by Cell Suspension Cultures of *Lithospermum erythrorhizon* I. Effect of Nitrogen Sources on the Production of Shikonin Derivatives. *Plant Cell Rep. 1*, 59-60.

Y. Fujita, Y. Hara, C. Suga, and T. Morimioto. 1981b. Production of Shikonin Derivatives by Cell Suspension Cultures of *Lithospermum erythrorhizon* II. A New Medium for the Production of Shikonin Derivatives. *Plant Cell Rep. 1*, 61-63.

H. Fukui, N. Yoshikawa, and M. Tabata. 1983. Induction of Shikonin Formation by Agar in *Lithospermum erythrorhizon* Cell Suspension Cultures. *Phytochemistry 22*, 2451-2453.

C. Funk, K. Gugler, and P. Brodelius. 1987. Increased Secondary Product Formation in Plant Cell Suspension Cultures After Treatment with a Yeast Carbohydrate Preparation (Elicitor). *Phytochemistry 26*, 401-405.

K. Gugler, C. Funk, and P. Brodelius. 1988. Elicitor-Induced Tyrosine Decarboxylase in Berberine-Synthesizing Suspension Cultures of *Thalictrum rugosum*. *Eur. J. Biochem. 170*, 661-666.

Y. Hara, T. Yoshioka, T. Morimoto, Y. Fujita, and Y. Yamada. 1988. Enhancement of Berberine Production in Suspension Cultures of *Coptis japonica* by Gibberellic Acid Treatment. *J. Plant Physiol. 133*, 12-15.

T. Hashimoto and S. Azecki. 1988. Bioreactors for the Large Scale Culture of Plant Cells. In "Biotechnology in Agriculture and Forestry," Vol. 4, Y. P. S. Bajaj (ed.) Springer-Verlag, Berlin, pp. 104-122.

P. K. Hegarty, N. J. Smart, A. H. Scragg, and M. W. Fowler. 1986. The Aeration of *Catharanthus roseus* L. G. Don Suspension Cultures in Airlift Bioreactors: The Inhibitory Effect at High Aeration Rates on Culture Growth. *J. Exp. Bot. 37*, 1911-1920.

L. Heide and M. Tabata. 1987a. Enzyme Activities in Cell-Free Extracts of Shikonin-Producing *Lithospermum erythrorhizon* Cell Suspension Cultures. *Phytochemistry 26*, 1645-1650.

L. Heide and M. Tabata. 1987b. Geranylpyrophosphate:p-Hydroxybenzoate Geranyltransferase Activity in Extracts of *Lithospermum erythrorhizon* Cell Cultures. *Phytochemistry 26*, 1651-1655.

N. Hiraoka, M. Tabata, and M. Konoshima. 1973. Formation of Acetyltropine in *Datura* Callus Cultures. *Phytochemistry12* 795-799.

Y. C. Hong, T. P. Labuza, and S. K. Harlander. 1989. Growth Kinetics of Strawberry Cell Suspension Cultures in Shake Flask, Airlift, Stirred-Jar and Roller Bottle Bioreactors. *Biotechnol. Prog. 5*, 137-143.

W.-S. Hu and D. I. C. Wang. 1986. Mammalian Cell Culture Technology: A Review from an Engineering Perspective. In "Mammalian Cell Technology," W. G. Thilly (ed.), Butterworths, Boston, pp.167-197.

K. Kato, Y. Shiozawa, A. Yamada, K. Nishida, and M. Noguchi. 1972. A Jar Fermentor Culture of *Nicotiana Tabacum* L. Cell Suspensions. *Agric. Biol. Chem. 36*, 899-902.

A. Kato, Y. Shimizu, and S. Nagai. 1975. Effect of Initial K_La on the Growth of Tobacco Cells in Batch Culture. *J. Ferment. Technol. 53*, 744-751.

A. Kato, S. Kawazoe, M. Iijima, and Y. Shimizu. 1976. Continuous Culture of Tobacco Cells. *J. Ferment. Technol. 54*, 82-87.

A. Kato, S. Kawazoe, and Y. Soh. 1978. Viscosity of the Broth of Tobacco Cells in Suspension Culture. *J. Ferment. Technol. 56*, 224-228.

Y. Kawase and M. Moo-Young. 1986. Mixing and Mass Transfer in Concentric-Tube Airlift Fermentors: Newtonian and Non-Newtonian Media. *J. Chem. Tech. Biotechnol. 36*, 527-538.

D.-I. Kim, H. Pedersen, and C.-K. Chin. 1988. Effects of Light on Berberine Production in Cell Suspension Cultures of *Thalictrum rugosum*. *Biotechnol. Lett. 10*, 709-712.

D. J. Kim and H. N. Chang. 1990. Enhanced Shikonin Production from *Lithospermum erythrorhizon* by *In Situ* Extraction and Calcium Alginate Immobilization. *Biotechnol. Bioeng. 36*, 460-466.

K.-H. Knobloch and J. Berlin. 1983. Influence of Phosphate on the Formation of the Indole Alkaloids and Phenolic Compounds in Cell Suspension Cultures of *Catharanthus roseus*. I. Comparison of Enzyme Activities and Product Accumulation. *Plant Cell Tissue Organ Culture* 2, 333-340.

K.-H. Knobloch and K. Hahlbrock. 1977. 4-Coumarate:CoA Ligase from Cell Suspension Cultures of *Petroselinum hortense* Hoffm. *Arch. Biochem. Biophys. 184*, 237-248.

Y. Kobayashi, H. Fukui, and M. Tabata. 1988. Berberine Production by Batch and Semi-Continuous Cultures of Immobilized *Thalictrum* Cells in an Improved Bioreactor. *Plant Cell Rep.* 7, 249-252.

Y. Kobayashi, H. Fukui, and M. Tabata. 1989. Effect of Oxygen Supply on Berberine Production in Cell Suspension Cultures and Immobilized Cells of *Thalictrum minus*. *Plant Cell Rep. 8*, 255-258.

W. Kreis and E. Reinhard. 1989. The Production of Secondary Metabolites by Plant Cells Cultivated in Bioreactors. *Planta Med. 55*, 409-416.

T. Kuboi and Y. Yamada. 1976. Caffeic Acid-O-Methyltransferase in a Suspension of Cell Aggregates of Tobacco. *Phytochemistry 15*, 397-400.

G. Lilley, A. E. Clark, and G. C. Lawrence. 1981. Control of the Production of Cephamycin C and Thienamycin by *Streptomyces cattleya* NRRL 8057. *J. Chem. Tech. Biotechnol. 31*, 127-134.

S. H. Mantell and H. Smith. 1983. Cultural Factors That Influence Secondary Metabolite Accumulations in Plant Cell and Tissue Cultures. In "Plant Biotechnology," S. H. Mantell and H. Smith (eds.), Cambridge University Press, New York, pp. 75-108.

I. A. Marques and P. E. Brodelius. 1988. Elicitor-Induced L-Tyrosine Decarboxylase from Plant Cell Suspension Culture. I. Induction and Purification. *Plant Physiol. 88*, 46-51.

J. F. Martin and A. L. Demain. 1980. Control of Antibiotic Biosynthesis. *Microbiol. Rev. 44*, 230-251.

B. Maurel and A. Pareilleux. 1985. Effect of Carbon Dioxide on the Growth of Cell Suspensions of *Catharanthus roseus*. *Biotechnol. Lett.* 7, 313-318.

W. J. McManamey and D. A. J. Wase. 1986. Relationship between the Volumetric Mass Transfer Coefficient and Gas Holdup in Airlift Fermentors. *Biotechnol. Bioeng. 28*, 1446-1448.

M. Midler and R. K. Finn 1966. A Model System for Evaluating Shear in the Design of Stirred Fermentors. *Biotechnool. Bioeng. 8*, 71-84.

H. Mizukami, M. Konoshima, and M. Tabata. 1977. Effect of Nutritional Factors on Shikonin Derivative Formation in *Lithospermum* Callus Cultures. *Phytochemistry 16*, 1183-1186.

H. Mizukami, M. Konoshima, and M. Tabata. 1978. Variation in Pigment Production in *Lithospermum erythrorhizon* Callus Cultures. Phytochem*istry 17*, 95-97.

T. Morimoto, Y. Hara, Y. Kato, J. Hiratsuka, T. Yoshioka, Y. Fujita, and Y. Yamada. 1988. Berberine Production by Cultures of *Coptis japonica* Cells in a One-Stage Culture Using Medium with a High Copper Concentration. *Agric. Biol. Chem. 52*, 1835-1836.

P. Morris. 1986. Regulation of Product Synthesis in Cell Cultures of *Catharanthus roseus*. II Comparison of Production Media. *Planta Med. 52*, 121-126.

K. Nakagawa, H. Fukui, and M. Tabata. 1986. Hormonal Regulation of Berberine Production in Cell Suspension Cultures of *Thalictrum minus*. *Plant Cell Rep. 5*, 69-71.

J. Y. Oldshue. 1966. Fermentation Mixing Scale-up Techniques. *Biotechnol. Bioeng. 8*, 3-24.

A. Pareilleux and R. Vinas. 1984. A Study on the Alkaloid Production by Resting Cell Suspensions of *Catharanthus roseus* in a Continuous Flow Reactor. *Appl. Microbiol. Biotechnol. 19*, 316-320.

G. F. Payne, M. L. Shuler, and P. Brodelius. 1987. Large Scale Plant Cell Culture. In "Large Scale Cell Culture Technology," B. K. Lydersen (ed.), Hanser Publishers, New York, pp. 193-229.

G. F. Payne, N. N. Payne, and M. L. Shuler. 1988a. Bioreactor Considerations for Secondary Metabolite Production From Plant Cells: Indole Alkaloids from*Catharanthus roseus.. Biotechnol. Bioeng. 31*, 905-912.

G. F. Payne, N. N. Payne, M. L. Shuler, and M. Asada. 1988b. *In Situ* Adsorption for Enhanced Alkaloid Production by *Catharanthus roseus*. *Biotechnol. Lett. 10*, 187-192.

R. Phillips and G. G. Henshaw. 1977. The Regulation of Synthesis of Phenolics in Stationary Phase Cell Cultures of *Acer pseudoplatanus* L. *J. Exp. Bot. 28*, 785-794.

A. Prokop and M. Z. Rosenberg. 1989. Bioreactor for Mammalian Cell Culture. *Adv. Biochem. Eng. Biotechnol. 39*, 29-71.

A. Razzaque and B. E. Ellis. 1977. Rosmarinic Acid Production in *Coleus* Cell Cultures. *Planta 137*, 287-291.

C. W. Robinson. 1986. On Power Input and Liquid-Phase Mixing Times in Pneumatically-Agitated Contactors. *Can. J. Chem. Eng. 64*, 523-528.

J.-P. Roustan, C. Ambid and J. Fallot. 1982. Influence de l'acide 2,4-dichlorophenoxyacetique sur l'accumulation de certains alcaloides indoliques dans les cellules quiescentes de *Catharanthus roseus* Cultivees *in vitro*. *Physiol. Veg. 20*, 523-532.

O. P. Sahai and M. L. Shuler. 1984. Multistage Continuous Culture to Examine Secondary Metabolite Formation in Plant Cells: Phenolics from *Nicotiana tabacum*. *Biotechnol. Bioeng. 26*, 27-36.

M. Sakuta, T. Takagi, and A. Komamine. 1987. Effects of Nitrogen Source on Betacyanin Accumulation and Growth in Suspension Cultures of *Phytolacca americana*. *Physiol. Plant. 71*, 459-463.

F. Sato and Y. Yamada. 1984. High Berberine-Producing Cultures of *Coptis japonica* Cells. *Phytochemistry 23*, 281-285.

O. Schiel and J. Berlin. 1987. Large Scale Fermentation and Alkaloid Production of Cell Suspension Cultures of *Catharanthus roseus*. *Plant Cell Tissue Organ Culture 8*, 153-161.

O. Schiel, K. Jarchow-Redecker, G.-W. Piehl, J. Lehmann, and J. Berlin. 1984. Increased Formation of Cinnamoyl Putrescines by Fedbatch Fermentation of Cell Suspension Cultures of *Nicotiana tabacum*. *Plant Cell Rep. 3*, 18-20.

A. H. Scragg, P. Morris, E. J. Allan, P. Bond, and M. W. Fowler. 1987. Effect of Scale-Up on Serpentine Formation by *Catharanthus roseus* Suspension Cultures. *Enzyme Microb. Technol. 9*, 619-624.

K. Shetty, R. A. Korus, and D. L. Crawford. 1989. Growth Kinetics and Phenolics Production in *Glycine max* Cell Suspension Cultures. *Appl. Biochem. Biotechnol. 20/21*, 825-843.

M. L. Shuler, J. W. Pyne, and G. A. Hallsby. 1984. Prospects and Problems in the Large Scale Production of Metabolites from Plant Cell Tissue Cultures. *J. Am. Oil Chem. Soc. 61*, 1724-1728.

S. P. Sigurdson and C. W. Robinson. 1976. A Comparison of Oxygen Transfer Power Requirements and Economy in Nonmechanically and Mechanically Agitated Fermentors. *Dev. Ind. Microbiol. 18*, 529-547.

A. J. Sinskey, R. J. Fleischaker, M. A. Tyo, D. J. Giard, and D. I. C. Wang. 1981. Production of Cell Derived Products: Virus and Interferon. *Ann. N.Y. Acad. Sci. 369*, 47-59.

N. J. Smart and M. W. Fowler. 1984a. An Airlift Column Bioreactor Suitable for Large-Scale Cultivation of Plant Cell Suspensions. *J. Exp. Bot. 35*, 531-537.

N. J. Smart and M. W. Fowler. 1984b. Mass Cultivation of *Catharanthus roseus* Cells Using a Nonmechanically Agitated Bioreactor. *Appl. Biochem. Biotechnol. 9*, 209-216.

J. I. Smith, A. A. Quesnel, N. J. Smart, M. Misawa, and W. G. W. Kurz. 1987. The Development of a Single-Stage Growth and Indole Alkaloid Production Medium for *Catharanthus roseus* (L.) G. Don Suspension Cultures. *Enzyme Microb. Technol. 9*, 466-469.

J. B. Snape, N. H. Thomas, and J. A. Callow. 1989. How Suspension Cultures of *Catharanthus roseus* Respond to Oxygen Limitation: Small-Scale Tests with Applications to Large-Scale Cultures. *Biotechnol. Bioeng. 34*, 1058-1062.

D. J. Styer. 1985. Bioreactor Technology for Plant Propagation. In "Tissue Culture in Forestry and Agriculture," R. R. Henke, K. W. Hughes, M. P. Constantin, and A. Hollaender (eds.), Plenum, New York, pp. 117-130.

S. S. Sukan and F. Vardar-Sukan. 1987. Mixing Performance of Air-Lift Fermentors against Working Volume and Draft Tube Dimensions. *Bioprocess Eng. 2*, 33-38.

M. Tabata and Y. Fujita. 1985. Production of Shikonin by Plant Cell Cultures. In "Biotechnology in Plant Science: Relevance to Agriculture in the Eighties," M. Zaitlin, P. Day, and A. Hollaender (eds.), Academic Press, New York, pp. 207-218.

M. Tabata, H. Mizukami, N. Hiraoka, and M. Konoshima. 1974. Pigment Formation in Callus Cultures of *Lithospermum erythrorhizon*. *Phytochemistry 13*, 927-932.

B. Tal, J. S. Rokem, and I. Goldberg. 1983. Factors Affecting Growth and Product Formation in Plant Cells Grown in Continuous Culture. *Plant Cell Rep. 2*, 219-222.

H. Tanaka. 1981. Technological Problems in Cultivation of Plant Cells at High Density. *Biotechnol. Bioeng. 23*, 1203-1218.

H. Tanaka. 1987. Large-Scale Cultivation of Plant Cells at High Density: A Review. *Process Biochem.*, August, 106-113.

H. Tanaka, F. Nishijima, M. Suma, and T. Iwamota. 1983. Rotating Drum Fermentor for Plant Cell Suspension Culture. *Biotechnol. Bioeng. 10*, 2359-2370.

B. Ulbrich, W. Wiesner, and H. Arens 1985. Large Scale Production of Rosmarinic Acid from Plant Cell Cultures of *Coleus blumei* Benth. In "Primary and Secondary Metabolism of Plant Cell Cultures," K.-H. Neumann, W. Barz, and E. Reinhard (eds.), Springer-Verlag, New York, pp. 293-303.

F. Vardar and M. D. Lilly. 1982. Effect of Cycling Dissolved Oxygen Concentrations on Product Formation in Penicillin Fermentations. *Eur. J. Appl. Microb. Biotechnol. 14*, 203-211.

F. Wagner and H. Vogelmann. 1977. Cultivation of Plant Tissue Cultures in Bioreactors and Formation of Secondary Metabolites. In "Plant Tissue Culture and Its Biotechnological Applications," W. Barz, E. Reinhard, and M. H. Zenk (eds.), Springer-Verlag, New York, pp. 245-252.

D. A. J. Wase, W. J. McManamey, S. Raymahasay, and A. K. Vaid. 1985. Comparisons between Cellulase Production by *Aspergillus fumigatus* in Agitated Vessels and in an Air-Lift Fermentor. *Biotechnol. Bioeng. 28*, 1166-1172.

G. Wilson and P. Marron. 1978. Growth and Anthraquinone Biosynthesis by *Galium mollugo* L. Cells in Batch and Chemostat Culture. *J. Exp. Bot. 29*, 837-851.

Y. Yamada and Y. Fujita. 1983. Production of Useful Compounds in Culture. In "Handbook of Plant Cell Culture. Techniques for Propagation and Breeding," Vol. 1, D. A. Evans, W. R. Sharp. P. V. Ammirato, and Y. Yamada (eds.), Macmillan Publishing Co., New York, pp. 717-728.

Y. Yamada and H. Morikawa. 1985. Proptoplast Fusion of Secondary Metabolite-Producing Cells. In "Primary and Secondary Metabolism of Plant Cell Cultures," K.-H. Neumann, W. Barz, and E. Reinhard (eds.), Springer-Verlag, New York, pp. 255-217.

Y. Yamada and F. Sato. 1981. Production of Berberine in Cultured Cells of *Coptis japonica*. *Phytochemistry 20*, 545-547.

T. Yamakawa, S. Kato, K. Ishida, T. Kodama, and Y. Minoda. 1983. Production of Anthocyanins by *Vitis* Cells in Suspension Culture. *Agric. Biol. Chem. 47*, 2185-2191.

H. Yamamoto, K. Kakagawa, H. Fukui, and M. Tabata. 1986. Cytological Changes Associated with Alkaloid Production in Cultured Cells of *Coptis japonica* and *Thalictrum minus*. *Plant Cell Rep. 5*, 65-68.

K. Yazaki, H. Fukui, M. Kikuma, and M. Tabata. 1987. Regulation of Shikonin Production by Glutamine in *Lithospermum erythrorhizon* Cell Cultures. *Plant Cell Rep. 6*, 131-134.

N. Yoshikawa, H. Fukui, and M. Tabata. 1986. Effect of Gibberellin A3 on Shikonin Production in *Lithospermum* Callus Cultures. *Phytochemistry 25*, 621-622.

M. H. Zenk, H. El-Shagi, and B. Ulbrich. 1977. Production of Rosmarinic Acid by Cell Suspension Cultures of *Coleus blumei*. *Naturwissenschaften 64*, 585-586.

M. H. Zenk, M. Rueffer, M. Amann and B. Deus-Neumann. 1985. Benzylisoquinone Biosynthesis by Cultivated Plant Cells and Isolated Enzymes. *J. Nat. Prod. 48*, 725-738.

CHAPTER 7

IMMOBILIZED PLANT CELLS

A. INTRODUCTION

Immobilization refers to the confinement of cells within or on particles, as well as to the confinement of biocatalyst particles within a reactor, under conditions favoring product release and continuous cell reuse. Following successes with enzymatic and microbial processes, immobilization has been suggested as an engineering strategy to improve the overall volumetric productivity of secondary metabolite production by undifferentiated plant cells. In addition, immobilization can also have a direct positive or negative impact on plant cell physiology. Immobilized plant cells have been used for single- and multistep biotransformations of precursors to desired products, as well as for the *de novo* biosynthesis of secondary metabolites.

This chapter is written in the spirit of a workshop rather than a literature review. Extensive reviews on various aspects of plant cell immobilization are readily available in the literature (Hulst and Tramper 1989; Hall et al. 1988; Brodelius 1988a; Berlin 1985; Rosevear and Lambe 1985; Rhodes 1985; Brodelius 1983; Prenosil and Pedersen 1983; Shuler 1981). The purpose here is to critically appraise progress in the area and redefine challenges, to provide a framework for understanding the effects of immobilization, and, finally, to briefly review available bioreactor options.

Immobilized plant cell technology is relatively young, and much of the research is largely in the preliminary experimental stages, with little work having been devoted to process design. The resolution of issues at the level of laboratory experimentation is still needed before larger scale operation can be generally successful. Consequently, this chapter deals more directly with experimental research and details and, in that sense, requires a different orientation from the reader than does the chapter on suspension culture.

B. BACKGROUND

B.1. Rationale for Immobilizing Cells

B.1.1. Process Engineering Issues The engineering advantages and limitations of immobilized plant cell systems have been thoroughly discussed in several contexts (Shuler 1981; Bucke 1983; Rosevear and Lambe 1985; Payne et al. 1987). Though we restrict our discussion to structurally undifferentiated cells in this chapter, it should be noted that the use of structurally differentiated tissues such as root and shoot cultures may also be considered an immobilized cell process from a bioprocessing standpoint provided that the product is extracellular and the biomass is reused. Immobilization of cells has the following advantages:

- Immobilization is compatible with non-growth associated product formation. This is an important advantage for plant cells since many, though by no means all, secondary metabolites are produced after growth ceases in a typical batch operation.
- Growth and product formation phases are effectively decoupled. Since the biomass is reused continuously, the slow growth of plant cells becomes less of a problem.

- Immobilized cells are capable of maintaining viability and biosynthetic capacity for prolonged periods and are protected from mechanical stresses, including shear.
- The continuous and selective release of desired product into the extracellular medium may greatly facilitate product recovery and downstream processing compared to extraction from tissue.
- Other productivity-enhancing strategies are not only compatible but may act synergistically in association with immobilization. These strategies include production medium formulation, *in situ* product removal, elicitor treatment, and precursor feeding strategies.
- Processing steps which are facilitated by immobilization include rapid change of media, operation at dilution rates independent of growth rates, and simplified downstream processing because cell separation is not necessary. However, these steps are also possible with suspension cultures operated in the perfusion mode.

Some limitations are:

- The disadvantages of immobilization include the increased cost incurred in the immobilization procedure itself, the added complexity of process design and operation, and complications arising from the imposition of mass transfer limitations on the biocatalyst.
- The claim of higher cell densities in immobilized cell systems is not valid for plant cells since very high cell densities (perhaps higher than for immobilized cells in some cases) can be achieved in suspension culture.

The prerequisites for a successful immobilization process are as follows (see Figure 7.1):

- The product of interest should be synthesized by nongrowing or slowly growing cells; that is, the product should not be strictly growth-associated.
- Growth should be suppressed to prevent disintegration of the biocatalyst matrix and eventual disruption of the process.
- Immobilized cells should maintain prolonged viability and biosynthetic capacity with high rates of sustainable secondary metabolite production.

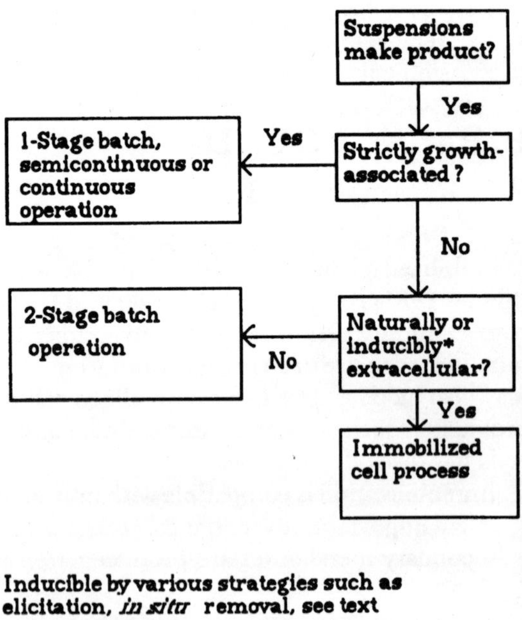

Figure 7.1 Flow diagram outlining the prerequisites for an immobilized cell process

Table 7.1. Dramatic effects of immobilization on secondary metabolism.

Species	Product	x-Fold change[a]	Type of immobilization	Reference
Capsicum frutescens	Capsaicin	> 100 (I)	Foam	Lindsey and Yeoman (1984a)
Capsicum annuum	Capsaicin	> 100 (I)	Gel	Ravishankar et al. (1988)
Tagetes patula	Thiophenes	ca. 20 (I)	Nat'l aggs	Hulst et al. (1989)
Coffea arabica	Methylxanthines	13 (I)	Gel	Haldimann and Brodelius (1987)
Catharanthus roseus	Ajmalicine	3.5 (I)	Gel	Asada and Shuler (1989)
Catharanthus roseus	Total alkaloids	No *de novo* synthesis	Membrane	Payne et al. (1988)
Mucuna pruriens	L-DOPA	No *de novo* synthesis	Gel	Wichers et al. (1983)
Salvia miltiorrhiza	Cryptotanshinone	2.5 (D)	Gel	Miyasaka et al. (1986)

[a] I = increase, D = decrease.

- At least partial product secretion is required. The process of immobilization itself, elicitor treatment, and the use of *in situ* product removal may cause increased product secretion.

B.1.2. Physiological Effects of Immobilization Though immobilization was originally proposed for its bioprocessing advantages, experimental evidence indicates that immobilization can have a dramatic impact on cellular physiology and secondary metabolism as well. Comparison to the literature on immobilized microbes must be done cautiously since such dramatic physiological responses are usually not observed with immobilized microbes. The more dramatic responses are summarized in Table 7.1. Note that both positive and negative effects are observed. Many attempts have been made to view immobilization in a larger biological context (Lindsey and Yeoman 1983a; Rosevear and Lambe 1985), and a collective viewpoint will be summarized here.

The term differentiation is repeatedly encountered in several contexts. Differentiation refers to metabolic or morphological specialization of cells. "Biochemical differentiation" refers to metabolic specialization such as the coordinated expression of specific enzymatic pathways to produce secondary metabolites. "Structural differentiation" refers to the specialization of cells in terms of form or morphology and can occur at various levels of increasing complexity, for example, cyto- (cellular), histo- (tissue level), or organ differentiation.

A range of extents of structural differentiation is encountered in plant cultures -

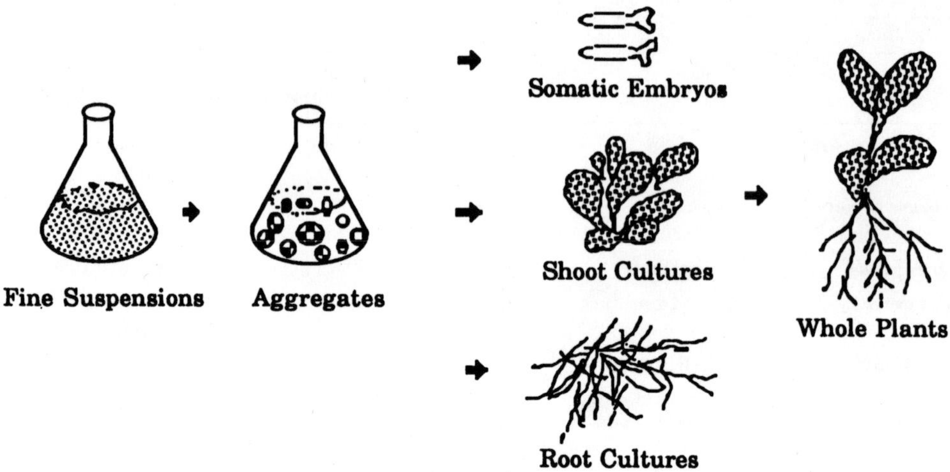

Figure 7.2 Range of extents of differentiation encountered in plant cell cultures

from fine suspensions to larger aggregates, partially organized tissues, mature organs, and finally whole plants (Figure 7.2). What characterizes this developmental sequence is the increasing level of cellular association and cooperation and the concomitant presence of spatial heterogeneities (such as concentration gradients of various metabolites) within the tissue.

In certain species, a high level of biochemical differentiation is expressed even in structurally undifferentiated fine suspensions, for example, shikonin, berberine, ajmalicine, rosmarinic acid, and others. However, in other species, secondary metabolism is "switched on" or enhanced only when accompanied by some level of structural differentiation (Collin 1987). Whereas simple aggregation enhances thiophene biosynthesis in *Tagetes* sp. (Table 7.1), a higher level of structural differentiation, viz., leaf organ cultures, is required for vindoline production by *Catharanthus roseus* (Chapter 9).

Immobilization can be viewed as a somewhat primitive attempt to induce cells to differentiate biochemically by imposing upon structurally undifferentiated cells some of the complexities inherent in the higher forms of differentiation. We shall see that different levels of complexity may be imposed by choosing the appropriate immobilization strategy and that immobilization can achieve the level of biochemical differentiation expressed by natural aggregates. Note that organs themselves may be considered to be "naturally" immobilized cells.

B.1.3. Comparison with an Alternative Technology The rationale for immobilization is perhaps best appreciated by comparison to an alternative technology. As an illustrative example, consider a hypothetical, non-growth-associated target compound that is intracellularly accumulated in a two-stage batch process and is secreted in an immobilized cell process (see Figure 7.3).

In the two-stage batch process, product accumulation takes place only in the stationary phase. The nonproductive growth phase thus forms a substantial portion of each batch production cycle. If immobilized cells are capable of prolonged biosynthesis, the growth phase becomes a less substantial portion of each production cycle.

The final process choice will, of course, be based on economic considerations. However, perhaps the first step in comparing the two processes would be to compare their respective volumetric productivities. The volumetric productivity is defined as the amount of secondary metabolite produced per unit volume per unit time. Note that the total volume and total time for a complete production cycle (i.e., including the growth phase) must be considered. From Figure 7.3, it is evident that for an immobilized cell process to be competitive with a two-stage batch process, the following criteria must be satisfied:

- The production phase for an immobilized cell process must be significantly longer than the stationary production phase of the two-stage batch.

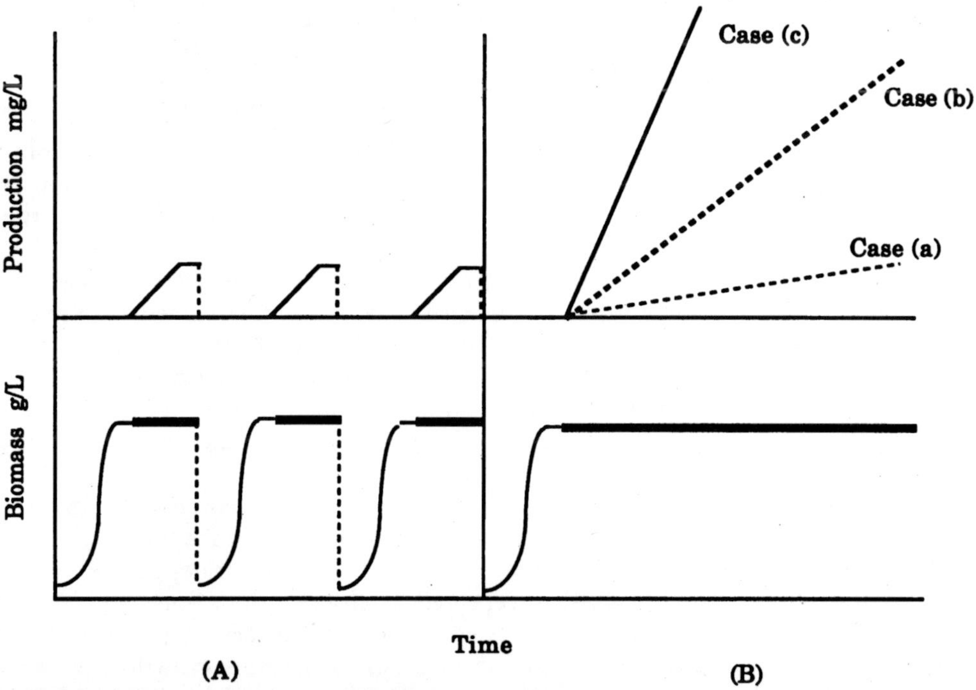

Figure 7.3. Comparison between an idealized two-stage batch (A) and an idealized immobilized cell (B) process. The lower panel shows biomass accumulation with time. The bold line signifies the portion of the production cycle when secondary metabolites are synthesized. Note that the reactor cell density for an immobilized cell culture may be less than for free cells. The upper panel shows secondary metabolite production with time. The batch process is terminated when the cells "fill up" with product, and a new cycle is started. Three cases are shown for immobilized cell extracellular product formation. In case (a), the specific productivity of free cells is greater than that for immobilized cells. In case (b), specific productivities of free and immobilized cells are equal. In case (c), the specific productivity of immobilized cells is greater than that of free cells.

- The density of immobilized cells per unit reactor volume must be maximized.
- The specific productivity (i.e., grams of product per gram of cells per unit time) for immobilized cells in the production phase must compare favorably to that of free cells. To elaborate, three cases are possible as shown in the upper right-hand panel of Figure 7.3.
- Curve (a): The specific productivity of immobilized cells is less than that for free cells. This may occur if the effectiveness factor (η, see Chapter 4, Section D) for secondary metabolite production is less than 1 due to mass transfer limitations, or if immobilization has a negative effect on physiology.
- Curve (b): The specific productivity of immobilized and free cells is equal. This may occur when immobilization does not affect cell physiology, and if mass transfer resistances are insignificant (η=1) .
- Curve (c): The specific productivity of immobilized cells is higher than that for free cells. This may occur when immobilization has a positive effect on cell physiology.

From the above comparison, it is clear that both engineering as well as physiological advantages should be exploited in developing an immobilized cell process.

B.2. Progress in Immobilized Cell Culture: Mini-Review

This section attempts to briefly summarize progress in the field of plant cell immobilization. It is by no means a comprehensive review, and readers wishing more details should consult the list of reviews in Section A. We will use this critical appraisal of the past in order to identify possible research focuses for the future.

Brodelius et al. (1979) first advocated immobilizing plant cells for secondary metabolite production and demonstrated the feasibility of gel-immobilized systems for *de novo* biosynthesis and for single- and multistep biotransformations. Since this report, immobilized plant cells have steadily attracted research interest. These studies may be conveniently classified into work on biotransformations and work addressing the physiological and engineering aspects of *de novo* secondary biosynthesis. Let us briefly consider each in turn.

Many of the engineering advantages of immobilization were demonstrated in early work on biotransformation. Alfermann et al. (1980, 1983) studied the biotransformation of cardiac glycosides by immobilized cells of *Digitalis lanata*. They demonstrated prolonged activity (up to 170 days) in repeated batch operation. Jones and Veliky (1981a) used immobilized cells of *Daucus carota* for biotransformation of cardenolides and demonstrated the use of a production medium to suppress growth while still supporting viability and biotransformation activity (Jones and Veliky 1981b). They suggested the use of periodic activation of biocatalyst particles with growth medium to sustain viability. Further work was devoted to the study of the mode of precursor addition. Study of the respiration characteristics of the biocatalyst particles suggested the presence of mass transfer limitations and pointed to the importance of adequate oxygen supply (Jones and Veliky 1981c; Veliky and Jones 1981). Galun et al. (1985) used gamma irradiation to arrest growth and perform biotransformations. The use of plant cells to perform biotransformations has since received steady interest, with recent work focusing on unique chemical capabilities such as stereoconversions (e.g., Hamada et al. 1988) and the use of unnatural precursors (e.g., Pras et al. 1988; Vanek et al. 1989).

The physiological effects of immobilization on *de novo* secondary biosynthesis attracted considerable research interest early on. Yeoman et al. (1980, 1982) first directed their investigations to the possible physiological mechanisms underlying non-growth-associated product formation. Lindsey and Yeoman (1983b) correlated higher secondary biosynthetic productivities with the extent of aggregation in several solanaceous species. The enhanced biosynthetic capabilities of natural aggregates as compared with fine suspensions has been subsequently reported by, among others, Fuller and Bartlett (1985), Watts et al. (1984), Dainty et al. (1985), McCown et al. (1988), Hulst et al. (1989) and Hoekstra et al. (1990). In an important paper, Haldimann and Brodelius (1987) showed that the higher alkaloid productivity of natural aggregates of *Coffea arabica* could be mimicked by gel-immobilized fine suspensions. The positive relationship between aggregation and secondary metabolism is obviously not universal for all systems, and examples to the contrary also exist (e.g., see Kinnersley and Dougall 1980; Ozeki and Komamine 1985).

Lindsey and Yeoman (1984a,b) themselves focused their efforts on capsaicin production by suspended and foam-immobilized cells of *Capsicum frutescens*. Suspensions produced only traces of the compound, whereas immobilized cells produced significant quantities of extracellular capsaicin. Prolonged viabilities and biosynthetic capabilities (up to 12 weeks) were reported in the foam-immobilized particles. The use of distant precursors and nutrient limitation resulted in growth suppression and dramatically enhanced productivities (by two to three orders of magnitude) compared to fine suspensions (Lindsey 1985). This is so far the most impressive positive effect of immobilization on cellular metabolism and secondary biosynthesis. The physiological effects of immobilization have been well documented in a number of systems, of which Table 7.1 provides a representative sample. Note that failures with immobilization, and negative effects on physiology, though less often reported, are also likely. The bases of these positive and negative effects still need to be elucidated.

A few efforts have been directed toward analyzing the microenvironment within biocatalyst particles. The issue of mass transfer resistances and resulting concentration gradients was addressed by Hulst et al. (1985) and Furusaki et al. (1988) and was experimentally confirmed by oxygen microprobe measurements by Robins et al. (1986). Vogel and Brodelius (1984) used ^{31}P-NMR spectroscopy to noninvasively monitor certain aspects of immobilized cell physiology. Bringi and Shuler (1990) have attempted to elucidate the differences in the microenvironment of suspended and immobilized cells, and the results of this investigation will be incorporated throughout this chapter.

Though the effects of immobilization on plant cell physiology are interesting in themselves from a fundamental point of view, it is clear that these advantages will not be technically relevant unless compelling engineering advantages are also concomitantly present. A steady stream of work has thus been devoted to examining the engineering issues pertaining to *de novo* secondary biosynthesis using immobilized plant cells.

Prolonged biosynthetic activity and viability have been demonstrated in many model systems. Nakajima et al. (1985) used gel-immobilized *Lavandula vera* for pigment production over a period of seven months. A production medium containing cysteine as an inducer was used to suppress growth. Recent work has resulted in an improved production medium that has accelerated pigment production, facilitated cell reuse, and circumvented the necessity of intermittent growth medium activation (Nakajima et al. 1989a,b). Kobayashi et al. (1988) demonstrated production of berberine by gel-immobilized *Thalictrum minus* cells at a constant rate for 60 days in repeated batch culture. Pareilleux and Majerus (1987) demonstrated stable and continuous production of indole alkaloids by gel-immobilized *Catharanthus roseus* for 60 days.

The next critical prerequisite for immobilization is that the product should be at least partially extracellular. Table 7.2 shows some cases in which the process of immobilization itself causes partial product secretion. A trivial explanation for this phenomenon would be cell lysis in a portion of the biocatalyst particle. Different methods of distinguishing between cell lysis and true secretion are discussed in Chapter 11. If immobilization does indeed cause product secretion in some cases, then an

Table 7. 2. Immobilization itself causes extracellular product accumulation.

Compound	Plant	% Product in the Extracellular Medium		Reference
		Free	Immobilized	
Echinatin	*Glycyrrhiza echinata*	20	67	Ayabe, *et al.* (1986)
Cryptotanshinone	*Salvia miltiorrhiza*	25	75	Miyasaka et al. (1986)
Ferruginol	*Salvia miltiorrhiza*	26	45	Miyasaka et al. (1986)
Solasodine	*Solanum surattense*	7	47-59	Barnabas and David (1988)
L-DOPA from tyrosine	*Mucuna pruriens*	(Suspensions accumulate product intracellularly by *de novo* biosynthesis)	90	Wichers et al. (1983)
Capsaicin	*Capsicum frutescens*	(Immobilized cells secrete product totally, suspensions only make traces)		Lindsey and Yeoman (1984a)
Capsaicin	*Capsicum annuum*	(Immobilized cells secrete product, suspensions do not synthesize it)		Ravishankar et al. (1988)
Ajmalicine	*Catharanthus roseus*	(Immobilized cells secrete 4 times as much as do suspension cultures)		Asada and Shuler (1989)
Limonene, α-pinene	*Pelargonium fragrans*	Not reported	80-90%	Charlwood et al. (1989)

Table 7.3 Elicitors cause extracellular product accumulation in suspension cultures.

Compound	Plant	x-Fold increase or de novo synthesis	Amount in medium (% of total product)	Reference
Sanguinarine	*Papaver somniferum*	26	43%	Eilert et al. (1985)
Glyceollin	*Glycine max*	*De novo* synthesis	"Major portion" (not quantified)	Funk et al. (1987)
Canthin-6-one	*Ailanthus altissima*	125	20-27%	Krauss et al. (1989)
Sanguinarine	*Papaver bracteatum*	500	75%	Cline and Coscia (1988)
Capsidiol	*Nicotiana tabacum*	*De novo* synthesis	> 90%	Chappell et al. (1987)
Alkaloid esters	*Cephalotaxus barringtonia*	*De novo* synthesis	up to 82%	Heinstein (1985)
Furoquinoline alkaloids	*Ruta graveolens*	*De novo* synthesis	40 - 60%	Eilert (1989)

important prerequisite will have been fulfilled. In other cases, however, immobilization does not cause any product release (e.g., Brodelius et al. 1979).

Elicitors have been successful in dramatically increasing rates of secondary biosynthesis in suspension cultures (DiCosmo and Misawa 1985; Eilert 1987). Interestingly, elicitors also seem to cause cells to partially secrete product as summarized for a few cases in Table 7.3. Examples to the contrary have been reported as well by, among others, Smith et al. (1987), Heinstein (1985), and Eilert et al. (1984). Again, if the appearance of product in the extracellular medium is not simply due to lysis, elicitors may well be advantageously incorporated into an immobilized cell process.

The incorporation of *in situ* product removal in an immobilized cell process may have significant advantages. *In situ* product removal may increase the rate and extent of product secretion, increase overall productivity, prevent product degradation, relieve feedback inhibition or toxicity, and simplify product separation. Some of these advantages have been well documented for suspension cultures (e.g., see Beiderbeck and Knoop 1988). Robins and Rhodes (1986) studied the stimulation of extracellular anthraquinone production by naturally aggregated cultures of *Cinchona ledgeriana*. Addition of adsorbent stimulated productivities 15-fold and increased the rate and extent of product secretion. The majority of the product was adsorbed on the resin, thus simplifying product separation. Asada and Shuler (1989) also showed that *in situ* adsorption increased the amount of extracellular ajmalicine 6-fold and roughly doubled total productivity compared to controls without product removal in suspension cultures of *Catharanthus roseus*. In their early report, Brodelius et al. (1979) found that immobilized *C. roseus* secreted significantly higher amounts of ajmalicine compared to values typical for suspension cultures. The extracellular medium in the immobilized cell reactor was constantly recirculated through a layer of chloroform. Brodelius et al. (1979) proposed that traces of chloroform in the medium could have permeabilized the cells, causing increased secretion. However, an additional or alternative explanation could be that continuous *in situ* removal of ajmalicine into the chloroform layer increased the extent of product secretion.

Other efforts have also been directed at inducing product secretion. The use of chemical permeabilizing agents such as DMSO for inducing excretion of metabolites from vacuoles has met with only limited success despite the promise of earlier studies (summarized in Brodelius 1988b). Other attempts at releasing predominantly intracellular products include cycling of the external pH (Renaudin and Guern 1982; Renaudin et al. 1985; Nef et al. 1987; Bouyssou et al. 1987), use of high-ionic-strength media (Tanaka et al. 1985), temperature-induced release (Cooke et al. 1986), use of high (Brodelius et al. 1988) and low electric fields (Pu et al. 1989), and ultrasound (Kilby and Hunter 1990).

The synergistic effects of immobilization, production medium formulation, elicitor treatment, and *in situ* product removal on ajmalicine production by gel-immobilized *Catharanthus roseus* was demonstrated by Asada and Shuler (1989). Figure 7.4 summarizes the results of various treatments on the formation and release of ajmalicine. This study shows how an integrated approach combining biological and engineering strategies can result in dramatic increases in productivity. However, a major technical hurdle in this study which re-

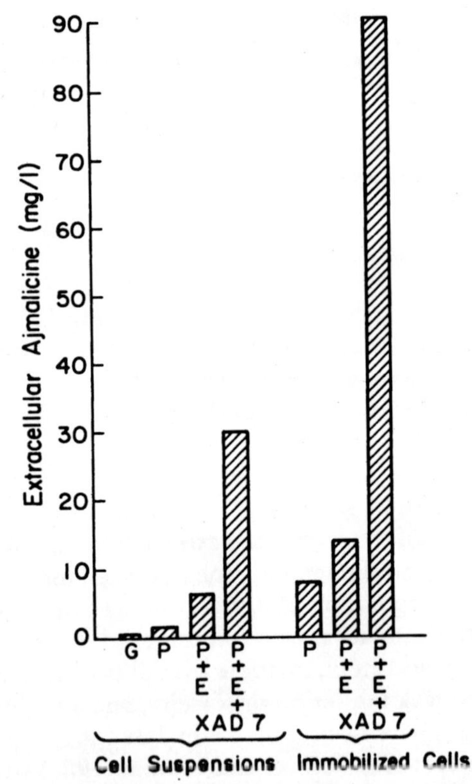

Fig 7.4 Summary of productivity enhancements for ajmalicine production by free and gel-immobilized cells of *Catharanthus roseus*. **G** stands for growth medium, and **P** for production medium. **E** indicates elicitor addition for a 2-day exposure time, and **XAD 7** indicates the addition of adsorbent resin for *in situ* product removal. Samples were analyzed 23 days after addition. After Asada and Shuler (1989).

mains to be resolved is the precipitous drop in productivity in the third batch cycle upon reuse of the biocatalyst particles. Unsuppressed cell growth also disrupted the biocatalyst particles upon repeated reuse. Improved conditions and further refinements in biocatalyst design will be necessary before a continuous system can be established.

Recent research has been focused toward refining biocatalyst design. Kobayashi et al. (1988) initially demonstrated the feasibility of the use of immobilized *Thalictrum minus* for berberine production. Kobayashi et al. (1989) subsequently identified intraparticle oxygen transport as a problem and redesigned their particles to improve productivities (see Example 3 in Chapter 4, page 100). Similarly, Furuya et al. (1984) used gel-immobilized *Papaver somniferum* for the biotransformation of codeinone to codeine. A kinetic study, prompted by the rapid loss of activity of the immobilized cell preparation, again identified intraparticle oxygen transport as a potential problem (Furusaki et al. 1988). We shall take a renewed look at this and other case studies from the literature in Section E.2.

There have been few reports of attempts at scale-up. Archambault et al. (1990a,b) have described a mechanically (2-L) and a pneumatically (7-L) agitated reactor for indole alkaloid production by cloth-immobilized *Catharanthus roseus*. Mavituna et al. (1987) have reported an attempt to immobilize *Capsicum frutescens* in polyurethane foam for capsaicin production in a 5-L bioreactor. Other smaller bioreactors have been used by Kobayashi et al. (1988), Shuler (1986), and Prenosil and Pedersen (1983). Small laboratory-scale packed beds have been routinely used in many laboratory studies by, among others, Brodelius et al. (1979), Jirku et al. (1981), Veliky and Jones (1981), and Furuya et al. (1984).

To summarize, immobilization can have positive effects such as prolonged viability, enhanced secondary biosynthesis, and partial product secretion. Immobilization may also have negative effects such as inhibition of secondary metabolism, reduced reaction rates compared to suspensions, and decline of activity of biocatalyst particles upon reuse. Clearly, for an immobilized process to be technically successful, the positive effects of immobilization must be completely exploited and the negative effects must be circumvented.

B.3. Phases in the Development of an Immobilized Cell Process

At least three phases in the development of an immobilized cell process can be identified:

(i) Preliminary Work

This is the first phase of laboratory-scale shake-flask experimentation. The objectives are usually to demonstrate the feasibility of immobilization and to check whether or not the prerequisites (see Section B.1.1) are fulfilled. Most of the work reported thus far falls into this category. As we have seen from the preceding mini-review, results at the end of this phase are often mixed; that is, both positive as well as negative effects of immobilization become apparent.

(ii) Biocatalyst Design

The second phase involves rational design of the biocatalyst particle and operating conditions for optimum performance. In particular, key variables to be determined include the history of cells prior to immobilization, the type of immobilization and matrix to be used, the size of the particle, and the number of cells per particle. Immobilization affects cells by various mechanisms which we shall discuss in detail in Section D. Before their contributions are understood, the effects must be isolated and individually studied. Such an understanding will enable more complete exploitation of immobilized cell technology.

This phase also involves determining the environment or series of environments to which the cells must be subjected for optimum performance. In particular, this includes the integrated use of enhancement strategies such as production medium formulation, elicitor treatment, *in situ* product removal, and precursor feeding. This phase is still carried out on a laboratory scale in shake flasks or small bioreactors and is an iteration of the preliminary work.

(iii) Bioprocess Design

The final phase includes scale-up of the immobilized cell process by providing optimally designed immobilized cell particles with the previously determined environment or sequence of environments in a pilot-scale and finally large-scale reactor. This phase also involves integration of productivity enhancement strategies at the process scale and downstream processing and hardware design.

B.4. Challenges

An appraisal of the progress made so far suggests that biocatalyst design is an important phase in the development of an immobilized cell process for at least two reasons:

- The merits and demerits of immobilization cannot be properly assessed and fairly compared to those of other technologies unless the immobilized cell process has been optimized at least to some extent.
- Biocatalyst design and determination of operating conditions is an essential step prior to scale-up. If this step is not performed, rational scale-up will not be possible since reactor options depend to some extent on the biocatalyst properties, such as size and density. Further, problems arising due to any uncertainties and ambiguities in the particle microenvironment at the small scale will only be exacerbated during scale-up.

The next focus of this chapter will be on issues that are relevant to biocatalyst design.

C. TYPES OF IMMOBILIZED CELLS

Before studying the possible mechanisms by which immobilization could affect cell performance, we first briefly describe the types of immobilized cells. Plant cell immobilization has more or less followed the protocols that have been developed for immobilization of enzymes and microbes (see Philips and Poon 1988). The three most commonly used methods will be described here, viz., natural aggregates and gel and foam immobilization. Immobilized cell performance is usually compared to that of undifferentiated fine suspension cultures.

The viability of an immobilized cell preparation is an important indicator of the efficacy of the immobilization procedure. It is also often valuable to observe whether there is positional dependence of viability in order to assess the adequacy of nutrient supply inside the particle. Most methods rely on staining of cross sections (e.g., Linsefors and Brodelius 1985) or on measuring respiration rates (Brodelius and Nilsson 1980). Parr et al. (1984) have described an elegant noninvasive method based on differential uptake of radioactive medium components. The ability of biocatalyst particles to show cell growth upon transfer to growth medium at the end of a study is not a good indicator of viability since it gives no positional information, nor is it a quantitative method. For example, if only 10% of the cells on the periphery of a biocatalyst particle remained viable, growth would still be visible when the particle was transferred to growth medium.

C.1. Natural Aggregates

Cells in a suspension culture rarely grow as single cells - typically, they grow as aggregates, with sizes ranging from fines as small as 100-500 μm (few cells) to aggregates measuring several millimeters in diameter (many thousands of cells). These large aggregates may be considered a form of "self-immobilized cells" which require no artificial matrix. Whereas natural aggregates are useful for shake-flask experiments, they may not always be suitable from the bioprocessing standpoint because of their susceptibility to mechanical damage and attrition.

The size of a natural aggregate is often critical in terms of its biosynthetic capability. Because the shear field in bioreactors is often not well characterized and may differ from shake-flask to bench to pilot scale, the size distribution of natural aggregates may vary considerably in the three cases. The difficulty in predicting and controlling the aggregate size distribution is primarily what precludes the use of natural aggregates for an immobilized cell process. It may, however, sometimes be possible to obtain mechanically robust large aggregates (see Fuller and Bartlett 1985; McCown et al. 1988; Hoekstra et al. 1990).

We believe that the use of natural aggregates as an experimental control during biocatalyst design at the shake-flask stage can provide valuable insights into how well other immobilization methods mimic "natural" immobilization.

C.2. Gel Immobilization

Gel immobilization consists of obtaining biomass from a rapidly growing suspension culture in growth medium, sieving fines, and entrapping them in a variety of polysaccharide gels prior to incubation under conditions favoring product formation and reduced growth.

Calcium alginate, κ-carrageenan, and agarose have been successfully used to immobilize plant cells. Detailed protocols are widely available from many sources (e.g.,

Brodelius and Nilsson 1980; Mosbach 1987). Briefly, the desired ratio of cells and gel solution is mixed and polymerized either by ionotropic gelation for alginate and k-carrageenan or by temperature-induced gelling for agarose. The method is simple to carry out aseptically and requires no elaborate apparatus. The mildness of the experimental conditions allows preparations with high retention of cell viability.

A few experimental details need to be emphasized:

- The history of cells prior to the immobilization step refers to whether they were in exponential, late-exponential, or in stationary phase and may affect the viability and biosynthetic capacity profoundly (e.g., see Jones and Veliky 1981c; Subramani et al. 1989).
- The size and geometry of the immobilized cell particle can be varied at will. Extrusion with a syringe normally yields spherical particles 3-5 mm in diameter depending on the viscosity of the gel solution. Smaller particles (less than 1 mm diameter) can be obtained by using a concentric flow of sterile air while extruding the particles (Rehg et al. 1986). Figure 7.5 shows a photograph of tobacco fines entrapped in 2% calcium alginate at two particle sizes. The size of the particle is a very important parameter since it affects mass transfer limitations and nutrient availability (e.g., see Robins et al. 1986).
- The initial cell loading (i.e., the number of cells per unit volume of support) may be altered over a wide range - from sparsely embedded cells to densely packed cells visually approaching tissue-like densities. The cell loading is important for gauging the presence of concentration gradients as discussed in Section D of Chapter 4. The maximum cell loading is usually limited by the viscosity required to produce spherical beads by extrusion. Practically, it is difficult to exceed cell loadings of 50-60% (grams fresh weight of cells/gram of gel solution). This is a disadvantage of gel immobilization since a significant portion of the reactor volume will eventually be occupied by gel.
- Another disadvantage with calcium alginate is its susceptibility to calcium chelators and its low mechanical strength. However, successful attempts have been made to improve the mechanical properties of the gel by using trivalent cations (Rochefort et al. 1986) and by blending silica gel with the alginate solution (Fukushima et al. 1988).

C.3. Foam Immobilizatiom

The reticulate polyurethane foams commonly used for plant cell immobilization consist of interconnected "pockets" of different sizes (typically 10 to 80 pockets or pores per inch). Cells invade pores, attach, and grow *in situ* to finally occupy the pore volume. Detailed protocols for foam immobilization have been described (Lindsey et al. 1983; Rhodes et al. 1987). Rhodes et al. (1987) have conducted a systematic study of the factors affecting plant cell immobilization in foam particles. The following factors have been found to be important:

- Different cell lines showed different propensities for attachment and growth.
- Foams from different commercial sources showed different phytotoxicities.

Figure 7.5 Photographs showing large (4.3 mm diameter) and small (ca. 1.5 mm diameter) alginate-immobilized beads of *Nicotiana tabacum* fines. The beads consist of ca. 50% by weight cells. Large beads were made by manual extrusion from a syringe; small ones were extruded in a concentric flow of air. Photograph by V. Bringi.

- The time of exposure of foam to cells was important, as were the volume fractions occupied by cells and foam particles during the loading step.
- Efficacy of cell invasion depended on the size of each pocket and on the size of the average aggregate in the suspension culture.

Some further general observations are:

- Since the porosity of foams can exceed 90%, very high cell densities may be achieved compared to those achieved with gel-immobilized particles.
- Foam-immobilized cells are different from gel immobilized cells and are closer to natural aggregates in that growth takes place *in situ*, whereas in gel immobilization, growth is accomplished prior to the immobilization step in suspension culture.
- A comparison of the extended viability of natural aggregates, gel-immobilized cells, and foam-immobilized cells has not been conducted but deserves further study.

C.4. Other Methods

Various other methods of immobilization have also been used with plant cells. The adsorption of cells onto cloth or other fibers has been reported by Payne et al. (1988) and Archambault et al. (1989; 1990a,b). The use of membranes to entrap plant cells has been investigated (Shuler et al. 1983; Prenosil and Pedersen 1983; Shuler 1986; Hallsby and Shuler 1986; Shuler 1987). Since adsorption and membrane entrapment both allow growth to occur *in situ*, they may be expected to have effects similar to those imposed on cells by foam immobilization.

Due to the convenience and ease of use, shake-flask studies are often conducted with foam-immobilized cells. Though we refer mostly to foams in this chapter, it is with the explicit understanding that other methods of immobilization may mimic foams and may, in some situations, be preferable to foams. For example, membrane entrapment provides a highly controllable system for fundamental studies on immobilization (Hallsby and Shuler 1986). Later, a new method of classifying immobilization methods will become evident - that is, one that is not based on physical method involved, but on level of complexity imposed on the immobilized cells.

D. POSSIBLE MECHANISMS BY WHICH IMMOBILIZATION COULD AFFECT CELL PERFORMANCE

Many hypotheses have been suggested to explain the dramatic positive and negative effects of immobilization on secondary metabolism, but few studies have been performed to test them. A thorough understanding of the effects of immobilization on cellular physiology is a prerequisite to rational design of the biocatalyst particle and operating conditions. An understanding of the negative effects is important since they may possibly be circumvented, and an understanding of the positive effects may lead to a more complete exploitation of this technology.

The first step is to identify and examine all the possible mechanisms by which immobilization could alter cellular physiology. In this section, an attempt has been made to integrate the results of various studies on biochemical and structural differentiation in plants and plant cell cultures. The advantages of using structural differentiation (e.g., cytodifferentiation of tracheary elements; see Fukuda and Komamine 1985), as a model system are that the position of the differentiated cell can be localized within a tissue and that differentiation is usually terminal; for example, a tracheary element does not revert to an undifferentiated state. Moreover, since structural differentiation culminates from a complex series of coordinated biochemical steps, it is likely that lessons learned from these model systems will be applicable to the simpler forms of biochemical differentiation which precede secondary metabolism in other immobilized plant cell cultures.

Clearly, not all effects will be equally important in every system. The purpose here is to anticipate the possible role of complexities, rather than to be confounded when confronted with them later. In Section E, we will see how this anticipation leads to a simple conceptual approach for separating effects and for studying them independently.

The five mechanisms by which immobilization may affect cellular physiology may be conveniently classified as follows:

1. Physical effects
2. Matrix effects
3. Spatial heterogeneity: Cell age distribution
4. Plasmodesmata
5. Spatial heterogeneity: Concentration gradients

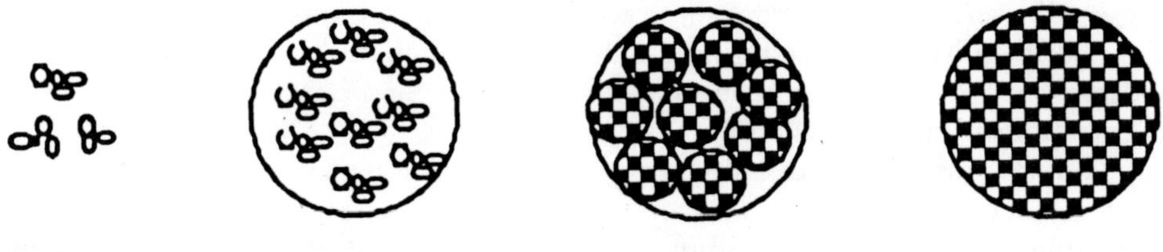

Figure 7.6 Sketch showing types of immobilized cells. (A) Nonimmobilized fines. (B) Gel-immobilized cells. (C) Foam immobilized cells. (D) Large natural aggregates.

The presence and relative severity of these effects on fine suspensions, gel-immobilized cells, foam-immobilized cells, and natural aggregates will be discussed in detail (refer to Figure 7.6).

D.1. Physical Effects

Immobilization, regardless of method employed, protects cells from applied mechanical stresses such as hydrodynamic fluid shear. There are two consequences of such physical protection on plant cell behavior:

(a) There is evidence that plant cells respond physiologically to applied mechanical stresses. Tanaka et al. (1988) reported that applied mechanical stresses affected not only the aggregate size distribution in a *Catharanthus roseus* suspension culture, but also the contents and relative fractions of pectin, hemicelluose, and cellulose in the cell wall. Immobilization, by conferring physical protection, may thereby alter cellular metabolism.

(b) A commonly observed phenomenon is that natural aggregates in a suspension culture need to reach a certain critical size before some aspect of differentiation is observed (Walker et al. 1979; Steward et al. 1958; Kuboi and Yamada 1978). Since the shear field in a cultivation vessel affects the aggregate size distributions profoundly, it can also have an indirect effect on differentiation. In such cases, immobilization may prove to have a positive effect simply because it allows clumps of cells to retain their critical size so that differentiation takes place.

An example of such an effect may explain the observations of Bringi (1991). "Nodules" which subsequently turned into shoot primordia were observed from the outer rim of large natural aggregates and gel-entrapped fines of a tobacco suspension culture. Fine suspension controls originally did not form these shoot primordia, perhaps because attrition prevented the nodules from forming and reaching a critical size. When the fine suspension was transferred to a lower-rpm shaker, consequently exposing cells to less shear, nodules and shoot primordia were observed. This suggests that gel immobilization may have been favoring structural differentiation primarily because cells were protected from shear and other stresses. However, this interpretation is not unequivocal since altering the shaking speed might have altered the rate of mass transfer between the gaseous and liquid phases, and hence the dissolved gas concentrations in the shake flask.

(c) Though immobilized cells are protected from certain mechanical stresses, they may experience other static mechanical stresses not experienced by fine suspensions - for example, when cells grow within the restricted volume of an immobilized cell support. Applied compressive stresses have been shown to cause alignment of cell divisions and to induce early signs of organization in tobacco callus (Lintilhac and Vesecky 1984). Since structural or morphological differentiation is often correlated with biochemical differentiation and secondary metabolism (Collin 1987), static mechanical stress may be another way by which immobilization could affect cell metabolism.

D.2. Matrix Effects

An artificial matrix used for immobilization may alter plant cell behavior either positively or negatively. Matrix effects are probably more likely in gel-immobilized systems since the ratio of gel to cell is typically high (>1:1), and the cells are in intimate contact with the gel. Some foam materials have been shown to have toxic properties, that is, negative matrix effects (Rhodes et al. 1987), but positive effects have not yet been documented, and natural aggregates do not use artificial matrices.

Since an artificial matrix provides a surface onto which medium components may adsorb or partition, the concentrations of these components will be decreased in the liquid medium and increased on the matrix. It has previously been suggested that an immobilized cell experiences an altered microenvironment because of the presence of locally high concentrations of the adsorbed components in the cell's proximity. However, Rouxhet and Mozes (1990) have pointed out that since the adsorbed component rapidly equilibrates with its concentration in the liquid medium, the chemical potentials (the true indicators of bioavailability) of the free and adsorbed components are the same. Furthermore, the dimensions of a typical cell are orders of magnitude greater than those of the adsorbed layer. These two facts imply that when a component adsorbs or partitions onto the matrix, the cell essentially experiences a concentration corresponding to that in the liquid medium (Rouxhet and Mozes 1990).

Cells in a gel may be exposed to a higher concentration of gel-stabilizing counterion in the liquid medium. The ion may have been leached from the gel, or may have been deliberately added to preserve gel stability. Thomasset et al. (1990) found that added calcium enhanced alkaloid metabolism in suspension cultures of *Datura innoxia*. Alternatively, Wichers et al. (1983) found that addition of calcium to the medium prevented biotransformation of tyrosine and release of L-DOPA by immobilized cells of *Mucuna pruriens*.

Concentrations of medium components such as hormones, phenolics, or even the secondary metabolite of interest may be altered by partitioning or adsorption onto the matrix. The physiology of cells exposed to these altered medium concentrations may differ from that of suspensions not exposed to any matrix effects. The ability of compounds such as activated charcoal to adsorb inhibitory or beneficial components from the medium is well known in plant cell culture (e.g., Fridborg et al. 1978; Beiderbeck and Knoop 1988). Presumably, an artificial matrix may behave in a similar manner.

The matrix itself may release inhibitory or stimulatory compounds into the medium. Beaumont and Knorr (1987) have noted the inhibitory effect of chitosan on immobilized celery cultures. The stimulatory effect of an acidic polysaccharide component of agar on shikonin production by *Lithospermum erythrorhizon* has been reported by Fukui et al. (1983). Ayabe et al. (1986) noticed an increased accumulation of the retrochalcone echinatin in the cells and medium of immobilized *Glycyrrhiza echinata* cultures. Fine suspension controls with 1% added sodium alginate also showed increased accumulation, suggesting that the cells were responding to the alginate, rather than to being immobilized *per se*. It should be noted, however, that dissolved alginate thickens the medium and thus probably reduces the rate of gas-liquid mass transfer. The alteration of dissolved gas concentrations could have played a role in the reported response. An alternative control might have involved adding empty polymerized beads of matrix to a fine suspension. Eilert et al. (1984) were able to partly mimic the increased production of acridone alkaloid epoxides by alginate-immobilized *Ruta graveolens* cultures simply by adding empty alginate beads to suspended cells.

Another type of matrix effect may involve more intimate contact between the cell and the gel. Schnabl et al. (1983) reported that alginate-immobilized protoplasts of *Vicia faba* showed lower rates of ethane production (an indicator of membrane turnover and senescence) and more prolonged viability than did nonimmobilized protoplasts. Similarly, Linsefors and Brodelius (1985) reported the enhanced viability of immobilized protoplasts of *Daucus carota* as compared to that of nonimmobilized protoplasts. It is not clear whether the above responses were due to a chemical interaction of the cell membrane with the gel matrix or due to physical protection as described in Section D.1. The two effects may be very difficult to separate from each other experimentally and will be lumped together in further discussions.

A set of experimental controls that can account for matrix effects is summarized in the Appendix.

D.3. Spatial Heterogeneity: Cell Age Distribution

Since natural aggregates grow by successive deposition of cell layers at the periphery, there exists a radial distribution of cell ages through the cross section of the aggregate. Foam-immobilized cells also exhibit cell age distribution at the start of the secondary metabolite production phase, since growth takes place *in situ* prior to the production phase. Since the cells constituting a gel bead are grown in suspension and immobilized just prior to production, no spatial cell age distribution is present in gel beads.

Cells of different ages may respond differently to the same environmental cues. For example, if only older vacuolated cells in the center of a large aggregate were particularly responsive to the signals promoting differentiation, then gel immobilization of fines would not be as successful. The importance of the history of the cells and tissue prior to the experimental treatment has not been adequately appreciated.

It is difficult to separate experimentally the effects of spatial cell age distribution from those arising due to plasmodesmatal connections; therefore, the two effects will be considered together for further discussions.

D.4. Plasmodesmata

D.4.1 Occurrence in Immobilized Cell Cultures Plasmodesmata are cytoplasmic channels between plant cells which allow for the transport of molecules (<700-900 Da) much like gap junctions between animal cells (Meiners et al. 1988). They are believed to play a role in intercellular communication and in integrating cellular and tissue function in the whole plant (e.g., see Gunning and Overall 1983; Gamalei 1985).

Plasmodesmata generally form during mitosis when a new cell plate is laid down between daughter cells. Cells in natural aggregates are mitotically related and, as expected, have been shown to possess plasmodesmata (Halperin and Jensen 1967; Sussex and Clutter 1967). Cells in the pockets of a foam grow and undergo mitosis *in situ*, and probably exhibit plasmodesmata, at least within the confines of each pocket. However, it is not clear whether the extent of plasmodesmatal contact in foam-immobilized cells is similar to that in a single natural aggregate.

Another situation in which plasmodesmata form may be relevant to immobilized cell cultures. Kollmann and Glockmann (1985) conducted an ultrastructural study of the interface of an inter-specific graft union between *Vicia faba* and *Helianthus annuus* plants. Within 6 to 20 days, some cell proliferation (termed "callus") occurred at the graft interface, and the presence of plasmodesmata between clearly identifiable *Vicia* and *Helianthus* cells was established. Plasmodesmata can thus form *de novo* between mitotically unrelated cells. However, for complete connections to be established, close contact and proper alignment between cells were found to be necessary (Kollmann et al. 1985). Thus, it is a matter of conjecture whether or not gel-immobilized cells form plasmodesmata after immobilization even in the absence of substantial growth in the gel. We believe that the likelihood is small since the gel may prevent adequate contact and proper alignment. If significant growth occurs within the gel, say, upon prolonged operation, then plasmodesmatal connections would be expected to form.

D.4.2. Possible Significance of Plasmodesmata for Cellular Metabolism There have been few studies on the role of plasmodesmata in biochemical and structural differentiation in plant cultures. Perhaps the most compelling evidence for the role of plasmodesmata in development comes from studies on the fern gametophyte, *Dryptopteris pseudomas* (summarized in Burgess 1985). The tip of the axially growing filament of cytoplasmically connected cells is initially unidimensional. The transition to two-dimensional growth takes place when the cell division of one of the cells takes place at right angles to the filament axis in a very predictable manner. When the cells are subjected to temporary plasmolysis to disrupt plasmodesmata, the normal pattern of cell division is disrupted - any or all cells may undergo division perpendicular to the filament axis. Plasmodesmata clearly play a role in ensuring the proper pattern of morphogenesis and structural differentiation in this organism. Stange (1984) has emphasized the importance of metabolic cooperation between neighboring cells on the early stages of differentiation.

More indirect evidence for the importance of plasmodesmata in structural differentiation has been shown by Sussex and Clutter (1967). Cells in large natural aggregates differentiated to form tracheary elements whereas cells from a fine suspension immobilized (reaggregated) in dialysis tubing would not, suggesting that plasmodesmata/spatial cell age distribution were necessary for differentiation in their *Eucalyptus* suspension culture. Bringi (1991) initiated 16 independent tobacco cell cultures in which cells in large natural aggregates (>1 mm diameter) differentiated to form tracheary elements. When fines were immobilized in calcium alginate beads to mimic large natural aggregates, 13 out of the 16 cell lines could not be induced to differentiate. Cells in large aggregates differentiated in the presence of empty gel, diminishing the possible importance of negative matrix effects. In these 13 cell lines, plasmodesmata/spatial cell age distribution were necessary for differentiation. In the three remaining cell lines, gel-immobilized cells were induced to differentiate, although at lower levels than in natural aggregates. For these cell lines, plasmodesmata/spatial cell age distribution were not necessary for differentiation, but their presence allowed for a much higher percentage of cells to differentiate. In yet another system, single mesophyll cells of *Zinnia elegans* are capable of differentiating directly to tracheary elements in very high percentages (Fukuda and Komamine 1980), showing that the requirement of plasmodesmata for tracheary element differentiation is by no means absolute.

Even though the above studies provide insights into the possible importance of plasmodesmata, a question of more direct relevance to the biotechnologist may be, "If one cell can be induced to make a valuable secondary metabolite, will cells that are connected to it also do so?" Again, there is evidence to support both possibilities. In fine suspensions producing secondary metabolites, microscopic observations indicate that producing and non-producing cells are often present side-by-side in the same aggregate (Deus-Neumann et al. 1983), suggesting that the presence of plasmodesmata alone is not sufficient for biochemical differentiation. In contrast, when tracheary elements form in the natural aggregates of a suspension culture, they often appear in clumps [see photographs in Torrey (1975), Lindsey and Yeoman (1985), and Lang and Kohlenbach (1982)], suggesting that once a cell differentiates, plasmodesmatally connected adjacent cells may show a higher propensity for differentiation. Such simplistic interpretations are, however, confounded by the fact that tracheary element differentiation is terminal, whereas a secondary metabolite-producing cell may presumably revert to a nonproducing cell.

In summary, the role plasmodesmata play in differentiation is unclear. Since the choice of immobilization method may affect the degree of plasmodesmatal contact, the importance or irrelevance of plasmodesmata must be considered separately for each system during biocatalyst design.

D.5. Spatial Heterogeneity: Concentration Gradients

The role of spatial heterogeneities such as concentration and electrical gradients in the control of morphogenesis and pattern formation in multicellular organisms is a well-established theory in developmental biology (e.g., see Wolpert 1969; Slack 1983; Warren Wilson 1978; Jaffe and Nuccitelli 1977). Not surprisingly, there has been much speculation in the literature about the presence of concentration gradients of nutrients and metabolites in plant cultures, and about their possible role in biochemical and structural differentiation. In fact, what was once stated only as a possible hypothesis has more or less been accepted, without rigorous experimental confirmation, as a conclusion. In this section, we will attempt to critically assess the possible role that concentration gradients may play in enhancing secondary metabolism in immobilized cell biocatalyst particles. In particular, we will examine the following issues in detail:

(i) Evidence linking concentration gradients to differentiation.
(ii) How do concentration gradients form?
(iii) How could concentration gradients affect differentiation?

D.5.1. Evidence Linking Concentration Gradients to Differentiation The dramatic responses of plant cells to immobilization such as those listed in Table 7.1 indirectly suggest that concentration gradients, in addition to all the other effects listed so far, may be important in enhancing biochemical differentiation. Studies on the positional dependence of structural differentiation within tissue have provided more compelling evidence for a link between concentration gradients and differentiation:

1. Early studies with calli of *Phaseolus* sp. and *Syringa* sp. have shown the importance of auxin and sucrose gradients in the induction of vascular tissues. Local application of auxin with an agar wedge caused vascular differentiation in a ring-like locus at a certain distance away from the point of application. This distance depended on the auxin concentration in the wedge and the sucrose concentration in the medium (Wetmore and Rier 1963; Jeffs and Northcote 1966, 1967).

2. Studies on tracheary element cytodifferentiation in suspension cultures showed that natural aggregates had to reach a certain critical size (typically >1 mm diameter) before the onset of differentiation (Steward et al. 1958; Kuboi and Yamada 1978). Sussex and Clutter (1967) reported that tracheary elements differentiated in a ring-like locus ca. 1 mm inside from the edge in the compact natural aggregates of a *Eucalyptus* suspension culture. Wilbur and Riopel (1971a,b) immobilized cells of *Pelargonium* in cone-shaped membrane supports and allowed growth and differentiation to take place *in situ*. When transverse sections of the cone were examined, tracheary elements were observed in a ring-like locus 0.7 mm inside from the edge regardless of the longitudinal position from which the section was taken.

3. We have routinely observed that the locus of lignification within aggregates of garlic cells is size dependent. Smaller aggregates (ca. 1 mm diameter) have a central core of lignified cells whereas larger aggregates (ca. 3 mm diameter) show a concentric sphere of lignification ca. 1 mm inside the aggregate (unpublished observations, Bringi 1987).

4. Hallsby and Shuler (1986) used a membrane-entrapped reactor to impose mass transfer limitations on a tobacco cell layer of defined thickness. When nutrient medium was forced through the cell layer, concentration gradients were absent, and no signs of morphogenetic changes were visible. When medium was supplied parallel to the cell layer, concentration gradients of various nutrients and metabolites must have been present through the cell layer. Under such conditions, cells responded by differentiating into morphologically distinct "nodules."

5. Turning to examples of organ differentiation from undifferentiated cultures, it is a common observation that a critical aggregate size has to be reached before root organogenesis can be induced (Steward et al. 1958; Walker et al. 1979).

6. Histological studies on shoot initiation in tobacco callus cultures have shown that points of origin are located in the basal portion of the callus. Inversion of the callus during the first nine days of culture resulted in shoots forming from the new basal portion. The workers concluded that physiological gradients within the tissue were somehow responsible for shoot initiation at specific loci (Ross and Thorpe 1973; Maeda and Thorpe 1979).

D.5.2. How Are Concentration Gradients Formed? A concentration gradient exists within a particle if cells have the ability to take up nutrients or produce products faster than these nutrients and products can diffuse into or out of the particle. The shape of the gradient depends on the geometry and reaction kinetics. Because many authors have drawn erroneous conclusions from invalid or incomplete analyses of mass transfer, we have provided a separate chapter describing the intuitive and quantitative aspects of mass transfer, (chapter 4). Thus our subsequent discussion is brief.

Concentration gradients within particles are very difficult to measure directly. However, the rate of reaction is a more easily measured parameter. Mass transfer limitations are therefore quantified by their effect on the reaction rate, or, more precisely, on the effectiveness factor, η, defined as:

$$\eta = \frac{\text{Observed rate with mass transfer limitations}}{\text{Rate if mass transfer limitations were insignificant}}$$

If mass transfer limitations are insignificant, $\eta = 1$.
If mass transfer limitations are significant, $\eta < 1$.

The presence or absence of concentration gradients is then indirectly surmised from the value of the effectiveness factor. Section D.3 of Chapter 4 outlines some subtleties in the relationship between concentration gradients and effectiveness factors.

Two analyses may be used to gauge the presence and severity of intra-particle mass transfer resistances. Thiele's analysis, described in Section D.4.2 of Chapter 4, is based on free-cell kinetic parameters and can be used to predict the effectiveness factor *a priori*. Weisz's analysis is appropriate when observed immobilized cell reaction parameters are available and is easier to use (Weisz 1973; see Section D.2 of Chapter 4).

$$\Phi_w = (R_{obs} L^2) / (D_{eff} C_b)$$

where Φ_w is Weisz's modulus, a dimensionless quantity; R_{obs} is the observed reaction rate based on unit particle volume; D_{eff} is the effective diffusivity within the biocatalyst particle; C_b is the steady-state bulk liquid concentration; and L is the appropriate length scale for diffusion (defined as the ratio of the volume of the particle to the surface area of the particle). Intuitively, Weisz's modulus may be considered to be a ratio of characteristic times of diffusion and reaction, or, since the characteristic time is the reciprocal of the rate,

$$\Phi_w \, \alpha \, (\text{Reaction rate}) / (\text{Diffusion rate})$$

Weisz's criterion is:

If $\Phi_w << 1$, Mass transfer limitations are insignificant relative to reaction rate.
Effectiveness factor is equal to one.
Concentration gradients are unlikely.

If $\Phi_w >> 1$, Mass transfer limitations are significant relative to reaction rate.
Effectiveness factor is less than one.
Concentration gradients are likely.

An effectiveness factor versus Weisz's modulus chart is provided in Figure 4.6 of Chapter 4. As Φ_w approaches 1, the chart must be consulted since the η versus Φ_w relation depends on the geometry and kinetic regime, as discussed in Section D.3 of Chapter 4. Weisz's analysis is applicable to steady-state reactions (e.g., oxygen uptake and transport) and to those pseudo-steady-state batch situations in which media components are depleted very slowly (e.g., sugars, nitrate, ammonium).

Of the various media components, which ones are likely to be mass transfer limited? The answer is obtained by looking at the parameters which make up Weisz's modulus. Those components which are taken up quickly (high R_{obs}) or whose bulk concentrations are low (low C_b) tend to have high Φ_w and are therefore more likely to be mass transfer limited. This is true especially for dissolved gases such as oxygen in immobilized cell preparations of moderate particle size (4 - 5 mm diameter) and cell loading (>20% by wt. cells) (e.g., Hulst et al. 1985; Furusaki et al. 1988; Kobayashi et al. 1989). Indeed, microprobe measurements have confirmed the presence of oxygen gradients in foam-immobilized particles of *Daucus carota* (Robins et al. 1986). Though R_{obs} for sugar is similar to that for oxygen, sugar gradients are insignificant in most immobilized cell preparations because of the high medium concentration, C_b. Kobayashi et al. (1989) showed that sugar, nitrate, ammonium, and phosphate gradients were not important in their gel-immobilized *Thalictrum minus* system. Similarly,

Weisz's analysis can be used to confirm that gradients of the same four nutrients could not have been important in foam-immobilized *Capsicum frutescens* from the data presented in Lindsey and Yeoman (1984b).

Weisz's modulus also suggests that the best method of avoiding mass transfer limitations is to reduce the particle size (since $\Phi_w \alpha L^2$). Reducing the number of cells per bead, thereby reducing R_{obs}, and operating at the highest possible bulk concentrations (C_b) are other ways to reduce mass transfer limitations.

Finally, an unsteady-state analysis (Section D.4.3 of Chapter 4) is required to gauge the presence of concentration gradients for those components which are rapidly taken up from the medium in batch experiments. For some species, for example, tobacco (Knobloch et al. 1981; Bringi 1991) and *Catharanthus roseus* (Vogel and Brodelius 1984), depletion of phosphate from the medium is extremely rapid, whereas for *Thalictrum minus* (Kobayashi et al. 1989) and for *Capsicum frutescens* (Lindsey and Yeoman 1984b), it is not. Similarly, Bringi (1991) found that the cytokinin benzyladenine was significantly depleted from the medium within a day by tobacco suspensions. As shown in Section D.4.3 of Chapter 4, for such cases, different portions of the biocatalyst particle may be exposed to different levels of the component, with the core perhaps being completely deprived of the component when mass transfer resistances are severe.

Clearly, concentration gradients caused by mass transfer limitations may play an important role in creating spatial heterogeneity within biocatalyst particles.

D.5.3. How Could Concentration Gradients Affect Differentiation ?

D.5.3.1. Mechanism of the Stimulus Two hypotheses to explain the response of immobilized cells to concentration gradients may be summoned.

(i) One hypothesis is based on the fact that plant cells can respond to gradients of various environmental factors. The directional development of vascular tissue from leaf initials in response to polar auxin gradients is one example (summarized in Burgess 1985). The axis of polarity of the initially apolar fertilized egg of the brown seaweed *Fucus* can be influenced by gradients in various external factors, such as light intensity, temperature, pH, and osmotic activity. A possible mechanism by which a gradient across a single cell can cause a polar physiological response has been summarized by Jaffe and Nuccitelli (1977) and by Robinson (1985). The cell initially responds to an asymmetry in the environment by polar ion effluxes, thus creating an electrical gradient across itself. Charged surface cell membrane components move by electrophoresis in response to the electric field, thereby reinforcing polarity and allowing cells to respond to the directional stimulus. Since single cells and cytoplasmically connected groups of cells can respond to various types of gradients, it is conceivable that immobilized cells, subjected to concentration gradients, can respond in ways that may ultimately lead to differentiation.

(ii) Another hypothesis is based on the contention that the gradients are not important *per se*, but that they help to create the correct microenvironment by helping to establish the appropriate array of concentrations of various nutrients and by-products at a particular locus within the immobilized cell particle. In this case, the response is simply to a nondirectional stimulus. Though further investigations are necessary before any generalizations can be made, it appears from one initial report that at least some gel-immobilized systems may respond to altered concentrations rather than to gradients. Bringi and Shuler (1990) induced tracheary element differentiation in gel-immobilized fines of a tobacco suspension culture under low growth conditions. In that system, it was shown, by eliminating the other possibilities, that the positional dependence of tracheary element differentiation must have been due to concentration gradients of as yet unidentified components. Now, gradients of nutrients and products are always steepest near the edge of the bead and always flattest toward the core (they are zero at the center) due to symmetry considerations. Since tracheary elements consistently formed at the highest levels at the core and at the lowest levels toward the edge, it has been argued that the cells were probably not responding to gradients *per se* and were therefore probably responding to altered concentrations (Bringi 1991; Bringi and Shuler 1990; see Figure 7.7).

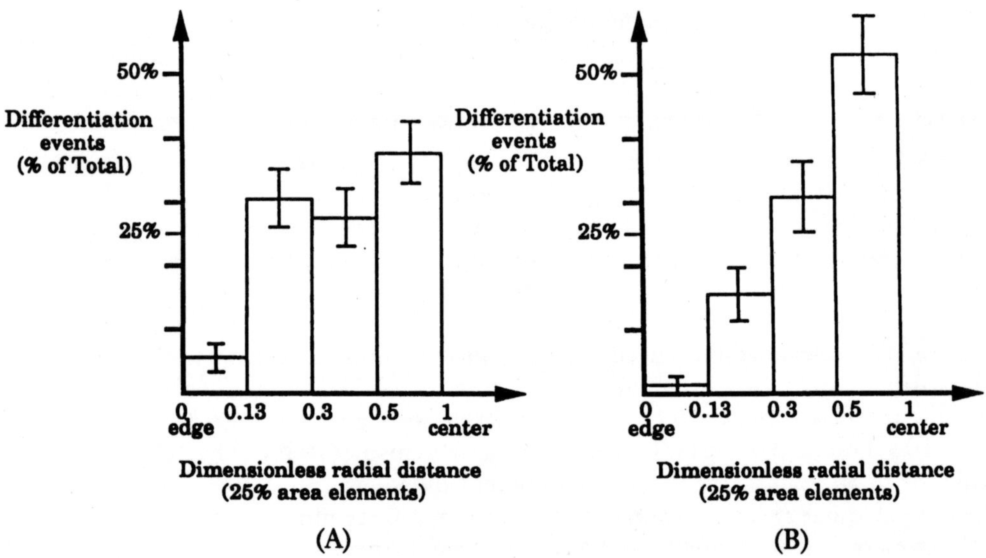

Figure 7.7. Positional dependence of tracheary element (TE) differentiation in 14day-old alginate-immobilized fines (125 - 500 μm) of *Nicotiana tabacum*. (A) 4 g (fresh weight) of cells per flask, total events = 96. (B) 1 g (fresh weight) of cells per flask, total events = 72. A thin center section of each bead was scanned microscopically, and each isolated TE or group of cytoplasmically connected TEs was considered a differentiation event. The frequency is plotted per unit area element, but the trends are reflective of events occurring on a unit volume basis. The error bars are standard deviation estimates based on the variance of a binomial distribution. Equal numbers of beads were counted in each case. After Bringi and Shuler (1990). Further experimental details are provided in Bringi (1991).

D.5.3.2. Consequences of Cells Responding to Altered Concentrations If, indeed, the speculation that cells respond to altered concentrations is borne out by further experiments with secondary metabolite-producing systems, then a typical gel immobilization experiment may be viewed as simply a convenient method for exposing cells to various microenvironments in a single bead. Two consequences become apparent:

1. If product formation is localized in a particular section of the biocatalyst particle, then it should be possible to predict the microenvironment favorable for product formation and reproduce it throughout a gradientless particle by manipulating bulk medium conditions.

2. If the cells are responding solely to altered concentrations of nutrients, then it may be possible to reoptimize medium conditions (including dissolved gas concentrations) and to induce suspension cultures to exhibit enhanced productivity. Immobilization may not be necessary for enhanced production. However, the "right" concentrations might include compounds generated by the cells themselves and compounds that are chemically unidentified. In such cases, recreation of the appropriate microenvironment in a suspension culture or in a gradientless bead will not be an easy task.

D.6. Summary of Possible Mechanisms

In summary, it should be recognized that different levels of complexity are superimposed when cells are gel-entrapped, foam-entrapped, or used as natural aggregates as compared to fine suspensions. These levels of complexity are summarized in Table 7.4.

All immobilization methods confer physical stability and shear protection to the majority of cells, though natural aggregates may be more susceptible to breakage upon exposure to mechanical stresses. Gel-entrapped cells are subjected to concentration gradients and matrix effects, but gel beads do not exhibit a positionally dependent cell age distribution, nor are all the cells symplastically interconnected by plasmodesmata. Foam-entrapped cells and natural aggregates exhibit plasmodesmatal connections, a spatially heterogeneous cell age distribution, and have concentration gradients imposed upon them in addition. The foam may have some minimal matrix effects, whereas natural aggregates of course do not use an artificial matrix.

Table 7.4. Summary of Levels of Imposed Complexity in Immobilized Cell Systems

Type of Immobilization	Physical Support	Matrix Effects	Concentration Gradients	Cell age distribution/ Plasmodesmata
Fine suspension	-	-	-	-
Gel-immobilized	+	+	+	-
Foam-immobilized	+	?*	+	+
Natural aggregates	+	-	+	+

* Only toxic effects have been reported so far with foam materials (Rhodes et al. 1987)

We have explored the various mechanisms by which immobilization could affect cellular metabolism. We have also recognized the levels of complexity imposed by each method of immobilization. Next, a conceptual framework is described for separating and studying the effects of immobilization independently. This framework will aid in experimental design and in the interpretation of experimental results and will also provide a basis for the choice of immobilization method and operating conditions.

E. SEPARATING THE EFFECTS OF IMMOBILIZATION

E.1. Concept

The outcome of an immobilization experiment will clearly depend on the level of complexity required to either switch on or enhance secondary metabolism. If simply protecting the cells from shear is sufficient, then all immobilization methods will be equally successful. If an altered microenvironment created by concentration gradients enhances secondary metabolism, then gel entrapment should be successful. Possible matrix effects should be accounted for by running proper control experiments. If a higher degree of complexity is necessary for secondary metabolism, then gel immobilization will be unsuccessful, and only foam-immobilized cells and natural aggregates will produce the secondary metabolite. Other methods such as adsorption onto cloth and membrane entrapment would also be expected to be successful in the absence of severe negative matrix effects. This simple concept logically leads to an experimental approach for separating the effects of immobilization (see the Appendix). Figure 7.8 summarizes the framework in the form of a flow diagram.

E.2. Case Studies: A Renewed Look at the Literature

Let us consider some examples from the literature and attempt to interpret results in the context of the conceptual framework outlined above.

E.2.1. Biotransformation of Codeinone The biotransformation of codeinone to codeine by ***Papaver somniferum*** is an example of a case in which immobilized cells have been investigated for their bioprocessing advantages even though fine suspensions are biosynthetically capable. This biotransformation has been studied in gel-immobilized (Furusaki et al. 1988; Furuya et al. 1984) and in foam-immobilized (Corchete and Yeoman 1989) systems. A comparison of the two sets of data with respect to the final conversion would suggest that the activity of foam-immobilized cells was superior to that of gel-immobilized cells. However, the two cell lines were different, and the two suspension cultures themselves showed different intrinsic biotransformation capabilities. Therefore, a better basis for a comparison would be the maximum specific biotransformation rate. The ratio of maximum specific biotransformation rates for immobilized to free cells was similar in the two reports and approximately equal to one in both cases. This indicates that immobilization did not dramatically alter intrinsic biocatalytic activity in either foam- or gel-immobilized cells. The choice of immo-

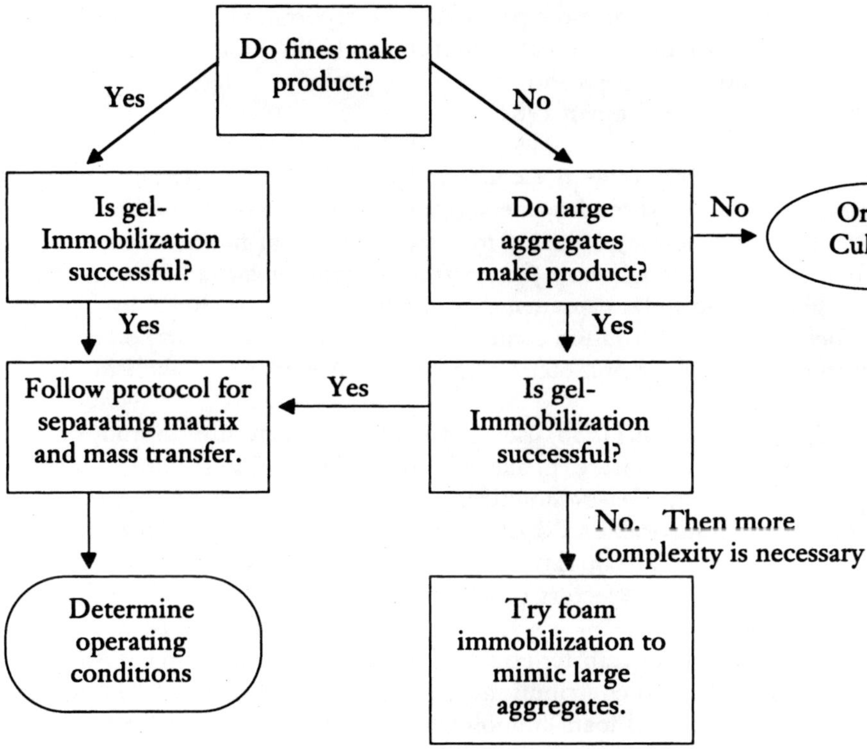

Figure 7.8. Framework for the separation of immobilization effects.

bilization method should be determined by other factors, for example, stability of activity upon reuse.

Furusaki et al. (1988) noted a rapid drop in activity upon reusing the gel beads for three cycles, whereas the fine suspensions could be reused without deterioration of activity. Clearly, this undesirable property of the gel-immobilized biocatalyst preparation had to be circumvented. Furusaki et al. (1988) conducted a kinetic analysis based on intrinsic free-cell uptake rates and concluded that concentration gradients of sugar and the precursor codeinone could not have been significant. They further recognized that for a zero-order reaction such as oxygen uptake, concentration gradients of oxygen could have been present in the particle even though the predicted effectiveness factor was 1 (see Section D.3 of Chapter 4). Furusaki et al. (1988) suggested that though cells could withstand lowered oxygen levels for short periods of time and transform codeinone at these low oxygen levels, perhaps prolonged exposure to depressed oxygen levels caused the inactivation of cells, resulting in rapid drop of biotransformation activity. Alternatively, a negative matrix effect could have been responsible, and the controls for mass transfer effects and for matrix effects outlined in the Appendix would likely provide the explanation for the loss of activity and would perhaps suggest a strategy to solve the problem. The repeated use of foam-immobilized cells was not reported by Corchete and Yeoman (1989). Other foam-immobilized systems have, however, been shown to retain prolonged viability and biosynthetic activity (Lindsey and Yeoman 1984b).

E.2.2. Capsaicin by Immobilized *Capsicum* spp. The most dramatic positive effect of immobilization on secondary metabolism was demonstrated by Lindsey and Yeoman (1984a,b) and Lindsey (1985). After manipulation of nutrient levels in the medium, foam-immobilized *Capsicum frutescens* cells produced two to three orders of magnitude higher extracellular capsaicin than the trace levels produced by fine suspensions. Since foam immobilization confers several levels of complexity on the cells, the exact mechanism by which immobilization affected secondary metabolism was not elucidated.

Recently, Ravishankar et al. (1988) used *Capsicum annuum* cultures to study capsaicin production. Suspension cultures produced no capsaicin within the limits of paper chromatography detection. Gel-immobilized cells, however, showed an increase in capsaicin production comparable to that observed by Lindsey and Yeoman (1984a). Since gel beads probably do not show a spatial cell age distribution or extensive plasmodesmatal contact, these factors could not have been necessary for enhancing capsaicin production in Ravishankar's system. The enhancement must have either been due to matrix or to mass transfer effects. The controls outlined in the Appendix would likely separate the two contributions. Further, since mass transfer limitations were common to the gel- and foam-immobilized systems, they could have played a role in Lindsey and Yeoman's study. A possible way of confirming this suggestion would be to follow the experimental approach of the Appendix with Lindsey and Yeoman's *Capsicum frutescens* system.

If gel- and foam-immobilized cells were equally capable of enhanced production, then the choice would depend on engineering issues such as viability of biocatalyst, ability to maintain prolonged biosynthesis, biocatalyst density in the bioreactor volume, and ease of scale-up of the immobilization procedure. If the immobilized cells were simply responding to altered concentrations of a particular nutrient, and if they did so when the appropriate microenvironment was reproduced in a suspension culture with reoptimized production medium (including dissolved gas concentrations), then suspension culture may prove to be a viable alternative. A comparison such as that in Section B.1.3 would then indicate whether immobilization was warranted or feasible.

E.2.3 Alkaloids from *Coffea arabica* Early studies of methylxanthine alkaloid production by *Coffea arabica* indicated that large natural aggregates were superior to fine suspensions (Frischknecht and Baumann 1985). Haldimann and Brodelius (1987), using the same system, showed that alginate-immobilized fines were also capable of increased alkaloid productivity, suggesting that spatial cell age distribution or plasmodesmatal contact were not necessary for enhanced secondary metabolism in this system. Interestingly, when the fines were released from the gel, they maintained higher productivity for a short time and then reverted to their original low-producing conditions. This example clearly demonstrates the reversible nature of the effect of immobilization on cellular physiology.

F. BIOPROCESS DESIGN FOR IMMOBILIZED CELLS

Once the biocatalyst has been designed and the operating conditions have been established at the bench scale, the next step is to provide the appropriate environmental conditions in a larger scale reactor. Before discussing the various bioreactor options, let us first briefly consider scale-up of the immobilization procedure itself.

F.1. Scale-up of the Immobilization Procedure

The first step in an immobilization process is the generation of biomass by conventional suspension culture. If mechanically robust natural aggregates are to be used as the immobilized cell particles, a separate immobilization step is not required, but a sieving or sedimentation step to separate aggregates of the desired size distribution may be necessary.

The technology developed for large-scale gel entrapment of enzymes and microbial cells should be largely transferable directly to plant cells. Nagashima et al. (1984) extruded gel-immobilized viable yeast directly into pilot-scale (4000-L) bioreactors with no specialized equipment and with high retention of viability, suggesting that plant cell immobilization should be similarly achievable. Further efforts have been directed toward larger scale production of small (<1 mm diameter) spherical gel beads (Siemann et al. 1990; Ogbonna et al. 1989).

The requirements for foam immobilization are somewhat different. The foam has to be seeded with cells and proliferation must be allowed to occur *in situ*. The factors affecting immobilization of plant cells in reticulate foams has been studied at the shake-flask level (Rhodes et al. 1987; see Section C.3), but rapid and efficient immobilization has been somewhat more difficult to achieve in a 5-L bioreactor (Mavituna et al. 1987). However, the ease and rapidity with which immobilization is achieved is dependent on a number of factors including the cell line itself, and certain lines may lend themselves to immobilization more readily than others (Archambault et al. 1990a,b).

F.2. Bioreactors for Immobilized Plant Cells

This section provides a brief overview of the bioreactor options for immobilized plant cell culture. The technology and know-how from chemical engineering experience with large-scale heterogeneous chemical catalysis and from enzymatic, microbial, and mammalian cell systems should be largely transferable to immobilized plant cell systems. Information on reactor design and operation can be found in many comprehensive texts (Satterfield 1970; Bailey and Ollis 1986; Wiseman 1985; Froment and Bischoff 1990).

Immobilized cell operation usually implies continuous flow through of liquid medium with total retention of biocatalyst particles inside the reactor. The classification of reactor choices is as follows:

1. Packed beds
2. Fluidized beds
3. Stirred tanks and other reactors

Other configurations such as membrane reactors have been discussed in literature reviews (Shuler 1988, 1987, 1986; Prenosil and Pedersen 1983).

F.2.1. Packed Beds A static bed of biocatalyst particles retained in a tower is called a packed bed or a fixed bed reactor. High cell densities and high volumetric productivities are achievable in packed beds, and therefore a smaller volume is required to perform a given conversion. Liquid medium is constantly recirculated to provide control over environmental conditions. Aeration can theoretically be achieved by: (i) concomitant exposure of the bed to liquid medium and gas (trickle bed reactor), (ii) alternate exposure of the bed to liquid and air (Kobayashi et al. 1988; see Chapter 9), and (iii) supply of dissolved gas by the recirculating medium, with gas replenishment being achieved outside the reactor. The latter option has generally been used on the laboratory scale.

The disadvantage of packed bed reactors stems from the critical dependence of operating conditions on the biocatalyst particle properties such as the size and compressibility. We have seen that a smaller particle diameter favors intraparticle mass transfer. Since the pressure required to pump liquid across the bed is inversely related to the particle diameter, a high pressure drop may be necessary to ensure the high recycle rates required for oxygen supply and for control over bulk medium conditions. These large pressure drops may cause problems with bed compaction (see Wiseman 1985), which would further aggravate the problem of increasing pressure drop. To operate at a lower pressure drop, larger particle diameters are necessary. However, internal and perhaps external mass transfer problems may arise with larger particles. To circumvent mass transfer limitations, larger particles with cells immobilized only on a surface layer [the so-called "pellicular" catalysts (Horvath and Engasser 1973)] may be used, but this reduces the high cell densities originally achievable. The particle size is usually a compromise between mass transfer and pressure drop considerations.

Though large-scale packed bed reactors have been successfully used for glucose isomerization and amino acid racemization with immobilized biocatalysts, further research is needed to confirm their suitability for immobilized plant cell and organ culture processes.

F.2.2. Fluidized Bed Reactors A fluidized bed refers to particles suspended by the flow of liquid or gas (or both) through the bed. Since pressure drop is not a consideration once the bed is fluidized, smaller particles can be used. The superior internal and external mass transfer characteristics due to the use of smaller particles and better mixing are the principal advantages of fluidized bed reactors. Aeration may be achieved outside the reactor with rapid recycle, or by introducing gas bubbles directly into the fluidized bed (a three-phase fluidized bed). Because of the complex fluid-flow phenomena, scale-up is often difficult, and particle attrition due to collisions may be severe. Moreover, cell densities will be lower than that achievable in a packed bed reactor.

The liquid fluidization velocity has to meet several constraints. The velocity should be high enough to suspend the particles. The volumetric flow rate should be high enough for dissolved oxygen supply to the bed if direct sparging is not used. The liquid residence time should be large enough for adequate conversions of substrate to take place. Consequently, high liquid recirculation rates are desirable. However, the density difference between the particles and fluid determines the maximum fluid velocity that can be used without washout of particles. Since the density difference between immobilized plant cell particles and the liquid medium is small, the particles could be weighted to allow a higher fluid recirculation velocity to be used.

At least two processes involving living immobilized cells have employed fluidized bed reactors. The Verax Corporation (Dean et al. 1987) has developed a fluidized bed system for immobilized mammalian cell culture. The process employs ca. 0.5-mm particles weighted with metal particles for an adequate density difference with the liquid medium. Gas transfer is achieved outside the reactor in the recycle loop via an artificial lung-type gas exchange device. However, the scale of operation (the largest reactor is 24 L) is small compared to the commercial requirements of most plant cell processes. The Kyowa Hakko Kogyo Co. in Japan reported the successful and stable operation of a fluidized bed reactor using alginate-immobilized yeast cells for the conversion of glucose to ethanol at a pilot scale of 4000 L (Nagashima et al. 1984). However, note that since the ethanol conversion is anaerobic, oxygen requirements were not a consideration for this design. Though these examples indicate that the technology exists for successful scale-up of immobilized cell processes, further research is necessary to confirm their applicability to plant cell processes.

F.2.3. Stirred Tank and Other Reactors A stirred tank reactor is a mechanically agitated, well-mixed vessel, similar to that used for suspension cultures. Its use with immobilized plant cells would be directly comparable to that of microcarrier cultures with animal cells. Aeration can be achieved inside the reactor by direct gas sparging. Mechanical agitation serves to suspend biocatalyst particles and to keep the reactor well mixed.

The principal advantage of a stirred tank reactor is the ability to exert careful control over the bulk medium conditions. Adequate intraparticle mass transfer may be achieved by using sufficiently small particles. When intraparticle limitations become insignificant, external mass transfer limitations become insignificant as well (see Chapter 4, page 84).

The disadvantages of a stirred tank are that mechanical agitation may cause particle damage due to collision and shear. The use of low-shear agitator designs may prove

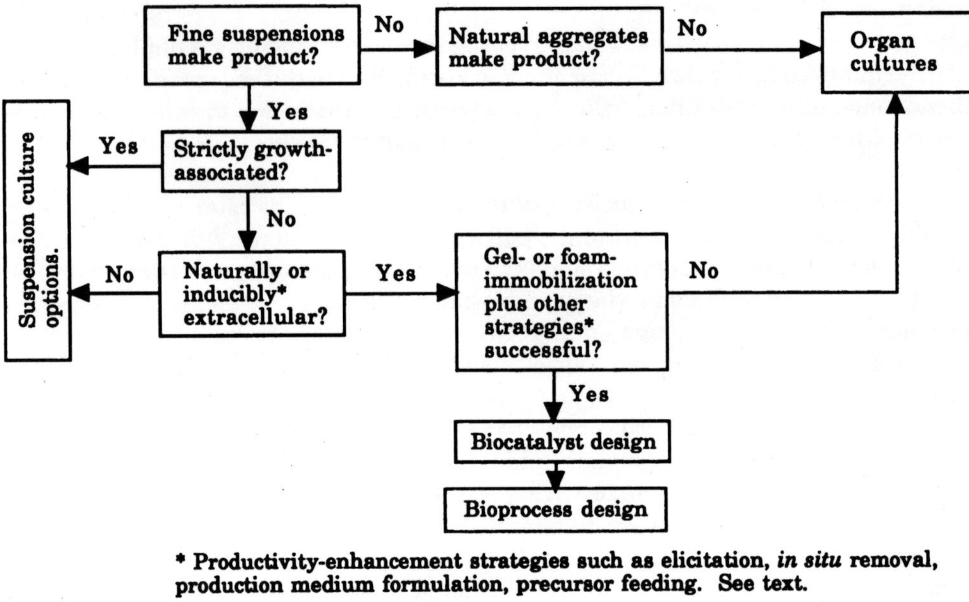

Figure 7.9. Immobilization in the context of suspension and organ culture options. This figure is in effect a synthesis of Figure 7.1 and Figure 7.8

to be useful in circumventing particle attrition, as will further improvements in the mechanical properties of immobilized cell particles. Note that the cells themselves are protected from mechanical stresses since they are entrapped within the particles. Another disadvantage of stirred tanks is that the maximum achievable cell density per unit reactor volume may not be as high as for other designs.

Other strategies such as the entrapment of cells in the shell side of hollow fiber membrane devices have proven successful for mammalian cell cultures (e.g., Hirschel and Gruenberg 1987). Though a few studies have been conducted with plant cell cultures (Shuler 1981; Hallsby and Shuler 1986; Prenosil and Pedersen 1983), these methods have not been as extensively studied as other immobilization methods. The possible advantages and disadvantages of membrane entrapment have been discussed (Shuler 1986; Payne et al. 1987).

G. CONCLUSIONS

An immobilized cell process is compatible with those secondary metabolites that are non-growth-associated and are at least partially extracellular. Figure 7.9 represents the immobilized cell process in the context of suspension culture and organ culture options.

Immobilization is a useful strategy from a bioprocessing point of view and, furthermore, can have dramatic effects on cellular physiology. Biocatalyst particle design is an important first step in assessing and exploiting the engineering and physiological benefits of immobilization.

Process scale-up involves the challenges of scaling up of the immobilization step, as well as of reactor design. Productivity enhancements such as *in situ* product removal and the use of elicitors must be integrated into the overall process. We believe that the technology developed for heterogeneous chemical catalysis and other microbial and mammalian cell processes should be directly applicable to plant cell processes as well.

The integrated use of engineering and biological knowledge at the laboratory as well as at the process scale will be an important factor in the successful design and operation of an immobilized cell process.

Appendix. A Suggested Experimental Approach for Separating the Effects of Immobilization

The conceptual idea embodied in Section E.1 logically leads to a simple experimental approach for independently studying the effects of immobilization on cell performance. This approach has been presented previously (Bringi and Shuler 1990) and is included here with some modifications.

Immobilized cell experiments can be broadly classified into two categories based on their objectives (see Figure 7.8):

(a) Fine suspensions already make significant amounts of product. Immobilization is used primarily for its engineering advantages and for any possible positive biological consequences that may accrue.

(b) Fine suspensions make little or no product whereas large natural aggregates do. Immobilization is used in an effort to mimic events occurring within natural aggregates and thus enhance productivity.

Consider a secondary metabolite producing system, such as an ajmalicine producing *Catharanthus roseus* suspension, corresponding to case (a) above, or the left-hand side of Figure 7.8. Fines from a suspension culture should be immobilized in moderately sized gel beads (4 mm diameter) with moderate cell loadings (20-30% by wt.) and subcultured under previously determined production conditions. A range of gels should be screened. If there is a complete shutoff of secondary metabolism, it is probably due to a negative matrix effect. If, however, some positive signs are observed, then the protocol for separating matrix and mass transfer effects should be followed.

Consider case (b), or the right-hand side of Figure 7.8. In this case, fine suspensions do not make appreciable levels of product, whereas large natural aggregates do. If cells within large aggregates make product simply because they are responding to imposed concentration gradients (or more precisely, to altered concentrations), then a gel-immobilized system with similarly imposed concentration gradients should work just as well, and an effort should be made to separate matrix and mass transfer effects. However, if gel immobilization is not successful in giving the expected levels of productivity, then a higher level of complexity, namely, plasmodesmata/spatial cell age distribution, may be necessary. Accordingly, foam immobilization, adsorption onto other matrices such as woven cloth, or membrane entrapment should be tried to mimic events occurring in large natural aggregates more accurately.

Foam immobilization may not always be successful because of differences between foam-immobilized cells and large natural aggregates. In foams, plasmodesmatal contact and spatial cell age distribution may perhaps be important only within each "pocket," whereas concentration gradients are imposed over the whole particle. In a natural aggregate, the length scale for concentration gradients, plasmodesmata, and spatial cell age distribution is the same, namely, the aggregate radius. Further studies are needed to examine whether such differences are important and to examine how closely the behavior of foam immobilized, cloth-immobilized, and membrane-entrapped cells resembles that of natural aggregates.

Protocol for Separating Matrix and Mass Transfer Effects

The following protocol allows the separation of matrix effects and allows the experimenter to chose a range of matrices that are compatible with secondary metabolism and prolonged cell viability. A study of mass transfer effects allows rational design of particle dimensions, and possibly provides insights into the proper environment or series of environments to which the cells must be subjected for high productivity.

The set of experimental controls which will account for the various types of matrix effects can be summarized. The following flasks should be tested for total productivity:

(i) Fine suspension
(ii) Fine suspension plus added soluble ion when appropriate
(iii) Fine suspension plus empty gel
(iv) Small beads of gel-immobilized cells (ca. 1 mm diameter) at high cell loading (ca. 33-50% fresh weight cells)
(v) Large beads at same cell loading

The total mass of cells and medium should be the same in all flasks, and the total amount of gel should be the same in flasks (iii) and (iv) and (v).

Possible Outcomes and Interpretations

Matrix Effects The positive or negative effect of the gel-stabilizing ion will be reflected in the flask with added dissolved ion when compared to the fine suspension. If the gel material is perfectly inert and does not affect cell metabolism, the "per cell" or specific productivity should be the same in fine suspension, fine suspension plus empty gel, and small beads. If the gel has a positive or negative effect on secondary metabolism, then small beads and the fine suspension plus empty gel will show higher or lower productivities, respectively, compared to the fine suspension control. If the gel needs to be in intimate contact with the cells to exert its positive or negative effect, then small beads will show higher or lower productivities, respectively, than either the fine suspension or the fine suspension plus empty gel. Small beads will also show higher productivities if physical protection is important. This interpretation is unequivocal since sufficiently small beads do not exhibit concentration gradients. A range of biologically compatible matrices should be chosen. The final choice will be influenced by processing constraints such as physical integrity, compressibility, and amenability to scale-up.

Mass Transfer Effects Once a biologically compatible matrix is chosen, a comparison of the total productivity of sufficiently small beads (Weisz's modulus << 1) and large beads will provide an insight into the effect of concentration gradients, or, more precisely, of altered concentrations, on secondary metabolism of gel-immobilized cells. Three outcomes are possible:

1. *Pronounced positive effect in small beads.* Concentration gradients are not significant in sufficiently small beads and cells essentially experience bulk medium conditions. Effectiveness factors for nutrients and products are equal to one, all cells are metabolically active, and most complete use is being made of the biomass. This is clearly the optimal condition for operation.

2. *Pronounced positive effect in large beads.* Since concentration gradients of nutrients such as oxygen can be significant in large beads, cells experience a range of different microenvironments depending on their position within the bead. It is possible that only a certain narrow range of microenvironments allows the favorable expression of secondary metabolism. Accordingly, an attempt should be made to localize the product-forming cells within the bead (e.g., by fluorescence microscopy or ELISA). The product may be difficult to localize if it is predominantly extracellular.

If spatial localization of secondary metabolite production is evident within the biocatalyst particle, then not all cells are equally active, and the biomass is not being used effectively. It should be possible to predict the microenvironment that favors secondary metabolism and reproduce it throughout small gradientless beads simply by manipulating the bulk concentrations. If such manipulations are unsuccessful, then perhaps the favorable microenvironment has not been understood correctly (e.g., consideration of the elevated levels of a product). Alternatively, the cells' response could be to a gradient rather than simply to an altered concentration.

3. *A mixed effect.* Small and large beads could possibly show different types of positive effects. We have observed that small beads of alginate-immobilized *Catharanthus roseus* produced more ajmalicine, per cell, than did large beads (Bringi 1989, unpublished observations). The large beads, however, secreted more of the ajmalicine they made into the medium. Immobilization has similarly been shown to cause secretion in a variety of other systems (see Table 7.2). Product secretion, if not simply due to lysis, is a critical prerequisite for a successful immobilized cell process.

The differences between the performances of small and large beads may be due to spatial heterogeneity within the large bead. Cells near the ***exterior*** of the large bead experience a microenvironment similar to that experienced by cells in small beads and may have accumulated product intracellularly as did the small beads. Cells in the ***interior*** of the large bead experience a very different microenvironment because of the presence of concentration gradients, perhaps one that favors secretion of the compound. It should be possible to replace the spatial heterogeneity by a temporal one, that is, make small gradientless beads, and cycle, say, the dissolved gas concentrations (O_2, or possibly CO_2 or the phytohormone C_2H_4) in the bulk medium such that cells first experience conditions typical of small beads and then experience the conditions typical of the interior of large beads. Both positive effects may possibly be achieved in the same system and efficient use is made of the biomass.

References

A. W. Alfermann, I. Schuller, and E. Reinhard. 1980. Biotransformation of Cardiac Glycosides by Immobilized Cells of *Digitalis lanata*. *Planta Med.* *40*, 218-223.

A. W. Alfermann, W. Bergmann, C. Figur, U. Helmbold, D. Schwantag, I. Schuller, and E. Reinhard. 1983. Biotransformation of b-Methyldigitoxin to b-Methyldigoxin by Cell Cultures of *Digitalis lanata*. In "Plant Biotechnology," S. H. Mantell and H. Smith (eds.), Cambridge University Press, Cambridge, U.K., pp. 67-74.

J. Archambault, B. Volesky, and W. G. W. Kurz. 1989. *Biotechnol. Bioeng.* *33*, 293-299.

J. Archambault, B. Volesky, and W. G. W. Kurz. 1990a. Development of Bioreactors for the Culture of Surface Immobilized Plant Cells. *Biotechnol. Bioeng.* *35*, 702-711.

J. Archambault, B. Volesky, and W. G. W. Kurz. 1990b. Production of Indole Alkaloids by Surface Immobilized *C. roseus* Cells. *Biotech. Bioeng.* *35*, 660-667.

M. Asada and M. L. Shuler. 1989. Stimulation of Ajmalicine Production and Excretion from *Catharanthus roseus*: Effects of Adsorption *In Situ*, Elicitors, and Alginate Immobilization. *Appl. Microbiol. Biotechnol.* *30*, 475-481.

S. Ayabe, K. Iida, and T. Furuya. 1986. Induction of Stress Metabolites in Immobilized *Glycyrrhiza echinata* Cultured Cells. *Plant Cell Rep.* *3*, 186-189.

J. E. Bailey and D. F. Ollis. 1986. "Biochemical Engineering Fundamentals," 2nd ed. McGraw-Hill, New York.

N. J. Barnabas and S. B. David. 1988. Solasodine Production by Immobilized Cells and Suspension Cultures of *Solanum surattense*. *Biotechnol. Lett.* *10*, 593-596.

M. D. Beaumont and D. Knorr. 1987. Effects of Immobilizing Agents and Procedures on Viability of Cultured Celery (*Apium graveolens*) Cells. *Biotechnol. Lett.* *9*, 377-382.

R. Beiderbeck and B. Knoop. 1988. Enhanced Production of Secondary Substances: Addition of Artificial Accumulation Sites to Cultures. In "Biotechnology in Agriculture And Forestry" Vol. 4, "Medicinal And Aromatic Plants I," Y. P. S. Bajaj (ed.), Springer-Verlag, Berlin, pp. 123-135.

J. Berlin 1985. The Use of Immobilized Plant Cells - an Evaluation. *International Association for Plant Tissue Culture Newsletter*, #46 , 8-14.

H. Bouyssou, A. Pareilleux and G. Marigo. 1987. The Role of pH Gradients across the Plasmalemma of *Catharanthus roseus* and Its Involvement in the Release of Alkaloids. *Plant Cell Tissue Organ Culture* *10*, 91-99.

V. Bringi. 1991. Differentiation in Plant Cultures: (1) Tracheary Element Differentiation in Immobilized Tobacco Cells; (2) Bioreactor Consideration for Shoot Culture Cultivation. Ph.D. Thesis, Cornell University, Ithaca, New York.

V. Bringi and M. L. Shuler. 1990. A Framework for Understanding the Effects of Immobilization of Plant Cells: Differentiation of Tracheary Elements in Tobacco. In "Physiology of Immobilized Cells," J. A. M. de Bont, J. Visser, B. Mattiasson, and J. Tramper (eds.), Elsevier Science Publishers, Amsterdam, pp. 161-172.

P. Brodelius. 1983. Immobilized Plant Cells. In "Immobilized Cells And Organelles," Vol 1, B. Mattiasson (ed.), CRC Press, Boca Raton, Florida, pp. 28-55.

P. Brodelius. 1988a. Immobilized Plant Cells as a Source of Biochemicals. In "Bioreactor Immobilized Enzymes And Cells," M. Moo-Young (ed.), Elsevier, Essex, pp. 167-196.

P. Brodelius. 1988b. Permeabilization of Plant Cells for Release of Intracellularly Stored Products: Viability Studies. *Appl. Microbiol. Biotechnol.* 27, 561-566.

P. Brodelius and K. Nilsson. 1980. Entrapment of Plant Cells in Different Matrices. *FEBS Lett. 122*, 312-316.

P. Brodelius, B. Deus, K. Mosbach, and M. H. Zenk. 1979. Immobilized Plant Cells for the Production and Transformation of Natural Products. *FEBS Lett. 103*, 93-97.

P. E. Brodelius, C. Funk, and R. D. Shillito. 1988. Permeabilization of Cultivated Plant Cells by Electroporation of Release of Intracellularly Stored Secondary Products. *Plant Cell Rep.* 7, 186-188.

C. Bucke. 1983. Immobilized Cells. *Phil. Trans. R. Soc. London Ser. B 300*, 369-389.

J. Burgess 1985. "An Introduction to Plant Cell Development," Cambridge University Press, Cambridge, U.K.

J. Chappell, R. Nable, P. Fleming, R. A. Andersen, and H. R. Burton. 1987. Accumulation of Capsidiol in Tobacco Cell Cultures Treated with Fungal Elicitor. *Phytochem. 26*, 2259-2260.

B. V. Charlwood, C. Moustou, J. T. Brown, P. K. Hegarty, and K. A. Charlwood. 1989. The Regulation of Accumulation of Lower Isoprenoids in Plant Cell Cultures. In "Primary and Secondary Metabolism of Plant Cell Cultures," W. G. W. Kurz (ed.), Springer-Verlag, Berlin, pp. 73-84.

S. D. Cline and C. J. Coscia. 1988. Stimulation of Sanguinarine Production by Combined Fungal Elicitation and Hormonal Deprivation in Cell Suspension Cultures of *Papaver bracteatum. Plant Physiol. 86*, 161-165.

H. A. Collin 1987. Determinants of Yield in Plant Tissue Cultures. *Adv. Bot. Res. 13*, 145-187.

A. Cooke, A. Cookson, and M. J. Earnshaw. 1986. The Mechanism of Action of Calcium in the Inhibition of High Temperature-Induced Leakage of Betacyanin from Beet Root Discs. *New Phytol. 102*, 491-497.

P. Corchete and M. M. Yeoman. 1989. Biotransformation of (-)-Codeinone to (-)-Codeine by *Papaver somniferum* Cells Immobilized in Reticulate Polyurethane Foam. *Plant Cell Rep. 8*, 128-131.

A. L. Dainty, K. H. Goulding, P. K. Robinson, I. Simpkins, and M. D. Trevan. 1985. Effect of Immobilization on Plant Cell Physiology - Real or Imaginary? *Trends Biotechnol. 3*, 59-60.

R. C. Dean, S. B. Karkare, P. G. Phillips, N. G. Ray, and P. W. Runstadler. 1987. Continuous Cell Culture with Fluidized Sponge Beads. In "Large Scale Cell Culture Technology," B. K. Lydersen (ed.), Hanser Publishers, Munich. pp. 145-167.

B. Deus-Neumann, G. Krauss, M. Hieke and D. Groger. 1983. Indole Alkaloid Formation and Storage in Cell Suspension Cultures of *Catharanthus roseus. Planta Med. 48*, 20-23.

F. DiCosmo and M. Misawa. 1985. Eliciting Secondary Metabolism in Plant Cell Cultures. *Trends Biotechnol. 3*, 318-322.

U. Eilert. 1987. Elicitation: Methodology and Aspects of Application. In "Cell Culture and Somatic Cell Genetics of Plants," Vol. 4, F. Constabel and I. K. Vasil (eds.), Academic Press, New York, pp. 153-196.

U. Eilert. 1989. Elicitor Induction of Secondary Metabolism in Dedifferentiated and Differentiated In Vitro Systems of *Ruta graveolens*. In "Primary and Secondary Metabolism of Plant Cell Cultures," W. G. W. Kurz (ed.), Springer-Verlag, Berlin. pp. 219-228.

U. Eilert, A. Ehmke, and B. Wolters. 1984. Elicitor-Induced Accumulation of Acridone Alkaloid Epoxides in *Ruta graveolens* Suspension Cultures. *Planta Med. 50*, 508-512.

U. Eilert, W. G. W. Kurz, and F. Constabel. 1985. Stimulation of Sanguinarine Accumulation in *Papaver somniferum* Cell Cultures by Fungal Elicitors. *J. Plant Physiol. 119*, 65-76.

G. Fridborg, M. Pedersen, L-E. Landstrom, and T. Eriksson. 1978. The Effect of Activated Charcoal on Tissue Cultures: Adsorption of Metabolites Inhibiting Morphogenesis. *Physiol. Plant. 43*, 104-106.

P. M. Frischknecht and T. W. Baumann. 1985. "Stress Induced Formation of Purine Alkaloids in Plant Tissue Culture of *Coffea arabica*." *Phytochem. 24*, 2255-2257.

G. F. Froment and K. B. Bischoff. 1990. "Chemical Reactor Analysis and Design," 2nd ed., John Wiley, New York.

H. Fukuda and A. Komamine. 1980. Direct Evidence for Cytodifferentiation to Tracheary Elements without Intervening Mitosis in a Culture of Single Cells Isolated from the Mesophyll of *Zinnia elegans*. *Plant Physiol. 65*, 61-64.

H. Fukuda and A. Komamine. 1985. Cytodifferentiation. In "Cell Culture and Somatic Cell Genetics of Plants", Vol 2, I. K. Vasil (ed.), Academic Press, New York, pp. 149-212.

H. Fukui, N. Yoshikawa, and M. Tabata. 1983. Induction of Shikonin Formation by Agar in *Lithospermum erythrorhizon* Cell Suspension Cultures. *Phytochem. 22*, 2451-2453.

Y. Fukushima, K. Okamura, K. Imai, and H. Motai. 1988. A New Immobilization Technique of Whole Cells and Enzymes with Colloidal Silica and Alginate. *Biotechnol. Bioeng. 32*, 584-594.

K. W. Fuller and D. J. Bartlett. 1985. The Chemosynthetic Potential of Plants and Its Realization by Immobilized Systems. In "Plant Products and The New Technology," K. W. Fuller and J. R. Gallon (eds.), *Ann. Proc. Phytochem. Soc. Eur. 26*, 229-247.

C. Funk, K. Gugler, and P. Brodelius. 1987. Increased Secondary Product Formation in Plant Cell Suspension Cultures after Treatment with a Yeast Carbohydrate Preparation (Elicitor). *Phytochem. 26*, 401-405.

S. Furusaki, T. Nozawa, T. Isohara, and T. Furuya. 1988. Influence of Substrate Transport on the Activity of Immobilized *Papaver somniferum* cells. *Appl. Microbiol. Biotechnol. 29*, 437-441.

T. Furuya, T. Yoshikawa, and M. Taira. 1984. Biotransformation of Codeinone to Codeine by Immobilized Cells of *Papaver somniferum*. *Phytochem. 23*, 999-1001.

E. Galun, D. Aviv, A. Dantes, and A. Freeman. 1985. Biotransformation by Division-Arrested and Immobilized Plant Cells: Bioconversion of Monoterpenes by Gamma-Irradiated Suspended and Entrapped Cells of *Mentha* and *Nicotiana*. *Planta Med. 51*, 511-514.

Y. V. Gamalei. 1985. Plasmodesmata: Intercellular Communication in Plants. *Soviet Plant Physiol. 32*, 134-148.

B. E. S. Gunning and R. L. Overall. 1983. Plasmodesmata and Cell-to-Cell Transport in Plants. *BioScience 33*, 260-265.

D. Haldimann and P. Brodelius. 1987. Redirecting Cellular Metabolism by Immobilization of Cultured Plant Cells: A Model Study with *Coffea arabica*. *Phytochem. 26*, 1431-1434.

R. D. Hall, M. A. Holden, and M. M. Yeoman. 1988. Immobilization of Higher Plant Cells. In "Biotechnology in Agriculture And Forestry," Vol. 4, "Medicinal and Aromatic Plants I," Y. P. S. Bajaj (ed.) , Springer-Verlag, Berlin, pp. 136-156.

G. A. Hallsby and M. L. Shuler. 1986. Altering Fluid Flow Patterns Changes Pattern of Cellular Associations in Immobilized Tobacco Tissue Cultures. *Biotechnol. Bioeng. Symp. Ser. 17*, 741-746.

W. Halperin and W. A. Jensen. 1967. Ultrastructural Changes during Growth and Embryogenesis in Carrot Cell Cultures. *J. Ultrastruct. Res. 18*, 428-443.

H. Hamada, N. Umeda, N. Otsuka and S. Kawabe. 1988. The Reductive Transformation of Decanones with Immobilized Cells of *Nicotiana tabacum*. *Plant Cell Rep.* 7, 493-494.

P. F. Heinstein. 1985. Future Approaches to the Formation of Secondary Natural Products in Plant Cell Suspension Cultures. *J. Nat. Prod. 48*, 1-9.

M. D. Hirschel and M. L. Gruenberg. 1987. An Automated Hollow Fiber System for the Large Scale Manufacture of Mammalian Cell Secreted Product. In "Large Scale Cell Culture Technology," B. K. Lydersen (ed.), Hanser Publishers, Munich, pp. 113-144.

S. S. Hoekstra, P. A. A. Harkes, R. Verpoorte, and K. R. Libbenga. 1990. Effect of Auxin on Cytodifferentiation and Production of Quinoline Alkaloids in Compact Globular Structures of *Cinchona ledgeriana*. *Plant Cell Rep. 8*, 571-574.

C. Horvath and J. M. Engasser. 1973. Pellicular Heterogeneous Catalysts. A Theoretical Study of the Advantage of Shell Structured Immobilized Enzyme Particles. *Ind. Eng. Chem. Fund. 12*, 229-235.

A. C. Hulst and J. Tramper. 1989. Immobilized Plant Cells: A Literature Survey. *Enzyme Microb. Technol. 11*, 546-558.

A. C. Hulst, J. Tramper, P. Brodelius, L. J. C. Eijkenboon, and K. Ch. A. M. Luyben. 1985. Immobilized Plant Cells: Respiration and Oxygen Transfer. *J. Chem. Tech. Biotechnol. 35B*, 198-204.

A. C. Hulst, M. M. T. Meyer, H. Breteler and J. Tramper. 1989. Effect of Aggregate Size in Cell Cultures of *Tagetes patula* on Thiophene Production and Cell Growth. *Appl. Microbiol. Biotechnol. 30*, 18-25.

B. K. Ishida 1988. Improved Diosgenin Production in *Dioscorea deltoidea* Cell Cultures by Immobilization in Polyurethane Foam. *Plant Cell Rep.* 7, 270-273.

L. F. Jaffe and R. Nuccitelli. 1977. Electrical Controls of Development. *Annu. Rev. Biophys. Bioeng. 6*, 445-467.

R. A. Jeffs and D. H. Northcote. 1966. Experimental Induction of Vascular Tissue in an Undifferentiated Plant Callus. *Biochem. J. 101*, 146-153.

R. A. Jeffs and D. H. Northcote. 1967. The Influence of Indol-3-yl Acetic Acid and Sugar on the Pattern of Induced Differentiation in Plant Tissue Culture. *J. Cell Sci. 2*, 77-88.

V. Jirku, T. Macek, T. Vanek, V. Krumphanzl, and V. Kubanek. 1981. Continuous Production of Steroid Glycoalkaloids by Immobilized Plant Cells. *Biotechnol. Lett. 3*, 447-450.

A. Jones and I. A. Veliky. 1981a. Biotransformation of Cardenolides by Plant Cell Cultures II. Metabolism of Gitoxigenin and Its Derivatives by Suspension Culture of *Daucus carota*. *Planta Med. 42*, 160-166.

A. Jones and I. A. Veliky. 1981b. Effect of Medium Constituents on the Viability of Immobilized Plant Cells. *Can. J. Bot. 59*, 2095-2101.

A. Jones and I. A. Veliky. 1981c. Examination of Parameters Affecting the 5ß-Hydroxylation of Digitoxigenin by Immobilised Cells of *Daucus carota*. *Eur. J. Appl. Microbiol. Biotechnol. 13*, 84-89.

N. J. Kilby and C. S. Hunter. 1990. Repeated Harvest of Vacuole-Located Secondary Product from In Vitro Grown Plant Cells Using 1.02 MHz Ultrasound. *Appl. Microbiol. Biotechnol. 33*, 448-451.

A. M. Kinnersley and D. K. Dougall. 1980. Increase in Anthocyanin Yield from Wild-Carrot Cell Cultures by a Selection System Based on a Cell-Aggregate Size. *Planta 149*, 200-204.

K. -H. Knobloch, G. Beutnagel, and J. Berlin. 1981. Influence of Accumulated Phosphate on Culture Growth and Formation of Cinnamoyl Putrescines in Medium-Induced Cell Suspension Cultures of *Nicotiana tabacum*. *Planta 153*, 582-585.

Y. Kobayashi, H. Fukui, and M. Tabata. 1988. Berberine Production by Batch and Semi-Continuous Cultures of Immobilized *Thalictrum* Cells in an Improved Bioreactor. *Plant Cell Rep.* 7, 249-252.

Y. Kobayashi, H. Fukui, and M. Tabata. 1989. Effect of Oxygen Supply on Berberine Production in Cell Suspension Cultures and Immobilized Cells Of *Thalictrum Minus*. *Plant Cell Rep. 8*, 225-258.

R. Kollmann and C. Glockmann. 1985. Studies of Graft Unions. I. Plasmodesmata between Cells of Plants Belonging to Different Unrelated Taxa. *Protoplasma 124*, 224-235.

R. Kollmann, S. Yang, and C. Glockmann. 1985. Studies on Graft Unions. II. Continuous and Half Plasmodesmata in Different Regions of the Graft Interface. *Protoplasma 126*, 19-29.

G. Krauss, G. J. Krauss, R. Baumbach, and D. Groger. 1989. Elicitation of Canthin-6-one Alkaloid Accumulation in Cell Suspension Cultures of *Ailanthus altissima* (Mill.) Swingle. *Z. Naturforsch 44C*, 712-714.

T. Kuboi and Y. Yamada. 1978. Changing Cell Aggregation and Lignification in Tobacco Suspension Cultures. *Plant Cell Physiol. 19*, 437-443.

H. Lang and H. W. Kohlenbach. 1982. Differentiation of Alkaloid Cells in Cultures of *Macleya* Mesophyll Protoplasts. *Planta Med. 46*, 78-81.

K. Lindsey. 1985. Manipulation, by Nutrient Limitation, of the Biosynthetic Activity of Immobilized Cells of *Capsicum frutescens* Mill. cv. annuum. *Planta 165*, 126-133.

K. Lindsey and M. M. Yeoman. 1983a. Novel Experimental Systems for Studying the Production of Secondary Metabolites by Plant Tissue Cultures. In "Plant Biotechnology," S. H. Mantell and H. Smith (eds.), Cambridge University Press, Cambridge, U.K., pp. 39-66.

K. Lindsey and M. M. Yeoman. 1983b. The Relationship between Growth Rate, Differentiation and Alkaloid Accumulation in Cell Cultures. *J. Exp. Bot. 34*, 1055-1065.

K. Lindsey and M. M. Yeoman. 1984a. The Synthetic Potential of Immobilised Cells of *Capsicum frutescens* Mill. cv. annuum. *Planta 162*, 495-501.

K. Lindsey and M. M. Yeoman. 1984b. The Viability and Biosynthetic Activity of Cells of *Capsicum frutescens* Mill. cv. annuum Immobilized in Reticulate Polyurethane. *J. Exp. Bot. 35*, 1684-1696.

K. Lindsey and M. M. Yeoman. 1985. Dynamics of Plant Cell Cultures. In "Cell Culture and Somatic Cell Genetics of Plants," Vol 32, I. K. Vasil (ed.), Academic Press, New York, pp. 61-101.

K. Lindsey, M. M. Yeoman, G. M. Black, and F. Mavituna. 1983. A Novel Method for the Immobilisation and Culture of Plant Cells. *FEBS Lett. 155*, 143-149.

L. Linsefors and P. Brodelius. 1985. Immobilization of Plant Protoplasts: Viability Studies. *Plant Cell Rep. 4*, 23-27.

P. M. Lintilhac and T. B. Vesecky. 1984. Stress-Induced Alignment of Division Plane in Plant Tissues Grown in Vitro. *Nature 307*, 363-364.

E. Maeda and T. A. Thorpe. 1979. Shoot Histogenesis in Tobacco Callus Cultures. *In Vitro 15*, 415-424.

F. Mavituna, J. M. Park, A. K. Wilkinson, and P. D. Williams. 1987. Characteristics of Immobilized Plant Cell Reactors. In "Plant and Animal Cells: Process Possibilities," C. Webb and F. Mavituna (eds.), Ellis Horwood, Chichester, U.K., pp. 92-115.

B. H. McCown, E. L. Zeldin, H. A. Pinkalla, and R. R. Dedolph. 1988. Nodule Culture: A Developmental Pathway with High Potential for Regeneration, Automated Micropropagation, and Plant Metabolite Production from Woody Plants. In "Genetic Manipulation of Woody Plants," J. W. Hanover and D. E. Keathley (eds.), Plenum, New York, pp. 149-166.

S. Meiners, O. Baron-Epel, and M. Schindler. 1988. Intercellular Communication - Filling in the Gaps. *Plant Physiol. 88*, 791-793.

H. Miyasaka, M. Masu, T. Yamamoto, Y. Endo, and K. Yoneda. 1986. Production of Cryptotanshinone and Ferruginol by Immobilized Cultured Cells of *Salvia miltiorrhiza*. *Phytochem. 25*, 1621-1624.

K. Mosbach (ed.). 1987. Immobilized Enzymes and Cells, Part B. In "Methods in Enzymology," Vol. 135, Academic Press, New York.

M. Nagashima, M. Azuma, S. Noguchi, K. Inuzuka, and H. Samejima. 1984. Continuous Ethanol Fermentation Using Immobilized Yeast Cells. *Biotechnol. Bioeng. 26*, 992-997.

H. Nakajima, K. Sonomoto, M. Usui, F. Sato, Y. Yamada, A. Tanaka, and S. Fukui. 1985. Entrapment of *Lavandula vera* Cells and Production of Pigments by Entrapped Cells. *J. Biotechnol. 2*, 107-117.

H. Nakajima, K. Sonomoto, F. Sato, K. Ichimura, Y. Yamada, and A. Tanaka. 1989a. Enhancement of Pigment Productivity of Immobilized Cultured *Lavandula vera* Cells by Limitation of Nitrogen Sources. *J. Ferment. Bioeng. 67*, 306-308.

H. Nakajima, K. Sonomoto, F. Sato, K. Ichimura, Y. Yamada, and A. Tanaka. 1989b. Influence of Carbon Source on Pigment Production by Immobilized Cultured Cells of *Lavandula vera*. *J. Ferment. Bioeng. 68*, 330-333.

C. Nef, C. Ambid, and J. Fallot. 1987. Influence of External pH on Alkaloid Production and Excretion by *Catharanthus roseus* Resting Cell Suspensions. In "Plant Vacuoles: Their Importance in Solute Compartmentalization in Cells and Their Application In Plant Biotechnology," B. Marin (ed.), *NATO ASI Ser., Ser. A: Life Sciences 134*, 529-533.

J. C. Ogbonna, M. Matsumura, T. Yamagata, H. Sakuma, and H. Kataoka. 1989. Production of Micro-Gel Beads by a Rotating Disk Atomizer. *J Ferment. Bioeng. 68*, 40-48.

Y. Ozeki and A. Komamine. 1985. Induction of Anthocyanin Synthesis in Relation to Embryogenesis in a Carrot Suspension Culture - A Model System for the Study of Expression and Repression of Secondary Metabolism. In "Primary and Secondary Metabolism of Plant Cell Cultures," K. H. Neumann, W. Barz, and E. Reinhard (eds.), Springer-Verlag, Berlin, pp. 99-106.

A. Pareilleux and F. Majerus. 1987. Continuous Production of Indole Alkaloids by Gel-Entrapped Cells of *Catharanthus roseus*. In "Plant Vacuoles: Their Importance in Solute Compartmentation in Cells and Their Applications in Plant Biotechnology," B. Marin (ed.), *NATO ASI Ser., Ser. A: Life Sciences 134*, 495-501.

A. J. Parr, J. I. Smith, R. J. Robins and J. C. Rhodes. 1984. Apparent Free Space and Cell Volume Estimation: A Non-Destructive Method for Assessing the Growth and Membrane Integrity/Viability of Immobilised Plant Cells. *Plant Cell Rep. 3*, 161-164.

G. F. Payne, M. L. Shuler, and P. Brodelius. 1987. Large Scale Plant Cell Culture. In "Large Scale Cell Culture Technology," B. K. Lydersen (ed.), Hanser Publishers, Munich, pp. 193-229.

G. F. Payne, N. N. Payne, and M. L. Shuler. 1988. Bioreactor Considerations for Secondary Metabolite Production from Plant Cell Tissue Culture: Indole Alkaloids from *Catharanthus roseus*. *Biotechnol. Bioeng. 31*, 905-912.

C. R. Phillips and Y. C. Poon. 1988. "Immobilization of Cells," Springer-Verlag, Berlin.

N. Pras, H. J. Wichers, A. P. Bruins, and T. M. Malingre. 1988. Bioconversion of Para-Substituted Monophenolic Compounds into Corresponding Catechols by Alginate Entrapped Cells of *Mucuna pruriens*. *Plant Cell Tissue Organ Culture 13*, 15-26.

J. E. Prenosil and H. Pedersen. 1983. Immobilized Plant Cell Reactors. *Enzyme Microb. Technol. 5*, 323-331.

H. T. Pu, R. Y. K. Yang, and F. L. Saus. 1989. Iontophoretic Release and Transport of Alkaloids from *Catharanthus roseus* Cells in a Ceramic Hollow Fiber Reactor. *Biotechnol. Lett. 11*, 83-86.

G. A. Ravishankar, K. S. Sarma, L. V. Venkataraman, and A. K. Kadyan. 1988. Effect of Nutritional Stress on Capsaicin Production in Immobilized Cell Cultures of *Capsicum annuum*. *Curr. Sci. 57*, 381-383.

T. Rehg, C. Dorger, and P. C. Chau. 1986. Trivalent Cation Stabilization of Alginate Gel for Cell Immobilization. *Biotechnol. Lett. 8*, 115-120.

J. Renaudin and J. Guern. 1982. Compartmentation Mechanisms of Indole Alkaloids in Cell Suspension Cultures of *Catharanthus roseus*. *Physiol. Veg. 20*, 533-547.

J. P. Renaudin, S. C. Brown, and J. Guern. 1985. Compartmentation of Alkaloids in a Cell Suspension of *Catharanthus roseus*: A Reappraisal of the Role of pH Gradients. In " Primary And Secondary Metabolism of Plant Cell Cultures," K. -H. Neumann, W. Barz, and E. Reinhard (eds.),

Springer-Verlag, Berlin, pp. 124-132.

M. J. C. Rhodes. 1985. Immobilized Plant Cell Cultures. *Top. Enzyme Ferment. Biotechnol. 10*, 51-87.

M. J. C. Rhodes, J. I. Smith, and R. J. Robins. 1987. Factors Affecting the Immobilisation of Plant Cells on Reticulated Polyurethane Foam Particles. *Appl. Microbiol. Biotechnol. 26*, 28-35.

R. J. Robins and M. J. C. Rhodes. 1986. The Stimulation of Anthraquinone Production by *Cinchona ledgeriana* Cultures with Polymeric Adsorbents. *Appl. Microbiol. Biotechnol. 24*, 35-41.

R. J. Robins, A. J. Parr, S. R. Richards, and M. J. C. Rhodes. 1986. Studies of Environmental Features of Immobilised Plant Cells. In "Secondary Metabolism in Plant Cell Cultures," P. Morris, A. H. Scragg, A. Stafford, and M. W. Fowler (eds.), Cambridge University Press, Cambridge, U.K., pp. 162-172.

K. R. Robinson. 1985. The Responses of Cells to Electrical Fields: A Review. *J. Cell Biol. 101*, 2023-2027.

W. E. Rochefort, T. Rehg, and P. C. Chau. 1986. Trivalent Cation Stabilization of Alginate Gel for Cell Immobilization. *Biotechnol. Lett. 8*, 115-120.

A. Rosevear and C. A. Lambe. 1985. Immobilized Plant Cells. *Adv. Biochem. Eng. 31*, 37-58.

M. K. Ross and T. A. Thorpe. 1973. Physiological Gradients and Shoot Initiation in Tobacco Callus Culture. *Plant Cell Physiol. 14*, 473-480.

P. G. Rouxhet and M. Mozes. 1990. The Micro-Environment of Immobilized Cells: Critical Assessment of the Influence of Surfaces and Local Concentrations. In "Physiology of Immobilized Cells," J. A. M. de Bont, J. Visser, B. Mattiasson, and J. Tramper (eds.), Elsevier, Amsterdam, pp. 343-354.

C. N. Satterfield. 1970. "Mass Transfer in Heterogeneous Catalysis," MIT Press, Cambridge, Massachusetts.

H. Schnabl, R. J. Youngman, and U. Zimmermann. 1983. Maintenance of Plant Cell Membrane Integrity and Function by the Immobilisation of Protoplasts in Alginate Matrices. *Planta 158*, 392-397.

M. L. Shuler. 1981. Production of Secondary Metabolites from Plant Tissue Culture - Problems and Prospects. In "Biochemical Engineering II," A. Constantinides, W. R. Vieth, and K. Venkatasubramanian (eds.), *Ann. N.Y. Acad. Sci.* 369, 65-79.

M. L. Shuler. 1986. Plant Tissue Culture Using Membranes. In "Membrane Separations in Biotechnology," W. C. McGregor (ed.), Marcel Dekker, New York, pp. 303-317.

M. L. Shuler. 1987. Immobilization of Cells by Entrapment in Membrane Reactors. In "Immobilized Enzymes and Cells, Vol. 135," K. Mosbach (ed.), Academic Press, New York, pp. 372-386.

M. L. Shuler. 1988. Bioreactors for Plant Cell Culture. In "Plant Cell Biotechnology," M. S. S. Pais, F. Mavituna, and J. M. Novais, (eds.), NATO ASI Series, Springer-Verlag, New York, pp. 329-342.

M. L. Shuler, O. P. Sahai, and G. A. Hallsby. 1983. Entrapped Plant Cell Cultures. *Ann. N.Y. Acad. Sci. 413*, 373-382.

M. Siemann, R. Muller-Hurtig, and F. Wagner. 1990. Characterization of the Rotating Nozzle-Ring Technique for the Production of Small Spherical Biocatalysts. In "Physiology of Immobilized Cells," J. A. M. de Bont, J. Visser, B. Mattiasson, and J. Tramper (eds.), Elsevier, Amsterdam, pp. 275-282.

J. M. W. Slack. 1983. "From Egg to Embryo," Cambridge University Press, Cambridge, U.K.

J. I. Smith, N. J. Smart, M. Misawa, W. G. W. Kurz, S. G. Tallevi, and F. DiCosmo. 1987. Increased Accumulation of Indole Alkaloids by Some Cell Lines of *Catharanthus roseus* in Response to Addition of Vanadyl Sulphate. *Plant Cell Rep. 6*, 142-145.

L. Stange. 1984. Cellular Interaction during Early Differentiation. In "Encyclopedia of Plant Physiology," Vol. 17, H. F. Linskens and J. Heslop-Harrison, Springer-Verlag, Berlin, pp. 424-452.

F. C. Steward, M. O. Mapes, and K. Mears. 1958. Growth and Organized Development of Cultured Cells. II. Organization in Cultures Grown from Freely Suspended Cells. *Am. J. Bot. 45*, 705-708.

J. Subramani, P. N. Bhatt, and A. R. Mehta. 1989. Alkaloid Production by the Immobilized Cells of *Solanum xanthocarpum*. *Curr. Sci. 58*, 510-512.

I. M. Sussex and M. E. Clutter. 1967. Differentiation in Tissues, Free Cells, and Reaggregated Plant Cells. *In Vitro 3*, 3-12.

H. Tanaka, C. Hirao, H. Semba, Y. Tozawa, and S. Ohmomo. 1985. Release of Intracellularly Stored 5'-Phosphodiesterase with Preserved Plant Cell Viability. *Biotechnol. Bioeng. 27*, 890-892.

H. Tanaka, H. Semba, T. Jitsufuchi, and H. Harada. 1988. The Effect of Physical Stress on Plant Cells in Suspension Cultures. *Biotechnol. Lett. 10*, 485-490.

B. Thomasset, E. Gontier, B. S. Sangwan-Norreel, and J. N. Barbotin. 1990. Influence of Stress Conditions (i.e., Immobilization) on Growth, Viability and Alkaloid Contents of *Coffea arabica* and *Datura innoxia* cells. In "Physiology Of Immobilized Cells," J. A. M. de Bont, J. Visser, B. Mattiasson, and J. Tramper (eds.), Elsevier, Amsterdam, pp. 597-602.

J. G. Torrey. 1975. Tracheary Element Formation from Single Isolated Cells in Culture. *Physiol. Plant. 35*, 158-165.

T. Vanek, V. V. Urmantseva, Z. Wimmer, and T. Macek. 1989. Biotransformation of 2-(4-Methoxybenzyl)-1-Cyclohexanone by *Dioscorea deltoidea* Free and Immobilized Plant Cells. *Biotechnol. Lett. 11*, 243-248.

I. A. Veliky and A. Jones. 1981. Bioconversion of Gitoxigenin by Immobilized Plant Cells in a Column Bioreactor. *Biotechnol. Lett. 3*, 551-554.

H. J. Vogel and P. Brodelius. 1984. An In Vivo ^{31}P NMR Comparison of Freely Suspended and Immobilized *Catharanthus roseus* Plant Cells. *J. Biotechnol. 1*, 159-170.

K. A. Walker, M. L. Wendeln, and E. G. Jaworski. 1979. Organogenesis in Callus Tissue of *Medicago sativa*: The Temporal Separation of Induction Processes from Differentiation Processes. *Plant Sci. Lett. 16*, 23-30.

J. Warren Wilson. 1978. The Position of Regeneration Cambia: Auxin/Sucrose Ratio and the Gradient Induction Hypothesis. *Proc. R. Soc. London, Ser. B 203*, 153-176.

M. J. Watts, J. J. Galpin, and H. A. Collin. 1984. The Effect of Growth Regulators, Light and Temperature on Flavour Production in Celery Tissue Cultures. *New Phytol. 98*, 583-591.

P. B. Weisz. 1973. Diffusion and Chemical Transformation: An Interdisciplinary Excursion. *Science 179*, 433-440.

R. H. Wetmore and J. P. Rier. 1963. Experimental Induction of Vascular Tissues in Callus of Angiosperms. *Am. J. Bot. 50*, 418-430.

H. J. Wichers, T. M. Malingre, and H. J. Huizing. 1983. The Effect of Some Environmental Factors on the Production of L-DOPA by Alginate-Entrapped Cells of *Mucuna pruriens*. *Planta 158*, 482-486.

F. H. Wilbur and J. L. Riopel. 1971a. The Role of Cell Interaction in the Growth and Differentiation of *Pelargonium hortorum* Cells In Vitro. I. Cell Interaction and Growth. *Bot. Gaz. 132*, 183-193.

F. H. Wilbur and J. L. Riopel. 1971b. The Role of Cell Interaction in the Growth and Differentiation of *Pelargonium hortorum* Cells In Vitro. II. Cell Interaction and Differentiation. *Bot. Gaz. 132*, 193-202.

A. Wiseman (ed.). 1985. "Handbook of Enzyme Biotechnology," 2nd ed., Ellis Horwood, Chichester, U.K.

L. Wolpert. 1969. Positional Information and the Spatial Pattern of Cellular Differentiation. *J. Theor. Biol. 25*, 1-47.

M. M. Yeoman, K. Lindsey, M. B. Miedzybrodzka, and W. R. McLauchlan. 1982. Accumulation of Secondary Products as a Facet of Differentiation in Plant Cell and Tissue Cultures. In "Differentiation In Vitro," M. M. Yeoman and D. E. S. Truman (eds.), Cambridge University Press, Cambridge, U.K, pp. 65-82.

M. M. Yeoman, M. B. Miedzybrodzka, K. Lindsey, and W. R. McLauchlan. 1980. The Synthetic Potential of Cultured Plant Cells. In "Plant Cell Cultures," F. Sala, B. Parisi, R. Cella, and O. Ciferri (eds.), Elsevier/North-Holland Biomedical Press, Amsterdam, The Netherlands, pp. 327-343.

CHAPTER 8

ROOT CULTURES

A. INTRODUCTION

A root culture consists of living root tissues growing in an organized differentiated form in the absence of other types of plant materials such as stem, leaf, or flower tissue. Root cultures are not new and were actually the first form of propagating plant culture (Robbins 1922). Earlier objectives were those of defining the growth requirements of isolated roots for indefinite growth in culture (White 1946). Once established, root cultures permitted study of a host of physiological questions (White 1946) including those involving root nutrition and hormone physiology (Torrey 1956; Butcher and Street 1964). With regard to secondary metabolism, Dawson (1942) used root cultures to confirm the hypothesis that nicotine synthesis in tobacco occurs in the roots followed by transport to the leaves.

Interest shifted to callus and suspension cultures, however, because of their potential for rapid growth and because, as is of value in many kinds of fundamental work, cells were observed to exist in a relatively homogeneous and undifferentiated state. For secondary metabolite production, not only was rapid growth attractive, but the apparent similarities to dispersed microbial cultures suggested that techniques employed in microbial fermentations might be applied directly. Nevertheless, frequent lack of secondary biosynthesis in suspensions, coupled with productive instability, has led to a renewed interest in both root and shoot cultures. Particularly with the availability of fast-growing and productive "hairy root" lines, there now exists a focus, in both academic and industrial laboratories, on root cultures for commercial production of useful products. The scale-up and commercialization of ginseng production by Nitto Denko Company (Ushiyama 1989) offers a recent example. Atropine and scopolamine have also been considered for commercial production (T. Hashimoto, personal communication) and a number of other prospects are currently under assessment.

Although large-scale production of secondary metabolites will be primarily emphasized in this chapter, other applied as well as fundamental objectives are of interest. Uniform clonal root cultures may be used for micropropagation of useful plant strains via shoot regeneration from rapidly propagating root cultures (Philip and Nainar 1988). Fundamental applications arise in studying the enzymology of primary and secondary metabolic pathways (Siebertz et al. 1989) and in studying shoot/root relations, nodulation, nutrient (including oxygen) uptake (Butcher and Street 1964), and host/pathogen interactions (Mugnier 1988).

In this chapter, considerable emphasis is given to issues which are peculiar to root (or organ) cultures and which are perceived as being critical to achieving commercial success for secondary metabolite production. With the exception of somewhat more complex product formation kinetics (Section G.2), productivity enhancement strategies are largely the same as those encountered in working with other types of plant cell cultures.

In contrast, two very important issues which must be discussed in a root culture-specific context are: (1) the improvement of growth rates using genetic engineering and physiological strategies, and (2) growth and production at high root density on a large scale. Growth rate enhancement, employing genetic, hormonal, nutritional, and other environmental strategies, requires an understanding of root anatomy and the mechanics of root growth (Section C). Genetic engineering strategies involving *Agrobacterium rhizogenes* are discussed in Section D.2. The influence of endogenously synthesized, as well as externally applied, plant growth hormones is discussed in some detail in Section E, while other nutritional and environmental factors are briefly considered in Section F.

Little work has been reported on successful culture of roots at high density in scalable growth or production systems, and this area is obviously of vital importance. Root cultures present altogether new growth morphologies and therefore also require different approaches to large-scale culture. In Section H, the difficulties of supplying nutrients, in particular, oxygen, are analyzed and promising root culture reactor options discussed.

B. RATIONALE AND ADVANTAGES

The possible advantages of working with root cultures instead of suspensions are that (1) secondary product formation is often associated only with structural differentiation to the level of organs, (2) roots tend to show greater genetic and productive stability over time, and (3) where relevant, root cultures may be intrinsically compatible with immobilized operation (engineering advantages of which have been listed in Chapter 7). These three advantages and the potential for high root culture growth rates need to be considered in detail before an accurate assessment of the potential of root cultures can be made.

Enhanced Biosynthetic Expression Root differentiation is required for enhanced secondary biosynthesis in a number of cases, some of which are listed in Table 8.1. For example, Flores et al. (1987) have demonstrated that cultures which are cycled back and forth between growth in callus versus root culture form show significant product accumulation only in the differentiated (root culture) state. In such cases, biosynthesis or accumulation is probably associated with the presence of certain organelles or cell

Table 8.1. Examples of cases in which root differentiation is necessary for enhanced secondary metabolism.

Plant	Compound	Reference
Senecio sp.	Pyrrolizidine alkaloids	Toppel et al. (1987)
Solanaceae	Tropane alkaloids	Hashimoto and Yamada (1983)
Linum flavum	β-Peltatin	Berlin et al. (1986)
Valerianeae	Valepotriates	Violon et al. (1984)

types, which form only in structurally differentiated root tissue (e.g., see Hashimoto and Yamada 1989). Alternatively, cell- or tissue-specific expression may result from direct regulatory controls over secondary metabolic gene expression such that, *in vivo*, secondary compounds are synthesized only in a temporally coordinated, tissue- and cell-specific manner. Given the complex ecological role of many secondary compounds, the existence of adaptive organ- and tissue-specific expression should perhaps not be surprising.

Genetic engineering approaches may succeed in decoupling secondary biosynthesis from various forms of differentiation, if the metabolic routes are well known. However, in complex cases involving cellular compartmentalization and highly toxic products, direct genetic manipulation may not be feasible and organ cultures may remain the tissue culture option of choice. Commercially, it will frequently be necessary for tissue culture strategies to demonstrate short-term feasibility over competing agricultural or synthetic approaches. For new products, biosynthetic pathways may not be understood in time for genetic engineering to contribute to the establishment of tissue cultures as the supply option of choice. The use of root cultures is an effective short-term method of productivity enhancement which is compatible with subsequent use of genetically improved root lines.

Genetic and Productive Stability Unstable culture characteristics present obvious problems for commercial process operation. Instability may even preclude large-scale production altogether if productivities cannot be maintained through 20 doublings or so, as may be needed for growth from shake flask to full-scale reactor. Suspensions frequently display unstable rates of growth and product formation (Zenk 1984; Dougall 1985). In contrast, root cultures appear to offer relatively stable growth and production for at least one year (Kamada et al. 1986; Mano et al. 1989; Aird et al. 1988a; Aird et al. 1988b), and may be stable for two, three, or more years (Aird et al. 1988b; Signs and Flores 1990).

Much of the difference between suspensions and root cultures may stem from different relative stabilities at the chromosome level. Suspensions and callus cultures are well known to give rise to mixed populations consisting of polyploid and aneuploid cells (D'Amato 1985). Root cultures apparently retain diploidy in all species so far studied (Aird et al. 1988a; Aird et al. 1988b; Banerjee-Chattopadhyay et al. 1985; Hanisch ten Cate et al. 1987). Interestingly, Aird et al. (1988b) reported that abnormal chromosome numbers can be induced by culturing transformed tissues as suspensions for a period of time; upon regeneration of root cultures, resulting aneuploids were suggested to be stable. Genetic stability in root culture may not be complete, however, and over prolonged culture, changes have been observed. These include morphological and productivity changes (Aird et al.1988a) as well as sharp changes in growth rate (Robbins and Stebbins 1949; Street and McGonagle 1953). Root cultures no doubt are susceptible to mutational alterations as are suspensions, though little is known of relative mutation rates.

Compatibility with Immobilization A number of engineering advantages are associated with the use of immobilized cells and continuous flow-through (perfusion) systems. In this case, cells are maintained within the reactor for extended biosynthesis of secreted products (see Chapter 7). By extending the production phase, we minimize the importance of a slightly long growth phase (as a percent of total reactor cycle time). Slow-growing and therefore expensive plant cell biomass is, in this way, effectively "reused." Additionally, with continuous removal of an extracellular product, specific rates of production may exceed those for an intracellular product in which, for example, feedback controls limit product formation.

Root cultures may be compatible with "immobilized" cell operation because (1) cells are "self-immobilized," in that no artificial matrix is required for cell aggregation to improve catalyst retention within the reactor, and (2) roots are frequently capable of secretion (see Section G.3) as is essential for immobilized operation. Roots may also display an active metabolism at low growth rates over extended time periods because they consist of primarily differentiated tissue.

Growth Rates It has previously been suggested that organ culture growth rates are generally too slow for economical production of chemicals. With the use of "hairy root" transformed root cultures (see below) and/or the judicious use of auxin, one may observe growth rates as high as or higher than those of the corresponding suspension. Examples of reported growth rates are given in Table 8.2. Of course, such comparisons may depend very much on the relative extents of environmental optimization in the two cases, and, in this regard, much less effort has been devoted to medium optimization for root cultures.

It is perhaps not too surprising that root cultures are capable of rapid growth rates. Cell cycle times (the average time between mitoses) in the highly organized and small-celled region of the meristem average less than 10 h (average for four reports in Table I of Gould 1982). In contrast, for suspensions, divisions occur in a much less organized fashion in cells of widely varying size; cycling times average about 35 h (for 17 reports in Table I of Gould 1982). Of course, the percentage of cells which actively divide ("growth fraction") is also important in determining the specific growth rate, and suspensions might be expected to have a higher fraction of growing cells (though still often substantially less than 1.0; King 1980). Factors affecting "growth fraction" in root cultures are discussed in Appendix 1.

C. ANATOMY AND PHYSIOLOGY

C.1. Root Structure and Function

Many growth and production issues can only be understood in the context of root culture anatomy. These issues include: (1) the overall growth rate as determined by tissue-specific growth events; (2) the source, transport, and site of action of hormones; (3) the reaction/diffusion problem of nutrient (especially oxygen) supply to rootlet tissues; and (4) secondary metabolite biosynthesis, transport, accumulation, and

Table 8.2. Growth rates of root cultures.

Species	Doubling time (h)	Average[a] or maximum?	Reference
Asteraceae	24	Ave (1 wk)	Flores et al.(1988)
Hyoscyamus muticus	41	Ave	Flores et al.(1987)
Datura leichardtii	30	Max	Endo and Yamada (1985)
Bidens alba	33	Max	Norton and Towers (1986)
Tagetes patula	33	Max	Croes et al. (1989)
Calystegia sepium	45	Ave	Jung and Tepfer (1987)
Hyoscyamus muticus	50	Ave	Oksman-Caldentey et al. (1989)
Beta vulgaris	55	Ave	Hamill et al. (1986)
Hyoscyamus albus	65	Ave	Hashimoto et al. (1986)
Fedia cornucopiae	89	Ave	Charlwood et al. (1989)
Coreopsis tinctoria	91.2	Ave	Thron et al. (1989)

[a]Maximum doubling times were estimated by drawing a tangent to the batch growth curve at the maximum rate of biomass increase. Average doubling times were estimated from the reported growth index (final/initial weight) by assuming exponential growth throughout the growth period. Since average doubling times do not account for the lag and stationary phases, the maximum doubling time may be considerably lower than the average doubling time.

modification. The following serves as a brief introduction; however, the reader is referred to other sources for more complete anatomical and physiological descriptions (Cutter 1978; Salisbury and Ross 1985).

Cultured roots fairly closely resemble young intact roots (Butcher and Street 1964), of the type shown schematically in Figure 8.1. The meristem consists of dividing cells which enlarge and mature in the zone of elongation, giving rise to the various tissues of Figure 8.1. Tissues emanating from the meristem may be subdivided into (1) an outer layer consisting of the epidermis and the cortex, and (2) an inner cylinder, the stele, made up of vascular and pericycle tissues (see below). Neighboring cells are generally cytoplasmically connected via plasmodesmata. A distinct set of cell initials dividing toward the tip gives rise to the root cap.

Meristem. Within the meristem, cell division occurs in small optically dense and metabolically active cells, which are highly organized into cell files. A little understood central region of nondividing cells, known as the quiescent center, occurs in mature intact roots but is not known to occur in cultured roots, where all cells in the meristem may divide actively (Butcher and Street 1964).

Epidermis The epidermis surrounds and protects the rootlet and is primarily involved in nutrient uptake. Root hairs protrude from epidermal cells and serve to enhance nutrient uptake but are generally observed only in cultured roots above the liquid level (i.e., in a primarily gaseous environment). Ethylene may be involved in root hair formation (Section E.2).

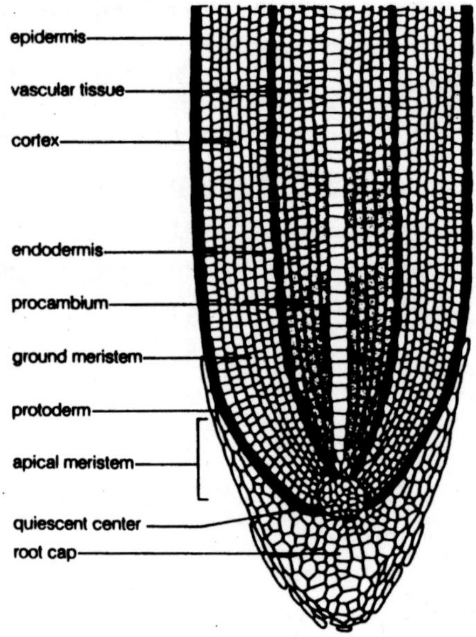

Figure 8.1. Young root anatomy in cross section. Adapted from Salisbury and Ross (1985). For many species, the stele is considerably narrower than that shown here.

Cortex The cortex is made up of large, vacuolated, thin-walled cells which are of a fairly generic type found throughout the plant, known as "parenchyma." In the intact plant, cell lysis in the cortex can give rise to a continuous system of gas spaces (aerenchyma) for rapid gas phase diffusion of oxygen down from the shoot (Jackson and Drew, 1984). Cortex lysigeny in response to ethylene (see below) may occur in cultured roots with, however, resultant spaces being filled with liquid in submerged systems.

Endodermis The endodermis is a cylindrical layer which comprises cells "mortared" together with a water-impermeable material known as suberin ("Casparian strips," shown in Figure 8.2). Free movement of water and other components to the stele is thereby restricted to a cytoplasmic (symplastic) mode in the intact plant, and an endodermis could conceivably affect movement of small polar molecules into or out of the stele in cultured roots.

Stele The stele is somewhat more compact than the cortex and consists of pericycle tissue and primary xylem and phloem. In the intact plant, xylem carries water and nutrients up from the roots while phloem brings carbohydrates and hormones down from the shoot. Vascular tissues are observed in cultured roots, though of course they do not function as they would in the intact plant (White 1936). Polar auxin transport toward the tip is thought to occur in cells around the vascular tissue (Torrey 1976), and, interestingly, auxin levels are observed to be much higher within the stele (Greenwood et al. 1973).

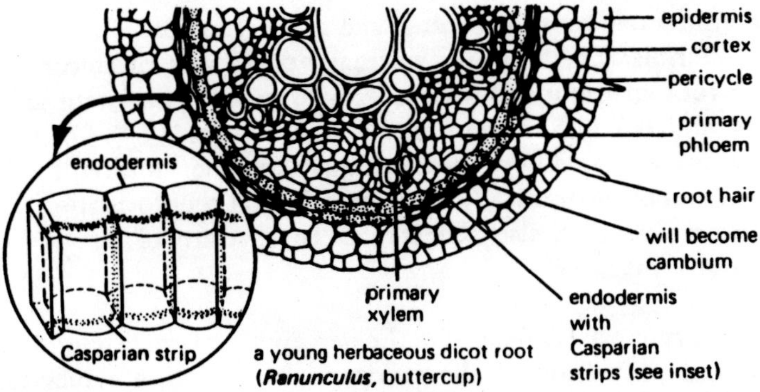

Figure 8.2. Young root cross section showing the stele and Casparian strips. Adapted from Salisbury and Ross (1985).

Most importantly for good growth, lateral roots are initiated in the endodermis or the pericycle, just inside the endodermis, as shown in Figure 8.3. Lateral roots reportedly form normally in cultured roots and are therefore not "adventitious," as is the case for shoot initiation in shoot cultures (Chapter 9).

In mature intact dicot and gymnosperm roots, two cylindrical layers of cell division, the vascular and cork cambium, develop inside the stele and result in cork

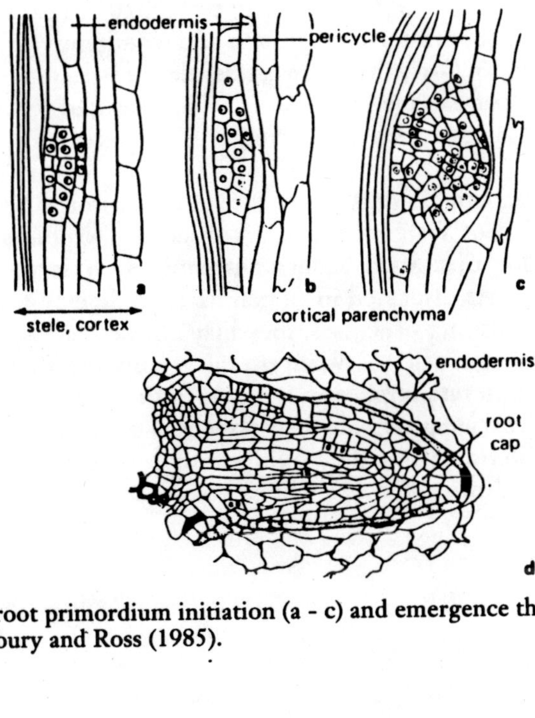

Figure 8.3. Lateral root primordium initiation (a - c) and emergence through the cortex (d). Adapted from Salisbury and Ross (1985).

formation, considerable root thickening, and a sloughing off of what was the original cortex and epidermis. This does not normally occur in root cultures, which retain typical young root anatomy and instead grow to a very consistent and narrow root diameter (Torrey 1956; White 1936; Butcher and Street 1964). Once primary tissue differentiation is completed beyond the elongation zone, cultured roots probably do not change very much anatomically, aside from lateral root formation. The immature nature of cultured roots is also reflected in the pattern of secondary metabolism (Hashimoto and Yamada 1983).

Root Cap The root cap is involved in the perception of gravity, light, and touch and provides physical protection to the meristem by secreting a lubricant known as "mucigel" (Feldman 1988) and by sloughing off at a fairly high rate, even in culture (White 1936). Mucigel consists of a complex mixture of organic components, including some secondary metabolites in some cases, and likely functions in the the intact plant in improving nutrient availability and to create a specific and favorable microbial ecology around the root (Feldman 1988). Certain secondary metabolites have been found to accumulate in the root cap in root cultures (Thron et al. 1989). It is reasonable to assume that the cap is involved in the synthesis or modification of some other secondary metabolites as well [e.g., gibberellins (Torrey 1976)].

C.2. Growth Kinetics

The growth of a root culture results from the processes of lateral root formation and rootlet elongation. Lateral root formation, in turn, may be resolved into (a) initiation of root primordia and (b) primordium growth and emergence (Wightman et al. 1980) as depicted in Figure 8.3. Growth of a tip is essentially restricted to one dimension and is linear with time (not exponential) and can be measured in units of length/time. Empirically, elongation can remain relatively constant over many days or weeks, even in batch culture.

Exponential increase in biomass is observed in root cultures (Mano et al. 1989) and results from lateral root formation and a consequent exponential increase in the number of elongating tips. Such behavior is similarly observed in fungal mycelial growth (Bader 1986). As discussed in Appendix 1, while elongation and lateral root formation represent distinct processes, they may not be independent: for example, if the distance (or number of cells) between laterals is constant (Barlow and Adam 1988), an increased elongation rate implies an increase in the lateral formation rate.

In terms of rootlet scale processes, in exponential growth, we expect meristems to elongate linearly with time while their numbers increase exponentially with time. On the scale of a large culture taken as a whole, growth appears simply exponential and "autocatalytic":

$$\frac{dx}{dt} = \mu x \qquad [8.1]$$

Such an expression should not be taken to mechanistically imply that all biomass contributes equally to the production of new biomass. Rather, as discussed in Appendix 1, μ may be shown to be proportional to the product of the elongation rate (cm/day) and the linear density of lateral roots (laterals/cm).

It is clear that equation [8.1] applies under conditions in which the elongation rate and the average distance between laterals are relatively constant. Less obviously, it can be shown that equation [8.1] applies to any chemical or morphological subcomponent, α, of the biomass such as tips, mature tissue filled with product, root cap tissue or phloem.

$$\frac{d\alpha}{dt} = \mu\alpha \qquad [8.2]$$

Intuitively, if the biomass is to retain its characteristic morphology and composition over time (as it certainly does, for example, from subculture to subculture), then chemical and morphological components must grow with the same exponential rate constant, μ. Such a concept is equivalent to the notion of "balanced" growth (Bailey and Ollis 1986), with the list of "balanced" components extended to include morphological units and tissues. The notion of balanced growth is considered in the analysis of growth and production kinetics, discussed further in Section G.2.3 and in Appendix 2.

Does the concept of balanced growth apply to real root cultures? Balanced growth should apply for suspended microbial cultures growing in a chemostat and may apply for periods of exponential growth in batch culture. For roots, exponential and balanced growth will often not be observed in batch culture because the environment tends to change rapidly relative to the rate of physiological phenomena such as the appearance of new lateral roots. However, continuous perfusion through a growing root culture is expected to minimize environmental changes and result in balanced growth over extended time periods.

A final point with regard to lateral root formation is that a certain delay will frequently be observed before root primordia develop and emerge as laterals. A time of between two and three days has been suggested by Hurren et al. (1988). Cultures without "reserve" root primordia will therefore always display a lag before exponential growth commences.

D. TYPES OF ROOT CULTURES

For the purposes of this chapter, a root culture is narrowly defined as a culture in which ongoing biomass generation occurs through the growth of organized root tissue: "roots from roots." Cell division occurs in organized root apical meristems, and new rootlets arise from root primordium initiation and emergence in a manner similar to that occurring in the intact plant.

There exist two principal types of "true" root cultures. Firstly, "untransformed" cultures (also termed "isolated," "ordinary," "excised," or "*in vitro*") are traditional root cultures which can be further classified according to whether or not an auxin growth regulator is exogenously supplied. Secondly, "transformed" or "hairy root" cultures result from the infection and subsequent genetic transformation of plant tissue by the pathogenic soil bacterium *Agrobacterium rhizogenes*. Transformation often predisposes cells to prolific root growth (Flores et al. 1987).

Other types of cultures involving roots have been reported. Hay et al. (1986) have discussed the production of quinine and quinidine from "root organ suspensions" - small clusters of differentiated rootlets. The generality of such an approach is not clear, however, and it is uncertain whether such a growth morphology could be maintained over a variety of nutritional, hormonal, and, importantly, hydrodynamic conditions. Another scheme involving the use of roots or "root-like-structures" relies on the large-scale, high-frequency redifferentiation of rootlets from suspension aggregates. This sometimes prolific emergence can produce quantities of biosynthetically active, differentiated tissue for secondary metabolite production. Biomass accumulation in this special case, however, occurs via suspension, and the relevant issues are therefore covered in the suspension and embryo culture chapters (Chapters 6 and 10).

D.1. Untransformed Root Cultures

Untransformed root cultures are obtained by inoculating a sterile root tip, millimeters to centimeters in length, into liquid medium. Inoculum-to-medium ratios must often be sufficiently high, and auxin levels (when used) relatively low, until root cultures are well established. Sterile root tips are typically obtained by excising the seedling radicle from surface-sterilized and germinated seeds. Alternatively, for species for which viable seeds are not available (e.g., wild or asexually propagating plants), shoot apical meristems can be excised, grown on solid medium, and induced to root (Borgman and Mudge 1986; Prince 1991). Occasionally, roots may be induced to form more or less directly from surface-sterilized explants (Norton and Towers 1986). Finally, redifferentiated rootlets from callus or suspension aggregates can be used to initiate a root culture proper.

Experimental work is normally carried out on a small scale in liquid medium. One frequently encounters considerable flask-to-flask variability in growth and productivity (Butcher and Street 1964); high numbers of replicates - in the range of 5 to 10 flasks or more - are sometimes necessary for statistically meaningful comparisons between treatments.

D.2. *Agrobacterium rhizogenes* and Hairy Roots

Hairy root disease is a rooty tumorous condition associated with infection by the pathogenic soil bacterium *Agrobacterium rhizogenes*. This species is closely related to *A. tumefaciens*, which similarly induces tumor formation ("crown gall" in this case).

The capacity for tumor induction is encoded by large virulence plasmids in each: those of *A. rhizogenes* strains are referred to as Ri for "root inducing" and those of *A. tumefaciens*, Ti, for "tumor inducing." On infection, a portion of the plasmid, the T-DNA, is more or less randomly integrated into the host plant genome (Ambros et al. 1986; Weissing et al. 1988). The process is thought to resemble bacterial conjugation, and details have been reviewed (Zambryski et al. 1989).

Confusion may arise over the apparently large variety of bacteria and plasmids used to induce hairy root formation. Though different bacterial strains have been identified, of primary interest is the type of Ri plasmid originating from them. There have been very few actual isolates of wild *A. rhizogenes*, and most of those studied are probably the same or are derived from either the A^4 type (probably the same as ATCC 15834, 1855, TR105; White and Sinkar 1987) or the 8196 type (TR7 and TR101 may be same isolate; White and Sinkar 1987). Type A^4, originally isolated from a rose in California, is strongly virulent on a variety of plants, and, on the basis of the opines (see below) synthesized, its Ri plasmid is referred to as "agropine type." Type 8196 was originally isolated from apple in California and, because of its weaker virulence, has somewhat less frequently been used for root culture initiation (Jung and Tepfer 1987; Knopp et al. 1988; White and Sinkar 1987). The 8196 plasmid is commonly referred to as "mannopine type," stemming from its inability to synthesize agropine. A third type, the "cucumopine type," has been described, and the T-DNA of this type shows substantial homology with that of mannopine and agropine types (Filetici et al. 1987; Peterson et al. 1989).

T-DNA contains a variety of genes, most of which are involved in tumor formation or in the synthesis of opines. Opines are unusual amino acids or sugars which are secreted by the plant tissue and utilized by the bacterium possessing genes for opine catabolism (Petit et al. 1983).

Several genes are involved in root formation. First, in agropine (A^4) type plasmids only, T-DNA contains genes for synthesis of the auxin indole-3-acetic acid (IAA) (Cardarelli et al. 1985). Given the important role of auxin in root formation (Section E.1), it is not surprising that the IAA synthase genes play a significant role in determining tumor (root) growth. The effects of transformation, however, are not explainable purely on the basis of elevated auxin levels in transformed tissues (Cardarelli et al. 1987; Spano et al. 1988), and, for example, mannopine and cucumopine strains are capable of inducing the hairy root phenotype in the absence of IAA synthase genes.

Importantly, a group of seven T-DNA genes are now being studied which, aside from IAA synthase genes, determine the root growth and morphological characteristics of hairy roots. These include the "rol" genes A, B, C, and D (Schmulling et al. 1988; Spena et al. 1987) as well as open reading frames (ORF) 8, 13, and 14 (Capone et al. 1989). Their functions are reportedly synergistic in stimulating root formation and operate via distinct mechanisms which are, as yet, not understood (Schmulling et al. 1988, 1989; Capone et al. 1989). With the possible exception of rol D (McInnes et al. 1989), metabolic effects of such genes are thought to be restricted only to cells in which the genes are actually expressed; effects are not interpretable simply in terms of the synthesis of diffusable hormones such as IAA or isopentenyl adenosine (IPA) (Schmulling et al. 1988).

Transformation with a desirable combination of T-DNA genes with promoters of desired strength (Schmulling et al. 1988) and cell-specific patterns of expression (Schmulling et al. 1989; Matsuki et al. 1989) could lead to intriguing possibilities for enhancing root culture growth rates. In this connection, it has been shown that genes which are shown to be important for root formation are not necessarily the same as genes which are useful for increasing root culture growth rates (Schmulling et al. 1988). The considerable fundamental and applied work now under way should soon lead to a much clearer understanding of T-DNA functions.

In contrast to the mannopine and cucumopine type T-DNAs, which consist of a single piece (Peterson et al. 1989), the T-DNA of the agropine type consists of two nonhomologous portions: TL-DNA and TR-DNA (for left and right, respectively). Schematics of agropine and mannopine plasmids are shown in Figure 8.4. The TR-DNA contains the IAA synthase genes, as well as genes for agropine synthesis. Rol genes are located on the TL-DNA. Mannopine and cucumopine type T-DNA, though not exactly the same, do display considerable homology with agropine type TL-DNA (Filetici et al. 1987). Moreover, mannopine and cucumopine type T-DNAs do not contain auxin synthase genes and rely instead on endogenous plant IAA levels for development of the hairy root phenotype. Transformants of agropine strains lacking auxin synthetic functions (via insertional mutagenesis) closely resemble mannopine type transformants (White and Sinkar 1987).

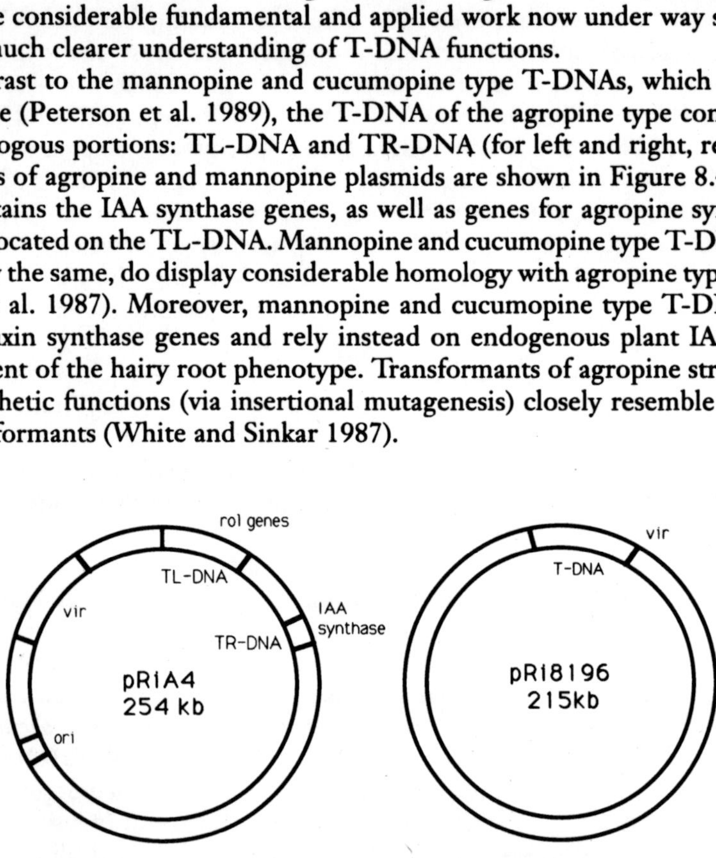

Figure 8.4. Simplified schematics of agropine (pRiA4) and mannopine (pRi8196) type Ri plasmids.

Agrobacterium-mediated transformation may be used for more than hairy root induction. The primary objective of *Agrobacterium*-related research has been to improve methods of inserting foreign genes into plants which, upon regeneration, display otherwise normal morphology. A variety of plasmid vectors are now available for genetic engineering, in which some or all of the tumor-inducing genes have been rendered inactive, or been eliminated, and various selectable markers incorporated (Weissing et al. 1988). For secondary metabolite production, the potential now exists for transforming plant tissue with T-DNA still containing genes for good root growth (e.g., specific TL-DNA genes), and which may further include genes coding for products which enhance or direct secondary metabolic pathways. One may envision, for example, elevating the enzyme levels of key rate-limiting steps through transformation with the relevant gene coupled to an appropriately expressive promoter.

The high degree of variability among transformed root lines (Mano et al. 1989), can be attributed, in part, to the manner of T-DNA integration. T-DNA border regions have been shown to be quite variable (Jouanin et al. 1987; Byrne et al. 1983; Weissing et al. 1988). Copy number is also variable and, for example, TL-DNA may occur independently of, or at different copy number than, TR-DNA (Jouanin et al. 1987). Expression of T-DNA sequences may be affected by position within the host genome, and, of course, plant sequences may be modified or inactivated during integration (Weissing et al. 1988). Based on such considerations, it is perhaps not surprising that considerable variability is observed when screening for useful transformed root lines. Much effort may be required, in some cases, to obtain productive and rapidly growing hairy roots (Mano et al. 1989).

In general, the host range for *Agrobacterium rhizogenes* is not well known (De Cleene and De Ley 1981; Mugnier 1988). Ri vir genes lying outside of the T-DNA regions, as well as genes on the bacterial chromosome, can be important in determining pathogenicity. The use of different plasmids and bacteria can therefore be important, even if the T-DNAs are identical. Generally, monocots have not been susceptible; however, successes have been reported with *Allium cepa* (Dommisse et al. 1990) and with maize and asparagus using *A. tumefaciens* (Weissing et al. 1988). It would appear that direct gene transfer methods will be successful in transforming monocots (Tepfer 1989), as has recently been demonstrated for the rol C gene in rice using electroporation (Matsuki et al. 1989). Difficulties may be encountered in achieving a good growth response with woody species (Hamill et al. 1989); however, this may be related to the difficulties of obtaining root cultures generally for woody species (Torrey 1988).

E. PHYSIOLOGY: HORMONES AND GROWTH REGULATORS

E.1. Auxin

A knowledge of the effects of auxin, with the possible involvement of ethylene, can be important for understanding and controlling root growth and development. The endogenous auxin levels of many cultured roots are suboptimal for maximum growth rates (Street 1957; Butcher and Street 1964; Mitra 1968), and the same has recently been demonstrated for a number of transformed lines as well (Yoshikawa and Furuya 1989; Mano et al. 1989; Knopp et al. 1988). Moreover, many monocot cultures cannot normally be established at all without exogenous addition of an auxin (Butcher and Street 1964). Though transformation with various TL-DNA genes may lessen the need for auxin, it is still unclear whether, in a given case, the strategy of choice will be (1) the use of untransformed root cultures with exogenous addition of auxin, or (2) the use of root cultures transformed with various complements of TL-DNA and, if necessary, with endogenous IAA synthesis (IAA synthase genes) or with exogenous auxin addition.

The following discussion is intended as an introduction to some of the applied issues in the auxin physiology of root cultures. Responses of root cultures to auxin are discussed. Consideration is then given to the relative merits of different modes of auxin addition, including endogenous synthesis. Auxin effects on secondary biosynthesis are further discussed in Section G.4.3.

E.1.1. Auxins of Interest Several auxins have been used in root culture work. Indolebutyric acid (IBA) has been repeatedly shown to be effective for use with root cultures, though others, for example, 4-chloro-IAA or 5,6-dichloro-IAA, may be even more effective (Prince 1991). Different auxins no doubt possess different relative affinities for each of the auxin transport, receptor, and auxin-conjugating enzyme sites as well as different membrane permeabilities and rates of degradation. Such differences may well give rise to the exploitable diversity in auxin pharmacology.

E.1.2. Responses to Auxin For root cultures, the three most important classes of response to auxin are: (1) the effects on elongation rate, (2) the stimulation of lateral root formation, and (3) the auxin-stimulated release of ethylene. The effects on elongation and lateral root formation are summarized schematically in Figure 8.5, while the effects of ethylene are discussed further in Section E.2.

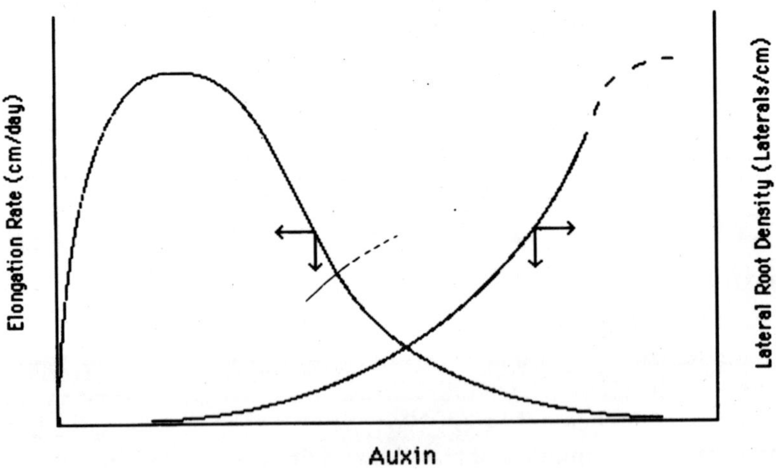

Figure 8.5. Schematic suggesting, qualitatively, the effect of auxin on root elongation rate and on lateral root density along a root axis. In particular instances, curves may differ from those shown, especially for transformed root lines.

It is generally acknowledged that auxin at low concentrations (probably less than 10^{-6} M endogenously) stimulates rootlet elongation (Salisbury and Ross 1985). At higher concentrations, elongation is inhibited (Mulkey et al. 1982; Salisbury and Ross 1985) as outlined in Figure 8.5.

The effects of ethylene on elongation (Section E.2) qualitatively resemble those of auxin, and because auxin stimulates ethylene formation, ambiguities often arise as to which hormone is fundamentally involved. Despite this, auxin's stimulatory and inhibitory effects are authentic and distinguishable from those of ethylene based on response and recovery times and the addition of ethylene inhibitors (Andreae et al. 1968; Rauser and Horton 1975; Torrey 1976; Mulkey et al. 1982).

In Section C.2, lateral root formation was stated to consist of primordium initiation followed by lateral root emergence; auxin may have independent effects on each process. At higher concentrations, auxin stimulates the initiation of lateral root primordia (Torrey 1976; Hurren et al. 1988). The total number of emerged laterals also increases, depending on the auxin employed (Wightman et al. 1980; Hashimoto et al. 1986). For this effect, the precise involvement of auxin, auxin-stimulated ethylene, or other factors is not understood at this time. Empirically, it has been observed in many systems that the linear density of laterals increases monotonically with increasing auxin levels as shown in Figure 8.5 (Mitra 1968; Hashimoto et al. 1986; Hurren et al. 1988; Croes et al. 1989).

As outlined in Appendix 1, during balanced exponential growth, the specific growth rate (μ) is approximately proportional to the product of the elongation rate and the lateral root linear density. The specific growth rate is therefore approximately represented by the product of the elongation rate and lateral density curves in Figure 8.5. Evidently an optimum in auxin concentration results from a compromise between stimulation of lateral root formation and inhibition of root elongation.

As discussed in Section D.2, several options will likely be available for transforming tissue with various complements of TL-DNA. Auxin may be used in conjunction; however, little or nothing is known of how the curves of Figure 8.5 will differ for transformed cultures. The preferred result would be for the optima for lateral root formation and elongation to move closer together such that the product of the two curves gives rise to a higher maximum growth rate. The work of Croes et al. (1989), however, does not support this preferred scenario.

E.1.3. Mode of Auxin Addition Whether due to normal endogenous auxin biosynthesis or to T-DNA IAA synthase genes, it seems reasonable to assume that endogenous biosynthesis gives rise to fairly steady free auxin levels. In contrast, quite a different pattern is likely for exogenous addition when, as in virtually all studies reported, auxin is added in one large "spike" at the start of each batch. It has been inadequately appreciated that auxin is very rapidly taken up (Prince 1991) and that large transients can be associated with "spike-wise" addition. "Spike-wise" feeding schemes likely result in swings from supra-optimal (inhibitory and ethylene-releasing) levels initially, to sub-optimal levels later in batch (Figure 8.6)—the initial spike

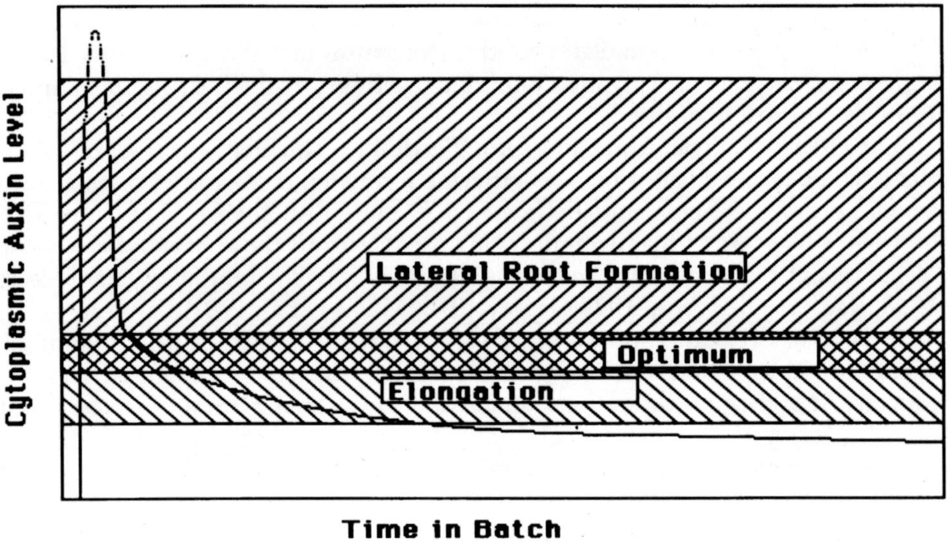

Figure 8.6. Schematic suggesting a possible scenario for the change in cytoplasmic auxin level with "spike-wise" auxin addition.

having been optimized as an unhappy compromise between the two. It is perhaps of little wonder that transformed root cultures, with TR-DNA synthase genes and increased auxin "sensitivity," have displayed superior growth rates to untransformed root cultures to which either auxin is added in "spike-wise" fashion or no auxin is added at all.

In what time-dependent manner should auxin be added? Steady fed batch addition seems a logical starting point. For large-scale root culture growth, external feeding of a concentrated auxin stream represents a trivial process elaboration. Operating in this mode is more difficult for shake-flask work; however, methods for chemical or physical slow release are possible (Reinecke and Bandurski 1987). In our laboratory, preliminary experiments with untransformed onion root cultures show that by adding IBA at 4 μM every 3 days instead of 30 μM every 14 days, the lag phase can be essentially eliminated and high sustainable growth rates achieved (Prince 1991).

More speculatively, given the level of control afforded by exogenous addition, it might even be possible to improve feeding programs relative to constant addition. For example, short elevations of auxin to enhance lateral root formation followed by longer periods of lower levels favoring elongation may be preferable to a single constant intracellular level. Gradients in the feed rate might also be programmed to account for the increase in root biomass. It is not clear, however, that in all cases an external source of auxin will offer the same growth-promoting effect as endogenous biosynthesis.

In addition to the possibility of time-varying rates of addition, other potential advantages of exogenous auxin addition may be cited:

1. Instead of relying strictly on the natural auxin, IAA, a wide selection of other growth regulators is available. Many of these perform much better than IAA when added exogenously (Wightman et al. 1980; Hashimoto et al. 1986). For food products, however, endogenous biosynthesis of the the natural auxin, IAA, may be preferable from a consumer/regulatory standpoint.
2. In immobilized root production reactors, simple hormone removal may help to slow growth.
3. In cases where auxin is truly shown to inhibit secondary biosynthesis, auxin removal from the external medium in production phase might help to reduce intracellular levels and improve productivity.
4. Transformation will seldom give rise to precisely optimal intracellular IAA concentrations, and therefore controlled external addition may prove beneficial.

E.2. Ethylene

The following responses to ethylene, potentially relevant to culturing roots, have been described: (1) concentration-dependent stimulation or inhibition of elongation (Konings and Jackson 1979), (2) effects on lateral root formation (Feldman 1984, Mudge 1988), (3) rootlet thickening (Whalen and Feldman 1988; Salisbury and Ross 1985), (4) cell lysis in the cortex (aerenchyma formation, Section C.1; Jackson and Drew 1984), (5) root hair formation (Reid 1987). Ethylene may also indirectly affect matters through a depression of auxin transport to the tip (Abeles 1973), or possibly through a reduction in auxin degradation (Moncousin et al. 1989).

E.2.1. Regulation of Ethylene Levels The relative rates of input and output govern the endogenous levels of ethylene within a root or mass of roots. Since ethylene has not been shown to be degraded within the plant, the sink for ethylene consists of gas or liquid phase diffusive and convective mass transfer away from root tissues. With regard to the ethylene source, roots endogenously synthesize ethylene at variable and species- or strain-specific rates of 1–30 nL/g fresh wt. per h (Feldman 1984; Abeles and Wydoski 1987).

Several factors have been shown to stimulate ethylene synthesis. Auxin above 10^{-6} M (for IAA) accelerates ethylene production in proportion to the amount added (Chadwick and Burg 1970). Wounding, or stress generally, is well known to stimulate ethylene formation (Abeles 1973). In relation to mechanical stresses, impeding root penetration or simply applying slight pressure to the root cap can result in manyfold increased rates of ethylene production (Feldman 1984). Growth depression as a consequence of root collisions with glass spheres has also been reported (Jackson and Gross 1978). Of relevance to reactor operation, ethylene levels would be expected to rise under conditions of high auxin concentration, under conditions of wounding or mechanical stress, and when poor liquid phase mass transfer restricts ethylene removal.

E.2.2. Responses to Ethylene

E.2.2.1. Effects on Elongation. The responses of root elongation to ethylene may be concentration dependent. Ethylene stimulates extension at low levels below about 0.1 ppm, depending on the rate of endogenous synthesis (Konings and Jackson 1978). Konings and Jackson (1978) also demonstrated a correlation between elongation rate and endogenous ethylene production rates though, of course, a causal relationship cannot be assumed. At higher concentrations, ethylene is inhibitory to root elongation (Konings and Jackson 1978). Importantly, though inhibition from short-term exposure to higher levels of ethylene may be reversible (Rauser and Horton 1975), longer term exposure may require long recovery periods or even be irreversible (Andreae et al. 1968; Chadwick and Burg 1970). Ethylene synthesized in response to spike-wise addition of auxin will certainly result in short-term depression of elongation and might result in prolonged growth suppression of emerged tips.

E.2.2.2. Effects on Lateral Root Formation. The effects of ethylene on lateral root initiation and emergence have been little studied and, as for adventitious root formation (Mudge 1988), both stimulatory and inhibitory responses have been reported (Zobel 1973; Graham and Linderman 1981; Radin and Loomis1969; Batten and Mullins 1978; Torrey 1976). It may well be that auxin is primarily responsible for root primordium initiation (Torrey 1976) whereas ethylene may facilitate lateral emergence through hydrolytic degradation of cortex ahead of the emerging rootlet (Linkins et al. 1973). From a practical standpoint, exogenous ethylene may well stimulate lateral formation in particular cases.

E.2.2.3. Effects on Radial Expansion. The commonly observed increase in root diameter in response to high auxin concentrations is likely due to ethylene. Exogenous ethylene induces radial cell enlargement primarily in the cortex (Whalen and Feldman 1988; Salisbury and Ross 1985). Thicker roots, while perhaps affording mechanical strength, also worsen oxygen mass transfer into the root and thereby place greater constraints on reactor design. In any case, such an effect probably occurs only when ethylene levels are inhibitory to elongation and may therefore not arise in an optimized system.

E.3. Other Hormones: Cytokinin and Gibberellic Acid

Compared to auxin or ethylene, the effects of cytokinin and gibberellic acid (GA) treatment appear to be less general and more system specific. Cytokinins are thought to be endogenously synthesized and required for cell division in root cultures (Torrey 1976). At 10^{-6} M or above, added cytokinins often result in shorter, thicker roots (Wightman et al. 1980; Norton and Towers 1986) and ultimately slower growth rates. Though cytokinins are generally quite inhibitory to lateral root formation, at very low concentrations (10^{-8} to 10^{-6} M) lateral formation may actually be stimulated (Wightman et al. 1980). Yoshikawa and Furuya (1987), for example, found kinetin (0.1 ppm) to be stimulatory to growth and saponin production for both transformed and untransformed root cultures.

Endogenous GA is thought to be required for growth (Torrey 1976). It has been shown to be stimulatory to root elongation, in some cases, apparently depending on the carbohydrate level (Butcher and Street 1964) and whether cultures are grown in the dark or light (Mitra 1968). GA may not generally play a role in lateral root formation (Torrey 1976; Wightman et al. 1980). Ohkawa et al. (1989) have recently reported, however, a 50% increase in growth rate, including a 40% increase in lateral root formation, with GA addition (0.1 mg/L) to transformed root cultures of *Datura innoxia*. Sugimoto et al. (1988) also reported growth improvements with GA addition to untransformed root cultures of *Stephania cepharantha*. Both GA and cytokinin likely act via interactions with other hormones (Torrey 1976).

F. OTHER MEDIUM AND ENVIRONMENTAL FACTORS

F.1. Magnetic and Electrical Effects on Growth

Magnetic and electrical effects on root growth have been little studied but are of potential interest for the external control and enhancement of root culture growth rates. Kato et al. (1989), working with transformed root cultures, have recently reported increases in root elongation rates under both increased and decreased magnetic field strengths relative to controls under Earth's magnetic field. Berry et al. (1947) observed an impressive 35- to 40-fold increase in lateral root formation in intact onion roots with applied electric fields of 15 mV/mm. Some inhibition of elongation might be expected, however, and media of lower ionic strength (conductivity) are required. It is not known whether such effects are mediated by hormone, for example, auxin, redistribution (Berry et al. 1947) or by some more fundamental reorientation of planes of cell division (Goldsworthy and Rathore 1985).

F.2. Other Nutritional and Environmental Factors

Root cultures, like callus and suspension cultures, grow on simple nutrient media. Standard plant cell media have been used effectively and serve as a good starting point in medium refinement for specific root cultures. Little work has been reported, however, on optimizing medium formulations specifically for root cultures.

Mano et al. (1989) examined the effects of 10 different basal media on growth and alkaloid production in *Duboisia leichhardtii* hairy root cultures. B5, SH, or LS basal media were found to induce callus formation whereas HF or HF-2 medium promoted good growth and alkaloid formation without callus formation. Additionally, it was shown that biomass yields could be improved about 3-fold if the inoculum was increased and the medium doubled in concentration (2XHF). Parr et al. (1988) found that *Catharanthus roseus* transformed root cultures grew only in half-strength B5 medium. Taya et al. (1989b) enhanced growth rates of transformed cultures through polypeptone addition at 1g/L. Hamill et al. (1989), in attempting to improve the growth rates of transformed *Cinchona* root cultures, could not identify an obvious nutritional deficiency.

One fairly general conclusion that can be drawn, with regard to carbohydrate source, is that dicotyledonous species prefer sucrose whereas monocots tend to grow better with glucose (Butcher and Street 1964; Burstrom 1941). Other carbohydrates have not been as effective (Butcher and Street 1964).

Temperature optima of about 26 to 30° C have generally been reported (Street 1957; Norton and Towers 1986; Burstrom 1941), though the drop-off may be somewhat steep above the optimum (Street 1957). Differences in optima for lateral root formation and elongation have been noted (Burstrom 1941).

Little generality emerges from reports on the effect of light on root culture growth rates: both stimulation (Flores et al. 1987) and inhibition (Street 1957; Norton and Towers 1986) have been reported. Effects on secondary metabolism may be more significant in particular cases; however, light will be somewhat difficult to supply to immobile root cultures on scale-up.

F.3. Oxygen and Anaerobic Metabolism

Oxygen is well known to be essential for the growth and prolonged viability of higher plants (Crawford 1982; Armstrong 1979). For example, anoxic periods of as little as 30 minutes can result in the death of soybean and cotton tap roots (Armstrong 1979). As outlined in Chapter 5, problems arise from the fact that the solubility of oxygen is quite low (ca. 8 mg/L) in the liquid medium surrounding submerged roots. Further, submerged roots probably do not contain gas spaces or aerenchyma, which normally serve, in the intact plant, to enhance internal oxygen diffusion and to reduce volumetric uptake rates (Armstrong 1979).

As shown in Figure 8.7, oxygen uptake rates are highest in the densely packed and metabolically active cells in the apex. Not surprisingly, root tips display the greatest sensitivity to oxygen deficits (Armstrong and Beckett 1987; Thomson et al. 1989a). The stele, important for lateral root generation, is also susceptible since it consists of more densely packed cells located along the central axis of the root.

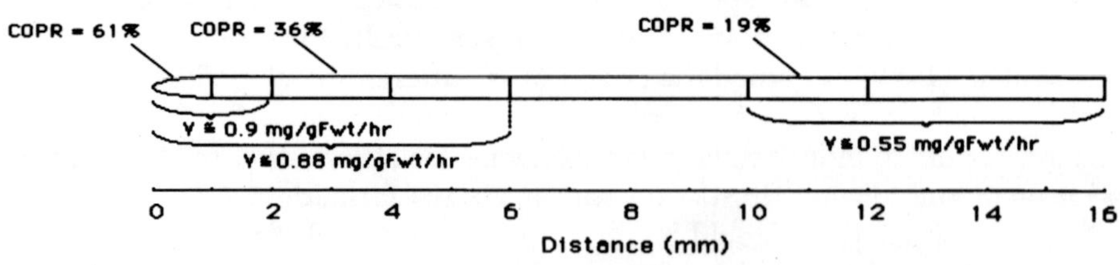

Figure 8.7. Variation along a rootlet of oxygen uptake rate and critical oxygen pressure for respiration (COPR). Data is for-7-day-old wheat seedling root tips. Data from Thomson et al. (1989a).

As the oxygen level drops, a critical oxygen partial pressure, COP, is reached at which oxygen deprivation limits various physiological phenomena. The intrinsic K_m (1/2 COP) for cytochrome oxidase-mediated respiration is probably quite low—less than about 0.01 mg/L (Jackson and Drew 1984). However, as discussed in Chapter 4, much higher oxygen concentrations are required on the surface of a root to drive radial oxygen diffusion into central regions. Moreover, other "alternative" oxidases, which may play important roles in biosynthesis and growth, can have much lower oxygen affinities (i.e., higher K_ms; Drew and Lynch 1980; Saglio et al. 1984; Thomson et al. 1989a). Observed COPs for respiration (COPRs) for roots are therefore often over 50% of air saturation and can be over 100% (Luxmoore et al. 1970; Saglio et al. 1984). Of course, an external mass transfer limitation around the root is often a factor (Armstrong 1979) and will only make matters worse (Section H). COPRs vary along the root (Figure 8.7) as is consistent with the notion that mass transfer limits uptake. The critical oxygen pressure for growth by elongation (COPE) is more difficult to determine than COPR and has been less well studied. To the extent that the functioning of alternative oxidases is important for growth, and because physiologically important regions generally possess the lowest oxygen levels, one might expect COPE to be greater than COPR (Saglio et al. 1984).

Thomson et al. (1989a, b) have described events following oxygen depression below COP in seedling wheat roots. Anaerobic metabolism is first observed in the tip. Ethanol, alanine, and lactate are accumulated in the first hour and, following a drop in cytoplasmic pH, only ethanol and alanine. Ethanol is the predominant species accumulated, but alanine production is significant and may be a more sensitive indicator of oxygen deficits. As anaerobiosis proceeds, ATP/ADP (Saglio et al. 1984) and K^+/Na^+ ratios drop (Thomson et al. 1989b) and growth is inhibited. Though anaerobic metabolism proceeds actively in higher plant roots, it probably serves only as a temporary means of maintaining viability (Drew and Lynch 1980). Recovery depends on the duration and extent of oxygen deprivation (Armstrong 1979); extended anoxia is probably fatal to all higher plants (Crawford 1982).

G. PRODUCT FORMATION IN ROOT CULTURES

The following section summarizes issues pertaining to secondary product formation using root cultures. First, differences in product levels between cultured and intact roots are considered in terms of more general differences between cultured and intact roots. Product formation kinetics in root cultures are somewhat more complex than for suspensions and, having received little previous attention, are discussed in some detail. Finally, prospects for product secretion are assessed and other productivity enhancement issues summarized.

G.1. Product Content of Cultured Versus Intact Roots

The comparison between root culture and plant product profiles may offer some suggestion of their relative capacities for biosynthesis. Parr and Hamill (1988) have suggested that root cultures in general resemble their intact counterparts reasonably well, in terms of both total and relative amounts of different compounds (compound profile). Similarity of product profile may be of particular desirability when producing natural flavors and aromas.

Despite gross similarities, it may not be correct to assume that a root culture will always produce the spectrum of compounds typical of the intact root. A variety of factors may be involved:

1. Root cultures are known to more closely resemble immature intact roots anatomically (Butcher and Street 1964), and their secondary biosynthetic pattern may reflect this as well (Hashimoto and Yamada 1983; Endo and Yamada 1985).
2. Lack of transport of secondary products from root to shoot (*Nicotiana:* Parr and Hamill 1987; *Tagetes*: Croes et al. 1989) or from the shoot down to the roots (Saito et al. 1989) could be another factor.
3. The history of the field-grown plant tissue, in response to changes in a variety of environmental factors over time, could play a role in influencing slow enzymatic or nonenzymatic transformations.
4. The biotic and chemical environment of the root culture obviously differs considerably from the complex ecology of the rhizosphere, and, in this regard, it is interesting to observe that alkaloid levels of aseptically grown plants also differ from those of soil grown plants (Kamada et al.1986; Parr and Hamill 1987).
5. Clone-to-clone variability in production can be substantial (Mano et al. 1989), and, moreover, reported concentrations are often for selected or otherwise manipulated cultures.
6. Product concentrations and profiles can also change dramatically over the course of a batch as a consequence of kinetic transients as discussed in more detail below.

G.2. Kinetics of Product Accumulation

An understanding of the kinetics of production, especially in relation to the kinetics of growth, is critical to (a) the interpretation and design of experiments and (b) selecting and designing a production process strategy. This section will first describe the variety of observed production kinetics in batch (Section G.2.1) and the spatial localization of sites of synthesis and accumulation within a rootlet (Section G.2.2). Section G.2.3 will then outline a simple conceptual model for understanding patterns of accumulation. Finally, other issues associated with understanding root culture kinetics will be mentioned (Section G.2.4).

G.2.1. Observed Overall Batch Accumulation Patterns On the scale of the culture as a whole, several different patterns of total product accumulation, relative to growth, are observed. The most common pattern is one of peak biosynthesis in late log or early stationary phase (Hashimoto et al. 1986; Mano et al. 1989; Norton and Towers 1986), biosynthesis coinciding roughly with a slowing of the growth rate (Figure 8.8). A similar pattern has been observed for thiophene accumulation in intact plant roots (Croes et al. 1989). Endo and Yamada (1985) observed biosynthesis once growth had begun, with the peak in specific productivity occurring in late log phase in some cases but as early as mid log phase for a culture which also possessed the slowest log phase growth rate. Other cases have been reported in which production fairly closely mirrored growth (Hashimoto and Yamada 1983; Violon et al. 1984); however, this pattern may be typical of more slowly growing cultures (see discussion below). An exception is the very close correspondence between phenylpropanoid accumulation and growth in reasonably rapidly growing transformed roots of *Coreopsis tinctoria* (Thron et al. 1989). Generally speaking, transitions in productivity do not appear to closely coincide with transitions in growth rate in batch culture.

G.2.2. Localization of Sites of Synthesis and of Accumulation An issue important to understanding patterns of synthesis and accumulation pertains to the actual site of synthesis and whether products are stored in the same cells or transported elsewhere for storage. For beet roots, pigments accumulate in more mature regions further from the tip (Hamill et al. 1986). Norton and Towers (1986) found higher polyacetylene levels in root tips of *Bidens alba*. Croes et al. (1989) likewise found higher levels of thiophenes in root tips of *Tagetes patula* root cultures. Intact plant roots, however, had lower levels in the tips than in more basal regions, suggesting that translocation can take place following synthesis (Croes et al. 1989). Recently, Hashimoto and Yamada (1989) have localized a key secondary metabolic enzyme, which converts hyoscyamine to scopolamine, within the phloem of root tissue. Clearly, spatial distribution of secondary product concentrations cannot, in general, be taken to reflect the localization of biosynthetic reactions.

G.2.3. A Simple Conceptual Model As shown in Figure 8.8, root culture growth and production curves resemble those generated by suspension cultures. The similarity is deceptive, however, because in root cultures an integrated combination of many different cell types gives rise to observed culture-scale kinetic behavior. Both the apparent non-growth-associated nature of biosynthesis and the influence of growth rate on the kinetic behavior may be understood on the basis of reasonable mechanistic assumptions about growth, biosynthesis and accumulation on the rootlet scale. The following qualitative discussion serves as an aid to intuitively understanding kinetic behavior and is outlined in more detail in Appendix 2. It will be shown that plots of batch growth and total product versus time are largely unsuitable for describing production kinetics, because they generally fail to distinguish between completely different kinetic patterns.

Looking at a single rootlet, new cells are produced by cell division in the tip while older cells move further and further from the tip as they age. As is observed for beet roots accumulating betacyanin pigments (Hamill et al. 1986), it may be that in many cases only mature cells further from the tip are capable of product biosynthesis or accumulation. A certain period of time will therefore be required for cells, produced in the tip, to mature sufficiently to achieve competence for product accumulation. A

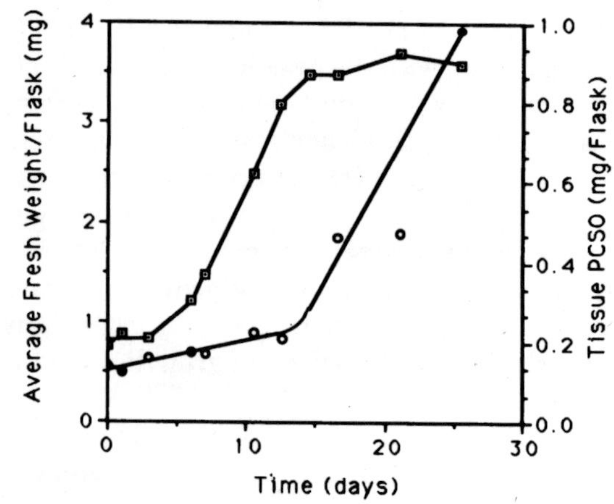

Figure 8.8. Typical batch culture growth (□) and product (○) accumulation curves (PCSO is propenyl cysteine sulfoxide). Data are for untransformed *Alliam cepa* (onion) root cultures with exogenous IBA addition (Prince 1991).

further period of time will be required for competent cells to accumulate product. Thus, subsequent to cell formation in the tip, a certain overall lag will be needed before such new cells accumulate product to some "final" level.

G.2.3.1. Case I: Apparent Non-Growth-Associated Product Formation We may consider what will happen, on the scale of the culture as a whole, when a rapid transition from the lag phase (in growth) to exponential growth occurs. New root tissue will be formed in the root tips with very little product being formed until the required lag for cell maturation and product "filling" has elapsed. During this period, the average amount of product per unit biomass will fall, and this may be termed product "dilution" (Section G.2.4). After a period of time corresponding to the rootlet level lag in production, product will begin to appear in a way that will precisely parallel the growth curve. This is shown in Figure 8.9, where the production curve is offset from the growth curve by the lag time.

If exponential growth is extended, for example, by perfusion with fresh medium, the growth and production curves are expected to extend in parallel indefinitely. The specific product content per unit biomass will remain constant as expected for balanced exponential growth (Section C.2). At the onset of stationary phase, when cells stop growing, the remaining pool of competent yet unfilled cells continues to accumulate product. Product formation then ceases after the characteristic lag time following the stoppage in growth. The specific product level should rise to its original level as shown in Figure 8.9.

Importantly, the lag is determined by intrinsic physiological factors which will differ for different root cultures and for different compounds. This may help to explain the considerable variety of kinetic behavior observed, from growth-associated to non-growth-associated, depending on the length of the lag. As well, slower exponential growth rates result in less noticeable or imperceptible deviations of production from growth curves because the lag in production becomes small relative to the total time in batch. Such a mechanism could explain some of the examples of growth-associated kinetics mentioned in Section G.2.1. In conclusion, the lag might represent a more fundamental descriptor of kinetic behavior, in some cases, than do terms referring to the relationship of production to transitions in the growth rate.

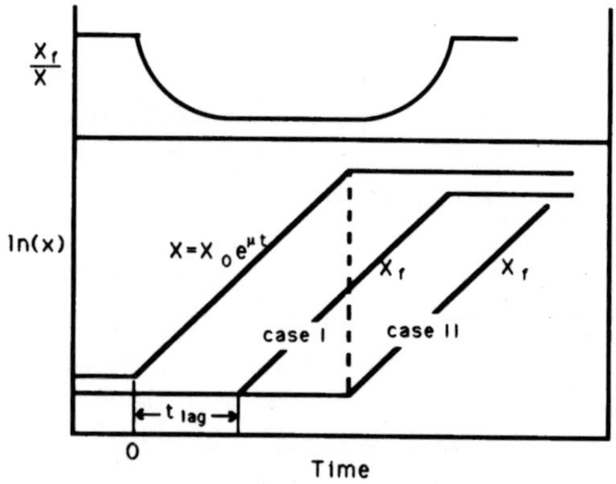

Figure 8.9. Schematic showing simplified case I and case II kinetics in batch culture. X_f represents biomass "filled" with product for case I.

Importantly, while this mechanism accounts for the non-growth-associated *appearance* of batch data, the appearance is artifactual because the ultimate rate of production is still tightly coupled to the growth rate. It should be noted, however, that although production ultimately parallels growth, volumetric productivity, for a given time in batch, is less for a longer lag (Appendix 2).

G.2.3.2. Case II: True Non-Growth-Associated Product Formation On the rootlet scale, the processes of cells developing competence or of accumulating product may be incompatible with conditions favoring growth of the culture as a whole. In this case, product formation will be truly non-growth-associated on the culture level as well. This is shown in Figure 8.9 in which case II is contrasted with case I. The physiological basis for the incompatibility of simultaneous growth and secondary metabolism has been dealt with by Yeoman et al. (1982). Possible reasons include (1) the type and levels of growth-promoting hormone may depress secondary metabolism; (2) during growth, high levels of medium components such as nitrate or phosphate may, in parallel, both stimulate growth and inhibit secondary metabolism; and (3) a physiological incompatibility may exist due to reduced intracellular levels of precursors for secondary metabolism. In a growing culture, the lag time for cells to accumulate product may therefore be indefinitely extended, and very little product may accumulate until the culture as a whole stops growing. At that time, cells may begin accumulating product to some final level.

G.2.3.3. Distinguishing between Apparent and True Non-Growth-Associated Product Formation The problem is that experiments are normally carried out in short-term batch culture in which the proposed lag in production is of roughly the same length as the exponential growth phase. Under these circumstances, an observed rise in production in late log or stationary phase could result from either case I or case II kinetics. Clearly, it will be impossible to use such batch data to address even simple questions regarding the feasibility of batch over two-stage batch or continuous production. It may be possible to run systems in perfusion mode for continued root growth, and even fairly long lags (case I) might then be distinguished from true non-growth-associated product formation (case II).

G.2.4. Other Kinetic Considerations

G.2.4.1. Extension of Kinetic Analysis to Synthesis and Accumulation in Other Tissues With regard to sites of synthesis versus sites of storage, the analysis of Section G.2.3 can apply to either rate-limiting accumulation or rate-limiting biosynthesis in mature tissue. Additionally, the analysis may be extended to rate-limiting synthesis or accumulation in the root cap. In this case, production would be expected to correlate with the rate of lateral root formation since new caps are formed at the tip of each lateral root. Some slight lag following lateral formation may be required for cap maturation. Rate-limiting involvement of the root cap could well give rise to apparent growth-associated product formation, even when growth rates are quite high, because the lateral formation rate may match the total growth rate fairly closely.

G.2.4.2. Product Dilution One consequence of non-growth-associated behavior is that when the culture is growing rapidly and production does not keep pace, specific product levels drop. Loosely termed "dilution," this effect can be distinguished from product degradation or modification by plotting the total amount of product per vessel versus time (Figure 8.8). Clearly, there exists, in such cases, a trivial interaction between growth and concentration, and therefore careful distinction must be made between the direct effect of a treatment on productivity and an indirect effect based on growth and dilution. Growth should not necessarily be construed as a handicap! Frequently, cultures grown under different treatments are all recovered at the same time, and results are plotted as specific product concentration versus treatment. Unfortunately, different treatments may also affect differently the growth rate and/or the lag phase and the time for entry into the stationary phase. Such data are often difficult or impossible to interpret without more complete time-course information.

G.3. Product Secretion

Achieving product secretion from living cells is critical to efficient operation of immobilized-root perfusion systems. It is encouraging to note that secretion is also important in a natural setting because released organics will often define the nature of ecological interactions with microbes and other organisms in the rhizosphere (Bowen and Rovira 1976). For example, infection with mycorrhizal fungi has been shown to induce release of volatile terpenes, which inhibit the growth of other pathogens (Bowen and Rovira 1976). A wide variety of types of compounds are reportedly secreted from both intact roots and root cultures, including alkaloids (Hashimoto and Yamada 1983; Hamill et al. 1986; Mano et al. 1989), terpenes, anthraquinones (Wilson et al. 1987), opines (Petit et al. 1983), vitamins, nucleotides, flavones, sugars, enzymes, amino acids, and auxins (Butcher and Street 1964; Melhuish 1962; Borner 1960). Butcher and Street (1964) have reviewed earlier work; general issues associated with product secretion are dealt with in Chapter 11.

Several reports offer compelling evidence to show that product release is a common feature associated with living root tissue and not just the result of cell lysis (Lundegardh and Stenlid 1944; Melhuish 1962). In our work (Prince 1991), total extracellular protein relative to total intracellular protein has been used as a rough measure of the percentage of biomass that has undergone cell lysis. Medium/biomass product ratios in considerable excess of this level are suggestive of authentic product secretion. Secretion of cell contents during the lag phase (Figure 8.10), in the absence of visible cell debris, is also indicative of true secretion. Further evidence for secretion comes from experiments in which ***in situ*** removal of secreted product through the use of resin adsorbents or increased dilution rates during perfusion (Rhodes et al. 1986) results in higher total production of alkaloid. Product release via increased rates of cell lysis would be expected only to affect the relative distribution of product and not the total amount produced.

Several factors reportedly favor secretion in roots: increased soil resistance— perhaps the result of ethylene release, nonsterile conditions, hormones such as auxins and gibberellins (Bowen and Rovira 1976), low O_2 or other "unfavorable" conditions (Martin 1957), high temperature (Martin 1957; Weathers et al. 1989), increased ionic strength, increased calcium, decreased pH (Melhuish 1962), on-line removal (Rhodes et al. 1986), and elicitor treatment (Signs and Flores 1990).

What can be said about the nature of product secretion from root cultures? Kandler (1951) has concluded that for amino acid secretion, results are consistent with a simple diffusive mechanism modified by differential reabsorption of specific amino acids. Melhuish (1962) has additionally pointed out the likelihood that only recently synthesized amino acids are readily available for secretion and the same may be true of many secondary metabolites. Products must also display adequate membrane permeability and, for example, only less polar compounds may accumulate in the medium (Hartmann and Toppel 1987). Interestingly, it has been shown that different compounds may be secreted from different locations along a rootlet (Bowen and Rovira 1976).

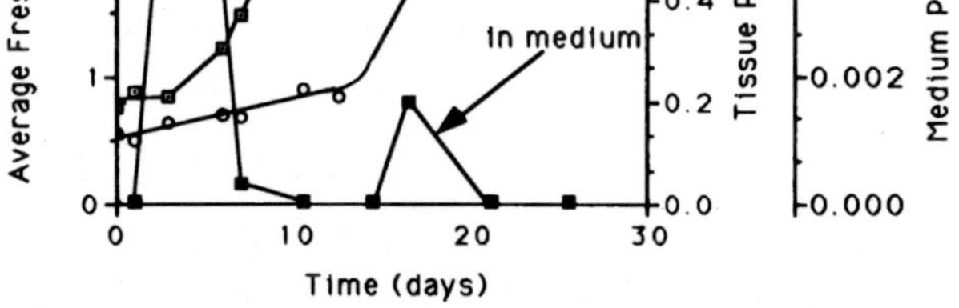

Figure 8.10. Batch culture growth (□) and product accumulation curves showing the appearance of extracellular product in the lag phase and at the onset of stationary biosynthesis (PCSO is propenyl cysteine sulfoxide).

Within a batch culture time course, secretion may be observed in the lag phase (especially with spike-wise auxin additions; Figure 8.10) or typically in late log phase when secondary product formation is active (Mano et al. 1989; Figure 8.10). As noted by Melhuish (1962), active product synthesis often favors higher rates of secretion. Products released in late stationary phase are often the result of cell lysis.

G.4. Productivity Enhancement Considerations

G.4.1. Root Line Selection *Screening* refers to the process of choosing promising root cultures from amongst many independently initiated lines. *Selection* is the technique of selecting improved variants, arising from spontaneous genetic change, within a suspension or organ culture line. Selection may be accomplished manually, by analyzing and subculturing only promising subclones, or through the use of selective media. Addition of a precursor at toxic levels, for example, might help to select for cultures with high rates of secondary metabolism capable of precursor detoxification (Rhodes et al. 1988).

Screening and selection strategies require an initial source of genetic variability from which useful root culture characteristics may be chosen. Sources of variability include different plant species for a given target compound; different strains; different seeds within a strain; irradiated seeds (Mitra 1972); somaclonal variation obtained through root regeneration from suspensions (Aird et al. 1988b) or protoplasts (Furze et al. 1987); and variability associated with Ri transformation (Mano et al. 1989).

An initial round of screening is frequently employed using unoptimized medium conditions, and this may be of value. Care must be taken, however, to ensure that cultures are old enough and have stabilized (dilution of parent plant material, loss of epigenetic changes), that the kinetics of production have been considered (Section G.2), and that statistical variability has been specifically estimated ahead of time. It is, of course, preferable to screen under optimized process conditions, though the fact that the "optimum" conditions may be, to some extent, root line specific may necessitate an iterative approach.

Although cell line selection has been an effective technique for enhancing suspension culture productivities, little work has been reported on selecting root cultures directly. This may be due to the perception that root cultures are more stable and therefore less likely to generate somaclonal variants. In one report, Endo and Yamada (1985) described 10 successive rounds of *Duboisia* root culture selection for higher alkaloid production. They reported some success in enhancing productivity, which, if due to genetic factors, would imply a fairly high frequency of gene-level alterations as a source of variability for subsequent selection. Others have employed, with marginal success, suspension- or callus-based screening and selection followed by root regeneration (Hashimoto and Yamada 1983; Furuya 1988); however, selected characteristics, especially those associated with a given ploidy, may or may not appear in the redifferentiated material (D'Amato 1985).

G.4.2. Production Medium Development Few reports exist on the development of root culture production media. Moreover, studies have mainly focused on demonstrating biosynthetic potential without specific consideration of production kinetics and operational mode. Variables examined include the effects of different basal media on growth and production (Mano et al. 1989; Endo and Yamada 1985; Hashimoto et al. 1986), sucrose levels (Endo and Yamada 1985; Norton and Towers 1986), sucrose/nitrate ratios (Norton and Towers 1986), and gibberellic acid (Sugimoto et al. 1988;

Ohkawa et al. 1989). A two-step approach, with enhanced production in the second step, has been proposed for the production of peroxidase, a primary metabolite (Taya et al. 1989b). Though some generalities are likely to emerge, medium optimization will continue to be carried out on a case-by-case basis.

G.4.3. Auxin Effects on Secondary Biosynthesis It has been suggested that exogenously added auxin depresses secondary product formation (Hamill et al. 1987; Flores et al. 1987). Certainly, when callus forms in response to higher levels of, for example, 2,4-D, inhibition is expected (Berlin et al.1986; Flores et al. 1987). That dedifferentiation is responsible, and not the auxin itself, is indicated by the fact that callus formed upon kinetin addition to transformed root cultures also results in a drop in product levels—polyacetylenes in this case (Norton and Towers 1986).

Other evidence suggests that auxin may directly inhibit product formation in root cultures (Norton and Towers 1986; Hashimoto et al. 1986). Caution must be exercised, however, in interpreting results comprised of single data points for each treatment, with all flasks having been recovered at the same time. Kinetically, not all of the important variables are maintained constant in such an experiment. Specifically, for auxin, two dramatic changes are expected in the growth curve: (1) auxin will usually enhance the growth rate, and (2) spike-wise auxin addition will extend the lag phase (Wijnsma et al. 1988). For non-growth-associated products, if experiments are recovered in advance of stationary phase production, cultures with higher growth rates will possess lower specific product levels. This issue of product "dilution" is discussed in Section G.2.4 and in Chapter 11. Stating that auxin stimulates growth is not equivalent to stating that auxin inhibits secondary metabolism.

Moreover, if production is characterized by a specific lag following growth (see Section G.2.3), for a given increase in biomass, the extent of product dilution will depend on the length of the lag phase in growth. This is shown in Figure 8.11, where for two equivalent increases in biomass, but with two different lag phases, the specific and total amount of product accumulated is completely different. Clearly, single time-point data for a set of auxin treatments often cannot be simply interpreted, and some form of time-course data, however labor intensive it may be to obtain, is often needed.

Good evidence for auxin inhibition of a specific secondary metabolic step has been obtained. Hashimoto et al. (1986) have clearly demonstrated the inhibition of the 6-β-hydroxylase-catalyzed conversion of hyoscyamine to scopolamine throughout the time course in batch. Conversely, Yoshikawa and Furuya (1987) reported both improved growth and enhanced saponin production, on a dry weight basis, with IBA addition to *Panax ginseng* root cultures. In summary, auxin may well inhibit important steps in some secondary metabolic pathways, even at growth-promoting concentrations, depending on the mode of auxin addition. However, inhibition cannot yet be considered to be of widespread generality.

G.4.4. Other Productivity Enhancement Techniques

G.4.4.1. Precursor Feeding and Biotransformation As for suspensions, the feeding of biosynthetic intermediates is likely to be an effective technique when rate-limiting

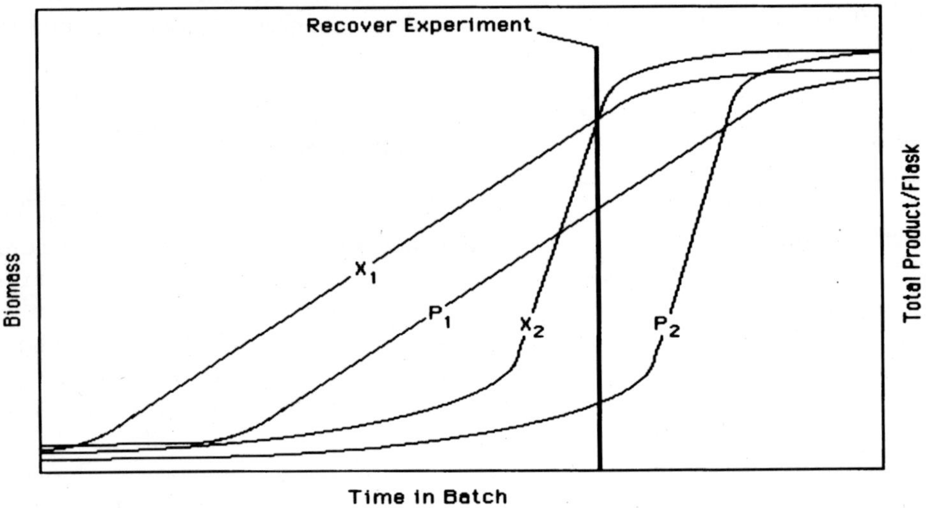

Figure 8.11. Two different growth and product accumulation curves with the same lag in production. For x1, a short lag phase for growth with a slower ultimate growth rate is assumed. For x2, a longer lag phase for growth with a high exponential growth rate is assumed. Recovering experiments as shown, x2 will contain much lower specific and total product simply due to alterations in the growth curve.

steps occur earlier in the biosynthetic pathway. The precursor chosen should, of course, not be overly toxic or expensive. Biotransformation involves a one- or two-step conversion of an advanced intermediate into final product. A few examples of precursor feeding and biotransformation using root cultures are given in Table 8.3.

In addition to enhancing productivities, precursor feeding may be used as a means of producing novel compounds and for modifying the relative levels of different compounds in a mixture. Such considerations can be important when producing flavors and fragrances.

As an example, we have examined the enhancement of onion flavor compounds, alkylcysteine sulfoxides, in root cultures (Prince 1991). Onion flavors are characterized by the relative contributions of methyl-, propyl-, and propenylcysteine sulfoxides. Garlic flavors are characterized by the profile of methyl-, propyl-, and allylcysteine sulfoxides. Through the addition of the amino acids cysteine, methionine, and glutathione, final product levels could be greatly enhanced; earlier steps in sulfate reduction are probably limiting in the absence of precursor addition. Importantly, different treatments elevate different flavor components to different extents (Figure 8.12), thereby enabling control over flavor profiles for specific commercial applications.

Through the addition and incorporation of thiol intermediates, it is possible to produce, in a controllable fashion, novel flavor composites with both garlic and onion

Table 8.3. Examples of precursor feeding and biotransformation using root cultures.

Products	Compounds added	Plant species	Reference
Scopolamine	Hyoscyamine	*Hyoscyamus niger*	Hashimoto and Yamada (1983)
Quinine, quinidine	Tryptophan	*Cinchona ledgeriana*	Hay et al. (1986)
Nicotine, anabasine	Lysine, cadaverine, putrescine, agmatine	*Nicotiana sp.*	Walton and Belshaw (1988) Walton et al. (1988)
Scopolamine, hyoscyamine	Putriscine, spermidine cadaverine, other amines	*Duboisia myoporoides*	Yoshioka et al. (1989)
Sugar esters	2-Phenylpropionic acid	*Panax ginseng*	Furuya et al. (1989)
Digitoxin	Digitoxigenin	*Panax ginseng*	Kawaguchi et al. (1990)

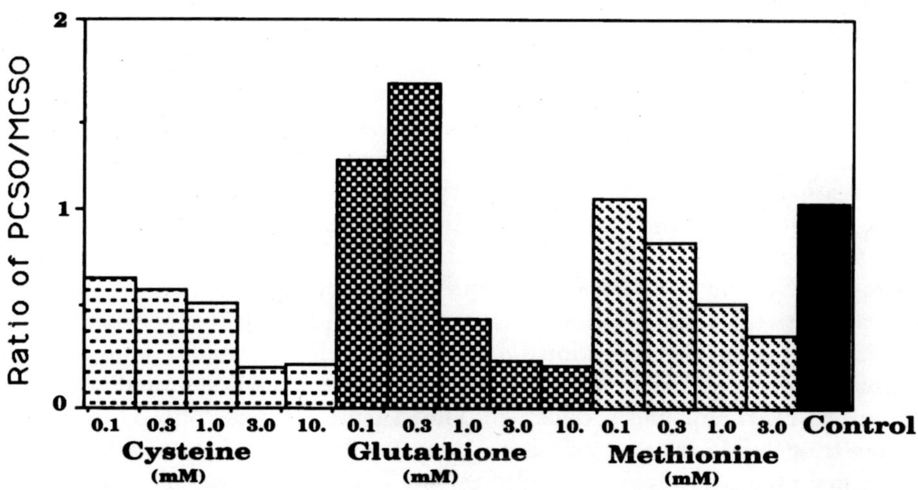

Figure 8.12. Effect of adding cysteine, glutathione and methionine on the relative levels of the flavor compounds S-methyl cysteine sulfoxide (MCSO) and S-propenyl cysteine sulfoxide (PCSO) (Prince 1991).

flavor components. Side chains that do not occur naturally may also be introduced; for example, we have produced onion cultures with high levels of ethylcysteine sulfoxide, which does not occur naturally in *Allium* species (Whitaker 1976). Such studies highlight the capabilities of plant cell cultures for flexible control and manipulation not easily achievable with agricultural methods.

G.4.4.2. Elicitor Treatment As discussed in Chapters 2 and 11, elicitor treatment has been demonstrated to be an effective method for enhancing secondary biosynthesis in suspensions and in immobilized cells. Though little work has as yet been reported, it is expected that elicitor treatment will be fully as effective with root cultures (Flores 1987). Flores et al. (1988) reported a 20-fold increase in polyacetylene levels upon treatment of *Bidens sulphureus*-transformed root cultures with fungal filtrates. Signs and Flores (1990) have also accomplished repeated elicitation of sesquiterpene biosynthesis in *Hyoscyamus muticus* root cultures. Secretion induced upon elicitation may prove beneficial for operation of immobilized cell processes.

G.4.4.3. In Situ *Product Removal.* Product removal has been demonstrated to improve alkaloid productivity in tobacco (Rhodes et al. 1986). Both removal through perfusion and with resin adsorption have been reported (Rhodes et al. 1986; Wilson et al. 1987). ***In situ*** product removal will likely be effective for enhancing productivity and achieving extracellular accumulation, as it has been for suspensions and immobilized cells (Chapter 6 and 7).

H. GROWTH AND PRODUCTION SYSTEM DESIGN

H.1. Introduction

Section B demonstrates that root cultures are capable of both rapid growth as well as superior rates of product formation in many cases. Most work so far, however, has been done in shake flasks, with the primary objective of simply demonstrating secondary biosynthetic potential. The next key step for process development is system design for large-scale growth and maintenance. As was the case for workers in the 1940s and 1950s trying to develop submerged growth systems for fungal antibiotic production, we similarly face the task of working with a new biological material with even greater constraints as regards mechanical stress sensitivity and nutrient supply difficulties.

In Chapter 6, various operational modes are described including growth and production in batch, fed batch, and perfusion. Choice of operational mode is primarily determined by kinetic and productivity enhancement considerations, and different options are possible for root cultures as they are for suspensions. Regardless of the operational mode, however, root cultures will have to be successfully grown to high density on a large scale. For non-growth-associated product formation, and especially for production with perfusion, roots must be maintained for prolonged periods beyond the growth phase. The following discussion therefore focuses specifically on factors associated with large-scale growth and maintenance which are common to all operational modes.

First, root culture characteristics significant for scale-up will be considered, followed by a review of reports of larger scale culture. Experimental evidence for the significance of convective mass transfer will then be described, and, based on this concept, an approach to understanding and designing submerged root culture processes will be outlined. The analysis illustrates, by example, the serious problem of supplying oxygen upon scaling up to larger clumps in suspended growth systems. Lastly, promising systems will be suggested which ensure good convective mass transfer and/or direct gas phase mass transfer to roots at higher average root densities.

H.2. Characteristics of Large-scale Root Culture Growth

Examining the characteristics of typical root cultures, we may anticipate some of the problems associated with larger scale culture. Morphologically, root cultures tend to grow as large clumps or root masses (Figure 8.13). Clumps typically grow larger and larger (roughly spherically) until the available reactor volume is filled. By this mechanism, root densities will tend to be higher in the central regions of a clump. Alternatively, on a smaller scale, and for cultures with lower linear lateral root densities (Section C.2), roots may tend to quickly fill the available reactor volume, with subsequent growth serving to increase homogeneously the average root density.

Root cultures are very metabolically active, and uptake rates for oxygen, hormone, and inorganic nutrients can be reasonably high. Root tips, for example, may respire at rates as high as 1.3 mg O_2/(mL×hr) (Luxmoore et al. 1970; Section F.3); at this rate, oxygen depletion occurs after about 22 seconds following air saturation. High uptake rates coupled with large clump sizes necessitate high rates of mass transfer into interior regions of root clumps (see Section H.4 below).

With regard to mechanical properties, the root mass or clump will normally remain stationary, and high relative liquid velocities can result in significant surface shear rates. Velocity gradients might also be expected to create critical torques over exposed rootlets.

Gravitropic bending of root tips, to the extent that it occurs, could result in nonhomogeneous biomass distributions within the reactor as tips seek the bottom of the growth vessel. Capone et al. (1989) report that roots transformed with any construction containing T-DNA open reading frames 13 and 14 (Section D.2) do not respond to gravity by orienting vertically downward. Similarly, elevated auxin in transformed or untransformed roots eliminates normal gravitropic bending (Capone et al. 1989). Untransformed root cultures, without added auxin, may or may not show significant gravitropic bending (White 1936; Capone et al. 1989).

H.3. Review of Published Studies

A handful of cases have been reported in which specific attempts to culture organs in vessels larger than shake flasks have been made. Rhodes et al. (1986) reported on cultured transformed roots of *Nicotiana rustica* in an 880-mL working volume fermenter. Initial batch growth was rapid and typical of growth in shake flasks. Following switching to perfusion mode at day 13, growth continued until approximately day 20

Figure 8.13. Root clumps of *Tagetes patula*, approximately 4 kg wet weight, obtained from a 20 l air-sparged reactor (photo courtesy of Dr. M.A. Hjortso).

when growth was suggested to have slowed, based on a leveling of the dissolved oxygen at about 30%. Final root density at day 30 was about 10 g dry wt./L. In subsequent work, bubble columns were used with perfusion for up to 86 days (Wilson et al. 1987). Despite the availability of fresh medium, nicotine production dropped precipitously at about 45 days, and overall growth rates were greatly depressed. Presumably, much of the biomass had stopped growing or had died due to oxygen limitation.

With regard to root clump size reduction, Hjortso and Flores (1987) observed what was presumed to be severe mass transfer problems in root masses and found that shear-induced size reduction of clumps led to severe root damage. Likewise, masses cut into pieces and impeller agitated gave rise only to a suspension (Hilton et al. 1988).

Several groups have referred to the effective use of impeller agitation in which roots or shoots are protected from direct impeller contact through use of a screen (Jung and Tepfer 1987; Hilton et al. 1988; Kondo et al. 1989; see Figure 9.5). Taya et al. (1989a) described operation of a bubble column with roots anchored to a polyurethane foam support, a similar column with trickling medium sprinkled from the top, and one in which the medium level is alternately raised and lowered through use of a siphon. Problems associated with operation of the medium cycling system are discussed in more detail in Section H.5 and in Chapter 9. Subsequently, Kondo et al. (1989) effectively accomplished alternate exposure to liquid and gas by anchoring roots to foam on the inner walls of a rotating drum. Both reports employing foam "immobilized" roots with good relative fluid velocity gave good results at the relatively low root densities employed (up to about 11 g dry wt./L).

H.4. Convective Mass Transfer Within the Root Mass

The challenge is clearly to achieve adequate mass transfer into the interior regions of large tortuous root masses. As outlined in Chapter 4, engineers working with porous catalysts traditionally calculate the Thiele modulus to obtain an impression of the relative rates of reaction and diffusion (and thus the sufficiency, or insufficiency, of substrate levels in the interior). If we do so for a small clump of 5 cm, with an average specific uptake of 0.3 mg O_2/g fresh wt. per h, we find that the zero-order Thiele modulus is about 11.6; less than 9% of the total mass should be oxygen sufficient with diffusion as the sole mode of mass transfer. In liquid systems, diffusion rates are extremely slow and, for typical reaction rates, operate only on length scales of millimeters at best.

What additional modes of mass transfer may be exploited? To the extent that any convection (fluid flow) occurs, large increases in effectiveness may be observed due to mass transfer enhancement (Nir and Pismen 1977; Rodrigues and Orfao 1984; Stephanopoulos and Tsiveriotis 1989; Bringi and Dale 1990). In dealing with porous organ culture clumps of many centimeters, convective mass transfer is expected to be the overwhelmingly dominant mode.

H.4.1. Experimental Evidence for Convection-Enhanced Mass Transport We have attempted to experimentally document the influence of intraclump convective mass transfer on rates of root culture respiration (Prince et al. 1991). As outlined in Section E of Chapter 4, circulating fluid around a root clump will give rise to a pressure distribution over the surface of the clump, which in turn may give rise to convective flows within the clump itself. The magnitude of this intraclump convective flow depends on the extraclump fluid velocity, via the induced pressure distribution, and on the hydraulic permeability of the porous root clump. For a given hydraulic permeability, the higher the pressure difference (external flow rate), the higher will be the flow velocity through the clump. Higher flow rates into a root clump imply an increase in the amount of oxygen available to respiring tissue on the interior of the root clump. If more cells are exposed to sufficiently high oxygen concentrations, rates of respiration, that is, the total oxygen uptake rate, will increase.

We have observed this phenomenon in spherical root clumps of known hydraulic permeability and known specific oxygen uptake rate, and in the characterizable flow field shown in Figure 8.14. Results displayed in Figure 8.15 show that a clump only 3.6 cm in diameter and of modest root density (0.32 g fresh wt./mL or 19 mg dry wt./mL), respires at maximal rates only when external velocities, in this relatively ideal flow field, reach about 1 cm/s. Convective mass transfer clearly plays a dominant role in supplying oxygen to even quite small root clumps. As well, given the velocity requirement of 1 cm/s in such an ideal flow field and for such a small clump size, adequate mass transfer may be very difficult to achieve in typical suspended growth systems—especially at higher root densities (see example below).

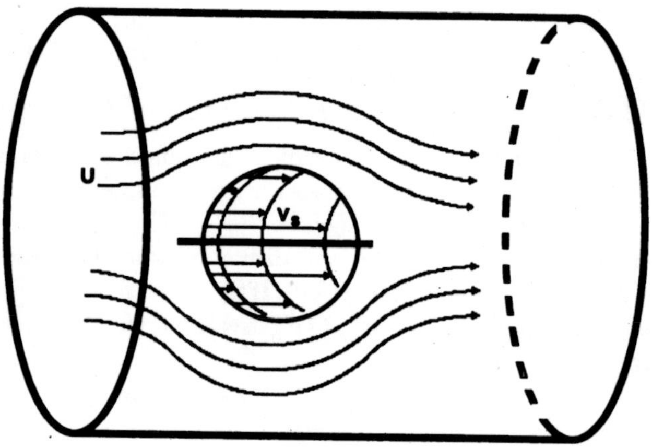

Figure 8.14. Suspended spherical root clump contained within a "tea ball," the stainless steel screen of which does not contribute significantly to convective flow resistance. Fluid flows past the root sphere at a known velocity and in a characterizable flow field. Fluid is recycled in a closed system and oxygen uptake is measured with time.

Since dissolved oxygen above the critical oxygen concentration is an absolute requirement for growth, systems which fail to induce good intraclump convective flows will fail to achieve good root culture growth rates and high final root densities. High cell densities are critical to operating any plant cell process efficiently, and, further, the period of most intense product formation often occurs only after high cell density has been achieved. Generous oxygen supply is then required throughout the culture volume, especially if the production phase is to be extended through perfusion and when oxygen requirements for product formation are greater than for growth. High root densities coupled with high oxygen requirements during production will frequently mean that design constraints apply primarily to the production phase.

H.4.2. Engineering Analysis Having clarified the problem, it is possible to develop equations describing the transfer and reaction of oxygen within a root culture system. In Chapter 4, the development of root culture design equations was outlined. As shown in Figure 4.11, the problem consists of three steps: (1) oxygen-saturated medium must flow into the root mass, (2) oxygen molecules must diffuse through the stagnant boundary layer surrounding the surface of each rootlet and (3) oxygen must simultaneously diffuse and react within the rootlet tissue itself. Following a fluid element along a streamline within the root mass, oxygen is depleted continuously (Figure 4.11). The design criterion is to prevent the dissolved oxygen in a fluid element from falling below a critical level before the fluid element exits on the other side of the root mass. Such an analysis can be used directly to design, for example, packed bed root culture reactors as discussed in Section H.5. It may also be used to shed light on the influence of scale on mixing velocities in suspended root clump growth reactors, as suggested in the following example.

Example Using the analysis detailed in Chapter 4, we may predict the effect of increasing the diameter of the 3.6-cm sphere depicted in Figure 8.14 by a factor of 5

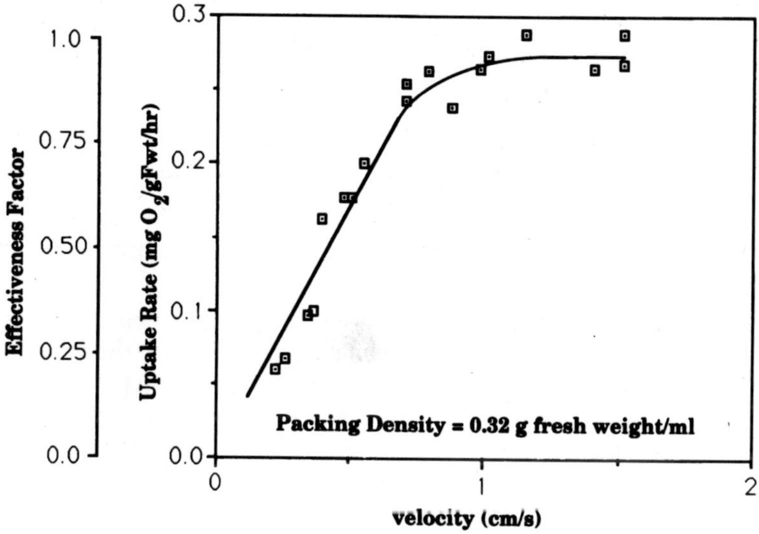

Figure 8.15. Results from operating the system shown schematically in Figure 8.14. As the external velocity increases, the oxygen uptake rate increases to the expected maximum for this onion root line.

to 18 cm. Root masses of at least 23 cm in diameter have been reported (Davioud et al. 1989). Roots are assumed to be 1 mm in diameter. Of interest is the effect on the minimum intraclump fluid velocity, the minimum dissolved oxygen concentration leaving the clump, and the external fluid velocity required for adequate internal convection and oxygen supply. Results are shown in Table 8.4 where we see that while the diameter increases by a factor of 5, the intraclump velocity increases only 3.6-fold (from 0.07 to 0.25 cm/s) because higher intraclump velocities reduce the boundary layer thickness and improve intraclump mass transfer rates to the rootlet surface. We may conclude that the intraclump velocity scales somewhat less than proportionally with the characteristic length of the root mass.

For suspended clumps, however, we must rely on external (extraclump) velocities to create pressure differences which then induce the required minimum intraclump velocity. As outlined in Section E.3 of Chapter 4, the external flow velocity, U, depends upon both the internal velocity, v_s, and the size of the porous object, d_{clump}. Depending on the flow regime, U scales at the very least as

$$U \propto (v_s d_{clump})^{1/2}$$

in intermediate-flow regimes and scales as

$$U \propto v_s d_{clump}^2$$

Table 8.4. Effect of increasing clump size from 3.6 cm to 18 cm.

Diameter (cm)	Min. O_2 conc. root surface (mg/L)	Min. O_2 conc. exiting clump (mg/L)	Min. internal velocity (cm/s)	Min. external velocity (estimated) (cm/s)
3.6	5.52	6.63	0.07	1
18	5.52	6.05	0.25	4.2 - 90

at lower "Reynolds numbers" (Chapter 4). In more turbulent flow fields, internal flow patterns are complex and probably result in stagnant regions within the clump which are difficult to supply with oxygen. We see therefore that to maintain oxygen sufficiency, on growing to an 18-cm clump, the internal velocity must increase 3.6-fold and the external velocity must increase at least $(3.6 \times 5)^{1/2} = 4.2$-fold (and likely 81-fold or more). Considering the fact that the geometry and flow field are idealized in this example and that pressure distributions are likely to be less favorable in suspended growth reactors, this example illustrates the problems and unpredictability associated with use of suspended growth reactors for root cultures.

Two factors may be considered which render the outlook slightly less bleak for suspended growth systems. First, sparging with oxygen-enriched air dramatically improves oxygen supply. Second, it should be mentioned that fluid convection will be induced when sparging directly through the porous root mass [mechanism described in Bringi and Dale (1990)], and this may help to expand oxygenated zones somewhat. Finally, until higher densities are reached, and for small clumps, it is possible for a high percentage of the growing tips to be located in the outer oxygen-rich regions of the clump. If growth proceeds outward from the surface of the clump, the root density should be lower, and the hydraulic permeability higher, in this outer region. These factors may help to explain why reasonably good growth is observed in a variety of systems up to about 10 g dry wt./L. A combination of strategies may keep suspended growth systems within the limits of aerobic operation up to moderate root densities.

At higher densities and for larger clumps, however, it will not be physically possible for a majority of tips to be inhomogeneously concentrated in oxygen-rich zones of the reactor. Although growth may continue in small oxygen-sufficient regions, the overall rate will drop. Densities of 30 g dry wt./L or more may be achievable under such circumstances over long periods of time, but only as a consequence of compacting dead tissue in oxygen-poor zones.

H.5. Alternative Design Approaches

In light of the special needs of organ cultures for good convective mass transfer, the suitability of a number of alternative approaches may be considered. These include packed bed with recycle, the medium and gas cycling approach, and the use of sprays or aerosols.

H.5.1. Packed Bed with Recycle One alternative design used in our laboratory has been the packed bed with fluid recycle. As shown in Figure 8.16, roots form a bed in the reactor portion through which medium is forced at a desired superficial velocity. Aeration is accomplished by air sparging in the recycle vessel. During continuous operation, because the recycle rate is rapid compared to feed/effluent flow rates, the system is well mixed and behaves like a "continuous stirred tank reactor" (Hill 1977).

The key advantage is that, rather than relying on extremely high velocities outside a clump, convection through the roots is achieved directly by minimizing short-circuiting around the outside. The relatively uniform velocity profile helps to reduce the possibility of shear damage and permits simple design through use of the analysis outlined above and in Chapter 4. Additionally, aeration is carried out in a separate vessel, which may be designed specifically for this purpose. High K_La values may be used, and the dissolved oxygen level can be maintained close to saturation. The trick to designing such systems is in simultaneously reducing fluid short-circuiting around root clumps while still offering ample space for root growth. In our experience, low aspect ratios (ratio of column height to diameter) permit unimpeded growth upward and yet induce good convective mass transfer with little compaction and low pressure drops across the bed.

H.5.2. Medium and Gas Cycling Another approach for obtaining good convective mass transfer and gas exchange is to simply alternate periods of exposure to liquid and gas phases. Several methods of accomplishing this are possible. The medium may be alternately pumped into and out of the culture chamber either using pumps (Bringi 1991) or a pump and siphon approach (Taya et al. 1989a). Organs may also be

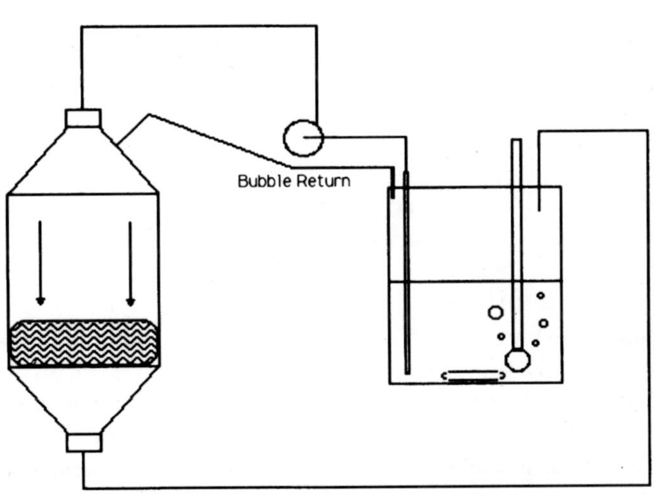

Figure 8.16. Packed bed root culture reactor with fluid recycle. Feed and effluent lines may be connected to the recycle vessel for operation in perfusion mode.

mechanically raised out of and lowered into the liquid medium, though this approach does not make efficient use of reactor volume. Alternatively, organs may be fixed within a rotating "drum" which is partially filled with medium (Kondo et al. 1989; see Chapter 9, Figure 9.6). If the hydraulic permeability is high enough, liquid will drain from the roots during gas phase exposure, permitting good convective mixing and gas exchange. However, this prerequisite will frequently not be met with root cultures, which tend to behave like sponges at higher densities. It is possible to estimate optimal times for exposure to the gas and liquid phases using a time constant analysis, and this is described in more detail in Section F.2.1 of Chapter 9. The "alternate exposure" technique is a low-shear technique but is not the simplest mechanically, and difficulties may be encountered on scale-up.

H.5.3. Spray and Aerosol Systems Since mass transfer tends to be a more difficult problem for oxygen compared to other nonvolatile components, another logical approach is to expose root masses to gas phase containing fine droplets of liquid medium. Oxygen concentrations are higher in the gas phase, and diffusivities are also much higher in gas than in liquid phases. Both sprays (Taya et al. 1989a) and aerosols (Weathers et al. 1989) have been used, although the latter may prove superior for liquid phase penetration into root masses.

The major advantage of working with aerosols appears to be the potential elimination of oxygen transfer problems. However, although oxygen supply is of central importance, nonvolatile components such as hormones and phosphate may also become limiting under some circumstances (see Chapter 7), and, in others, rapid exchange of medium components such as elicitors may be desirable. Supply or exchange of medium components will be less rapid using aerosol delivery, depending on the rate of droplet coalescence on root surfaces and on the homogeneity of aerosol distribution in the root mat (i.e., the residence time distribution). It appears likely that good convective distribution of aerosol through the root mat will be of benefit, as it is in packed bed submerged systems. Coalescence rates achieved in systems of the sort shown in Figure 8.17 appear to be satisfactory for supplying nonvolatile medium components at sufficient rates. Residence time distributions remain to be studied. Temperature control and heat exchange are not a significant problem in our experience. The performance of such systems on a large scale remains to be assessed.

H.6. Reactor Inoculation

A final important issue for root culture growth reactor design is that of inoculation strategy. In order to make efficient use of reactor volume, reasonably high inoculum levels are required so as to maintain high average biomass densities. Inoculum levels of 5 to 10% or more (volume of cells to total volume) are typical. It is not possible to pump or "pour" root culture inocula. Specialized feed reactor designs may be required which are physically integrated with the production-scale reactor and which are designed to supply roots in a desired shape or morphology. Further work is needed to develop feasible large-scale inoculation schemes.

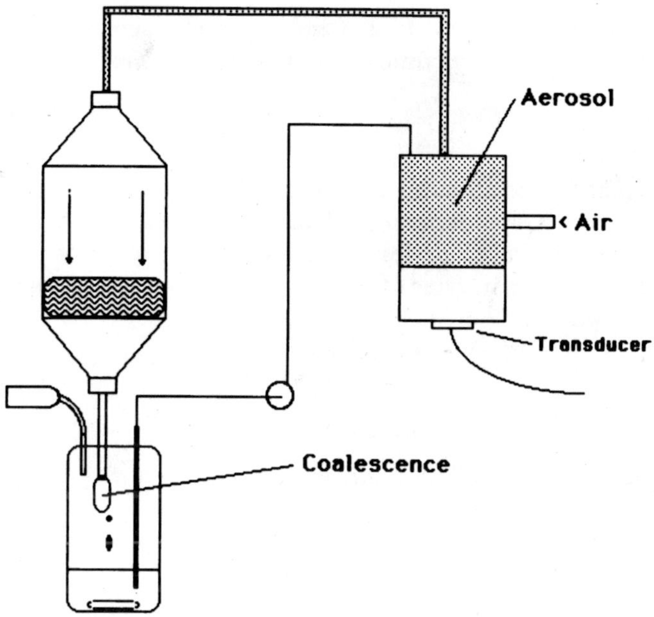

Figure 8.17. Aerosol system with medium recycle. Coalescence is achieved using a fritted glass dispersion tube. The system can be run continuously through inclusion of feed and effluent streams in the coalescence/recycle vessel.

Appendix 1: The Specific Growth Rate of Exponentially Growing Roots

It is of interest to express the specific growth rate of exponentially growing roots in terms of its constituent growth processes. As stated in the section on growth mechanics (Section C.2), growth can be resolved into (1) linear rootlet elongation coupled with (2) an exponential increase in the number of rootlets. In some species, the positioning of laterals can be quite precise relative to one another (Mallory et al. 1970). In other species, the distance between laterals is less regular but yet is still nonrandom, as might occur if, for example, an inhibitor is released from an established primordium, thereby preventing primordium formation for some critical distance within the inhibitor gradient. Barlow and Adam (1987) have recently suggested that, in tomato root cultures, a certain constant distance exists between prospective root primordia. Whether or not a lateral is actually formed at each possible site presumably depends on the hormonal and nutritional conditions. In any case, it is reasonable to define an average distance between laterals, L (cm), as the reciprocal of the linear root density, ρ (laterals/cm):

$$L = 1/\rho \quad [8.A1.1]$$

Given a steady rate of elongation, REL (cm/day), a characteristic time may be defined which represents the average time elapsed between lateral emergences:

$$t_l = (\rho \times \text{REL})^{-1} \qquad [8.A1.2]$$

We may assume that the positioning of laterals is fundamentally determined by processes governing the average distance of laterals from each other and that the average distance, L, will be more or less independent of the elongation rate. As discussed in Section C.2, constant rates of elongation and lateral root formation give rise to exponential growth,

$$\frac{dx}{dt} = \mu x \qquad [8.A1.3]$$

with a specific growth rate μ (h^{-1}). Upon integration, equation [8.A1.3] gives the familiar expression for the doubling time, t_d:

$$\ln\left(\frac{x}{x_o}\right)_{t_d} = \ln 2 = \mu t_d \qquad [8.A1.4]$$

Similarly, in the time t_l, there will be a certain constant increase in biomass:

$$\ln\left(\frac{x}{x_o}\right)_{t_l} = \mu t_l \qquad [8.A1.5]$$

where $\left(\frac{x}{x_o}\right)_{t_l}$ is a constant for a given culture growing under a given set of conditions and which can be shown to lie in the range $1 < \left(\frac{x}{x_o}\right)_{t_l} < 2$ (see below). Isolating μ in equation [8.A1.5] and using equation [8.A1.2], we have:

$$\mu = \frac{\ln\left(\frac{x}{x_o}\right)_{t_l}}{t_l} = \ln\left(\frac{x}{x_o}\right)_{t_l}(\text{REL})(\rho) \qquad [8.A1.6]$$

The specific growth rate is thus proportional to both the rate of elongation and the linear density of lateral roots. Intuitively, although lateral root formation is most important in determining growth, as is the division rate for microbial cells, elongation determines the rate of lateral formation by creating the backbone upon which to build new lateral roots.

The constant of proportionality in equation [8.A1.6], $\left(\frac{x}{x_o}\right)_{t_l}$, may be determinable geometrically. If an unusual rooting pattern were to occur in which the root meristem actually split into two equal meristems at each interval t_l, then t_l would be numerically equal to the doubling time, t_d. The constant of proportionality in equation [8.A1.6] would then be simply ln 2. Such an unusual dichotomous rooting pattern is observed naturally in conifers infected with a mycorrhizal fungus (Rupp and Mudge 1985). In other cases, the difference between x and x at a subsequent time t_l is determined by a Fibonacci sequence, which causes $\left(\frac{x}{x_o}\right)_{t_l}$ in equation [8.A1.6] to converge on a value something less than 2, depending on the geometry of the root culture in question (proof not shown). The ratio $\left(\frac{x}{x_o}\right)_{t_l} = 1$, of course, corresponds to a growth rate of zero.

A final note with regard to real root cultures is that the elongation rate of a given tip may change over time. Immediately following lateral root emergence, a length of time is likely required before elongation occurs at a normal rate. Such an effect is not inconsistent with the above analysis. Over longer periods of time, it is not known what the life span of a particular elongating tip will be. It has been assumed that all tips elongate indefinitely. A mechanism for limiting the growth of older tips has been described which involves the accumulation of supraoptimal levels of auxin in the oldest tip. Older work, especially with tomato (Street 1957), demonstrates this effect in the most apical root tip, probably as a consequence of high rates of endogenous auxin synthesis coupled with auxin transport along the longest section of continuous root to the oldest tip. Borgman and Mudge (1987) also have more recently demonstrated inhibition of the oldest main axis tip with applied auxin, while lateral roots were stimulated. Such effects may or may not be of wide generality, and, of course, some auxins are not transported very well.

Appendix 2: Analysis of Production Kinetics in Relation to Growth

The following analysis serves only to formalize the arguments of Section G.2.3, which are intended only as an aid to understanding kinetic behavior. The objective is not to initiate development of a root culture growth or production model for reactor design purposes. For example, no attempt is made to couple growth or production rates to levels of nutrients or hormones. To do so at this time would be of little value since far too little is yet known concerning, among other things, the regulation of intracellular auxin levels or of kinetic responses to changes in other important nutrients.

We may begin by defining some terms:

t_{cf}	time for cells to become competent and to fill with product
x	total biomass
x_u	new and "unfilled" biomass (biomass less than t_{cf} old and containing little or no product)
x_f	older biomass "filled" with product
x_o	initial total biomass
x_{-tcf}	total biomass at $t - t_{cf}$
f	unfilled fraction of the total biomass, i.e., x_u/x

$(1\text{-}f)$	fraction of the total biomass filled with product
f_o	f at $t = 0$

For simplicity, we will model early batch growth as consisting of an instantaneous transition from no growth in the lag phase to full exponential growth. Given the lag (t_{cf}) for new product to form, we may see what happens to specific and total product levels with time.

Looking at the pool of cells made up by x_u, we know that the input to this pool consists of all new growth because all new cells must pass through this immature phase. Output from this pool results from conversion of unfilled cells to filled cells exactly a period t_{cf} after they are formed. The rate of conversion is precisely determined by the rate of production of new cells that occurred at a time t_{cf} previous. It is helpful to acknowledge that cells of a given age proceed along the root axis together: "axial dispersion" or "cell displacement" is not substantial (Bertaud and Gandar 1985), and each rootlet may be considered to be a "plug flow" reactor (Hill 1977).

Restating,

$$\begin{pmatrix}\text{Rate of Change}\\ \text{of } x_u\end{pmatrix} = \begin{pmatrix}\text{Rate of Input From}\\ \text{New Biomass Formation}\end{pmatrix} - \begin{pmatrix}\text{Rate of Output from } x_u \text{ or}\\ \text{Rate of Biomass formation}\\ \text{at } t_{cf}\end{pmatrix}$$

$$\frac{dx_u}{dt} = \left(\frac{dx}{dt}\right)_t - \left(\frac{dx}{dt}\right)_{t - t_{cf}} \qquad [8.A2.1]$$

Two time periods may then be considered following the transition to exponential growth at $t = 0$: (1) the initial transient of length t_{cf} during which no new product is formed, and (2) the subsequent period during which product is formed and exponential and balanced growth proceed as described in Section C.2.

Initial Transient ($0 < t < t_{cf}$) When the culture just starts to grow at $t = 0$ (end of lag phase), all new growth contributes to an increase in x_u; however, there is no conversion of x_u to x_f, because this occurs only when $t \geq t_{cf}$. The second term on the right of equation [8.A2.1] is zero until $t = t_{cf}$ because no growth occurred before $t = 0$.

$$\frac{dx_u}{dt} = \left(\frac{dx}{dt}\right)_t = \mu x \qquad [8.A2.2]$$

Integrating this expression, we see that,

$$\begin{aligned} x_u(t) &= (\text{initial } x_u) + (\text{increase in } x) \\ &= f_o x_o + (x_o e^{\mu t} - x_o) \\ &= x_o(f_o - 1 + e^{\mu t}) \end{aligned} \qquad [8.A2.3]$$

The fraction of total biomass filled with product at any time during this initial phase is

$$f(t) = \frac{x_u}{x} = \frac{x_o(f_o - 1 + e^{\mu t})}{x_o e^{\mu t}} = (f_o - 1)\, e^{-\mu t} + 1 \qquad [8.A2.4]$$

It is evident that, throughout this period, the fraction consisting of unfilled biomass increases steadily, and conversely the filled fraction decreases. Product content (% dry wt.) also decreases, and product "dilution" results (Section G.2.4). At the end of this dilution phase, at $t = t_{cf}$, f will be:

$$f(t_{cf}) = (f_o - 1)\, e^{-\mu t} + 1$$

Balanced Exponential Growth and Production ($t > t_{cf}$) At time $t = t_{cf}$, output from the pool of x_u begins, as unfilled biomass, x_u, is converted to filled biomass, x_f, the second term in equation [8.A2.1] becomes nonzero. Under these circumstances,

$$\begin{aligned} \frac{dx_u}{dt} &= \left(\frac{dx}{dt}\right)_t - \left(\frac{dx}{dt}\right)_{t - t_{cf}} \\ &= \mu x - \mu x_{t - t_{cf}} \\ &= \mu(x - x_{t - t_{cf}}) \end{aligned} \qquad [8.A2.5]$$

We also know that

$$x_u = (x - x_{t - t_{cf}}) \qquad [8.A2.6]$$

Substituting into equation [8.A2.5] above,

$$\frac{dx_u}{dt} = \mu x_u \qquad [8.A2.7]$$

This represents a restatement of the concept of balanced growth (after all t_{cf}s) in the sense that all components should increase with the same exponential growth rate μ.

We may proceed in projecting the increase in x_u by simply integrating equation [8.A2.7] to give the familiar expression for exponential growth starting at $t = t_{cf}$:

$$x_u = \{(x_u)_{t_{cf}}\}\, e^{\mu(t - t_{cf})}$$

Substituting for $(x_u)_{t=t_{cf}}$ from equation [8.A2.3],

$$x_u = \{x_o(f_o - 1 + e^{\mu t_{cf}})\}\, e^{\mu(t - t_{cf})}$$

and therefore,

$$f = \frac{x_u}{x} = \left(f_o - 1 + e^{\mu t_{cf}}\right) e^{\mu(t - t_{cf})} e^{-\mu t}$$

$$= \left(f_o - 1 + e^{\mu t_{cf}}\right) e^{-\mu t_{cf}}$$

$$= \left(f_o - 1\right)\ e^{-\mu t_{cf} + 1} \qquad [8.A2.8]$$

Equation [8.A2.8] is seen to be independent of time, as expected for balanced growth, and is identical to the value given by equation [8.A2.4] at $t = t_{cf}$ at the end of the dilution phase. Likewise, the filled fraction, 1-f, after having declined exponentially in the dilution phase, is constant from t_{cf} on, and, therefore so is the product content (as a percent of dry wt.). The longer the lag, t_{cf}, the smaller is the filled fraction, and consequently also the volumetric productivity.

In the stationary phase, if growth stops instantaneously, the first term in equation [8.A2.1] becomes zero, and we expect an exponential increase in x_f at the expense of x_u. At a time t_{cf} after growth stops, we might also expect production to stop if t_{cf} is independent of the growth rate (Case I in Section G.2.3). The filled fraction, 1-f, increases correspondingly with increasing x_f, and the specific product content (% dry wt.) returns to its original level. Other behavior is, of course, possible in the stationary phase as outlined in Section G.2.3.

In conclusion, we have shown that the observed requirement of a cellular-level lag for biosynthesis necessitates the existence of product dilution. The fraction of biomass which is filled with product (and thus the product content as a percent of dry wt.) during exponential growth depends on the length of the lag, t_{cf}, and on the exponential growth rate. Since the rate of production of filled cells depends on their concentration,

$$\frac{dx_f}{dt} = \mu x_f, \text{ where } x_f = x\,(1 - f)$$

the volumetric productivity of the reactor depends on (1-f) and therefore the lag, t_{cf} (equation [8.A2.8]). Longer lags are generally bad for productivity, but nevertheless total production is still tightly coupled to growth.

References

F. B. Abeles. 1973. "Ethylene in Plant Biology," Academic Press, New York.

F. B. Abeles and S. G. Wydoski. 1987. Inhibitors of Ethylene Synthesis and Action: A Comparison of Their Activities in a Lettuce Root Growth Model System. *J. Am. Soc. Hort. Sci. 1*, 122-125.

E. L. H. Aird, J. D. Hamill, and M. J. C. Rhodes. 1988a. Cytogenetic Analysis of Hairy Root Cultures from a Number of Plant Species Transformed by *Agrobacterium rhizogenes. Plant Cell Tissue Organ Cult. 5*, 47-57.

E. L. H. Aird, J. D. Hamill, R. J. Robins, and M. J. C. Rhodes. 1988b. Chromosome Stability in Transformed Hairy Root Cultures and the Properties of Variant Lines of *Nicotiana rustica* Hairy Roots. In "Manipulating Secondary Metabolism in Culture," R. J. Robins and M. J. C. Rhodes (eds.), Cambridge University Press, Cambridge, U.K., pp. 137-144.

P. F. Ambros, A. J. M. Matzke, and M. A. Matzke. 1986. Localization of *Agrobacterium rhizogenes* T-DNA in Plant Chromosomes by *In Situ* Hybridization. *EMBO J. 5*, 2073-2077.

W. A. Andreae, M. A. Venis, F. Jursie, and T. Dumas. 1968. Does Ethylene Mediate Root Growth Inhibition by Indole-3-Acetic Acid? *Plant Physiol. 13*, 1375-1379.

W. Armstrong. 1979. Aeration in Higher Plants. *Adv. Bot. Res.* 7, 225-332.

W. Armstrong and P. M. Beckett. 1987. Internal Aeration and the Development of Stelar Anoxia in Submerged Roots. *New Phytol. 105*, 221-245.

F. G. Bader. 1986. Physiology and Fermentation Development. In "The Bacteria," I. C. Gunsalus, J. R. Sokatch, and L. N. Ornston (eds.), Academic Press, New York, pp. 311-321.

J. E. Bailey and D. F. Ollis. 1986. "Biochemical Engineering Fundamentals," McGraw-Hill, New York.

S. Bannerjee-Chattopadhyay, A. M. Schwemmin and D. J. Schwemmin. 1985. A Study of Karyotypes and Their Alterations in Cultured and *Agrobacterium* Transformed Roots of *Lycopersicon peruvianum* Mill. *Theor. Appl. Genet.* 71, 258-262.

P. W. Barlow and J. S. Adam. 1988. The Position and Growth of Lateral Roots on Cultured Root Axes of Tomato, *Lycopersicon esculentum (Solanaceae). Plant Syst. Evol. 158*, 141-154.

D. J. Batten and M. G. Mullins. 1978. Ethylene and Adventitious Root Formation in Hypocotyl Segments of Etiolated Mung-bean (*Vigna radiata* (L.) Wilczek) Seedlings. *Planta. 138*, 193-197.

J. Berlin, V. Wray, C. Mollenschott, and F. Sasse. 1986. Formation of ß-Peltatin-A Methyl Ether and Coniferin by Root Cultures of *Linum flavum. Nat. Prod. 49*, 435-439.

L. J. Berry, M. S. Gardiner and R. T. Gilmartin. 1947. Preliminary Studies of Atypical Growth in Onion Roots Subjected to Continuous Applied Electric Currents of Low Intensities. *Growth 11*, 155-175.

D. S. Bertaud and P. W. Gandar. 1985. Referential Descriptions of Cell Proliferation in Roots Illustrated Using *Phleum pratense* L. *Bot. Gaz. 146*, 275-287.

C. A. Borgman and K. W. Mudge. 1986. Factors Affecting the Establishment and Maintenance of 'Titan' Red Rasberry Root Organ Cultures. *Plant Cell Tissue Organ Cult. 6*, 127-137.

H. Borner. 1960. Liberation of Organic Substances from Higher Plants and Their Role in the Soil Sickness Problem. *Bot. Rev. 26*, 393-424.

G. D. Bowen and A. D. Rovira. 1976. Microbial Colonization of Plant Roots. *Annu. Rev. Phytopathol. 14*, 121-144.

V. Bringi. 1991. Differentiation in plant cultures: (1) tracheary element induction in immobilized tobacco cells; (2) bioreactor considerations for shoot culture cultivation. Ph.D. thesis, Cornell University, Ithaca, New York.

V. Bringi and B. E. Dale. 1990. Theoretical and Experimental Evidence for Convective Nutrient Transport in an Immobilized Cell Support. *Biotechnol. Prog. 6*, 205-209.

H. Burstrom. 1941. On Formative Effects of Carbohydrates on Root Growth. *Botaniska Notiser*. 310-334.

D. N. Butcher and H. E. Street. 1964. Excised Root Culture. *Bot. Rev. 30*, 513 -586.

M. C. Byrne, J. Koplow, C. David, J. Temp and M. D. Chilton. 1983. Structure of T-DNA in Roots Transformed by Agrobacterium rhizogenes. *J. Mol. Appl. Genet. 2*, 201-209.

I. Capone, L. Spano, M. Cardarelli, D. Bellincampi, A. Petit, and P. Costantino. 1989. Induction and Growth Properties of Carrot Roots with Different Complements of *Agrobacterium rhizogenes* T-DNA. *Plant Mol. Biol. 13*, 43-52.

M. Cardarelli, L. Spano, A. De Paolis, M. L. Mauro, G. Vitali, and P. Constantino. 1985. Identification of the Genetic Locus Responsible for Non-Polar Root Induction by *Agrobacterium rhizogenes* 1855. *Plant Mol. Biol. 5*, 385-391.

M. Cardarelli, L. Spano, D. Mariotti, M. L. Mauro, M. A. Van Sluys, and P. Constantino. 1987. The Role of Auxin in Hairy Root Induction. *Mol. Gen. Genet. 208*, 457-463.

A. V. Chadwick and S. P. Burg. 1970. Regulation of Root Growth by Auxin-Ethylene Interaction. *Plant Physiol. 45*, 192-200.

R. M. M. Crawford. 1982. Physiological Responses to Flooding. In "Encylopedia of Plant Physiology, New Series, Vol. 12B. Physiological Plant Ecology II. Water Relations and Carbon Assimilation," Springer-Verlag, Berlin, pp. 453-477.

A. F. Croes, A. J. R. van den Berg, M. Bosveld, H. Breteler and G. J. Wullems. 1989. Thiophene accumulation in relation to morphology in roots of *Tagetes patula*. *Planta. 179*, 43-50.

E. Cutter. 1978. "Plant Anatomy," Edward Arnold, New York.

F. D'Amato. 1985. Cytogenetics of Plant Cell and Tissue Cultures and Their Regenerates. *CRC Crit. Rev. Plant Sci. 3*, 73-112.

E. Davioud, C. Kan, J. Hamon, J. Tempe, and H. P. Husson. 1989. Production of Indole Alkaloids by *In Vitro* Root Cultures from *Catharanthus trichophyllus*. *Phytochemistry. 28*, 2675-2680.

R. F. Dawson. 1942. Nicotine Synthesis in Excised Tobacco Roots. *Am. J. Bot. 29*, 813-815.

M. De Cleene and J. De Ley. 1981. The Host Range of Infectious Hairy-Root. *Bot. Rev. 47*, 148-194.

B. Deus-Neumann and M. H. Zenk. 1984. Instability of Indole Alkaloid Production in *Catharanthus roseus* Cell Suspension Cultures. *Planta Med. 50*, 427-431.

E. M. Dommisse, D. W. M. Leung, M. L. Shaw, and A J. Conner. 1990. Onion is a Monocotyledonous Host for Agrobacterum. *Plant Sci. 69*, 249-257.

M. C. Drew and J. M. Lynch. 1980. Soil Anaerobiosis, Microorganisms, and Root Function. *Annu. Rev. Phytopathol. 18*, 37-66.

T. Endo and Y. Yamada. 1985. Alkaloid Production in Cultured Roots of Three Species of *Duboisia*. *Phytochemistry 24*, 1233-1236.

L. J. Feldman. 1984. Regulation of Root Development. *Annu. Rev. Plant Physiol. 35*, 223-242.

L.J. Feldman. 1988. The Habits of Roots - What's Up Down Under. *Bioscience 38*, 612-618.

P. Filetici, L. Spano, and P. Constantino. 1987. Conserved Regions in the T-DNA of Different *Agrobacterium rhizogenes* Root-Inducing Plasmids. *Plant Mol. Biol. 9*, 19-26.

H. E. Flores, M. W. Hoy, and J. J. Pickard. 1987. Secondary Metabolites from Root Cultures. *Trends biotechnol. 5*, 64-69.

H. E. Flores, J. J. Pickard, and M. W. Hoy. 1988. Production of Polyacetylenes and Thiophenes in Heterotrophic and Photosynthetic Root Cultures of Asteraceae. In "Naturally Occurring Acetylenes and Related Compounds," J. Lam, H. Breteler, and T. Arnason (eds.), Elsevier, Amsterdam.

T. Furuya. 1988. Saponins (Ginseng saponins). In "Cell Culture and Somatic Cell Genetics of Plants," Vol. 5, Academic Press, New York, pp. 213-234.

T. Furuya, M. Ushiyama, Y. Asada, and T. Yoshikawa. 1989. Biotransformation of 2-Phenylpropionic Acid in Root Culture of *Panax ginseng*. *Phytochemistry 28*, 483-487.

J. M. Furze, J. D. Hamill, A. J. Parr, R. J. Robins, and M. J. C. Rhodes. 1987. Variations in Morphology and Nicotine Alkaloid Accumulation in Protoplast Derived Hairy Root Cultures of *Nicotiana rustica*. *J. Plant Physiol. 131*, 237-246.

A. Goldsworthy and K. S. Rathore. 1985. The Electrical Control of Growth in Plant Tissue Cultures: The Polar Transport of Auxin. *J. Exp. Bot. 36*, 1134-1141.

A. R. Gould. 1982. Control of the Cell Cycle in Cultured Plant Cells. *CRC Crit. Rev. Plant Sci. 1*, 315-344.

J. H. Graham and R. G. Linderman. 1981. Effect of Ethylene on Root Growth, Ectomycorrhiza Formation and Fusarium Infection of Douglas Fir. *Can. J. Bot. 59*, 149-155.

M. S. Greenwood, J. R. Hillman, S. Shaw, and M. B. WIlkins. 1973. Localization and Identification of Auxin in Roots of *Zea mays*. *Planta. 109*, 369-374.

J. D. Hamill, A. J. Parr, M. J. Rhodes, R. J. Robins, and N. J. Walton. 1987. New Routes to Plant Secondary Products. *Bio/Technology 5*, 800-804.

J. D. Hamill, A. J. Parr, R. J. Robins, and M. J. C. Rhodes. 1986. Secondary Product Formation by Cultures of *Beta vulgaris* and *Nicotiana rustica* Transformed with *Agrobacterium rhizogenes. Plant Cell Rep. 5*, 111-114.

J. D. Hamill, R. J. Robins, and M. J. C. Rhodes. 1989. Alkaloid Production by Transformed Root Cultures of *Cinchona ledgeriana. Planta Med.55*, 354-357.

C.H. Hanisch ten Cate, K. Sree Ramulu, P. Dijkhuis, and B. de Groot. 1987. Genetic Stability of Cultured Roots Induced by *Agrobacterium rhizogenes* on Tubers of Potatoe cv. Bintje. *Plant Sci. 49*, 217-222.

T. Hartmann and G. Toppel. 1987. Senecionine N-Oxide, the Primary Product of Pyrrolizidine Alkaloid Biosynthesis in Root Cultures of *Senecio vulgaris. Phytochemistry 26*, 1639-1643.

T. Hashimoto and Y. Yamada. 1983. Scopolamine Production in Suspension Cultures and Redifferentiated Roots of *Hyoscyamus niger. J. Med. Plant Res.* 47, 195-199.

T. Hashimoto and Y. Yamada. 1989. Paper presented at International Chemical Congress of Pacific Basin Societies, Hawaii.

T. Hashimoto, Y. Yukimune, and Y. Yamada. 1986. Tropane Alkaloid Production in *Hyoscyamus* Root Cultures. *J. Plant Physiol. 124*, 61-75.

C. A. Hay, L. A. Anderson, M. F. Roberts, and J. D. Phillipson. 1986. *In Vitro* cultures of *Cinchona* Species. *Plant Cell Rep. 5*, 1-4.

C.J. Hill. 1977. "An Introduction to Chemical Engineering Kinetics and Reactor Design," John Wiley & Sons, New York.

M. G. Hilton, P. D. G. Wilson, R. J. Robins and M. J. C. Rhodes. 1988. Transformed Root Cultures - Fermentation Aspects. In "Manipulating Secondary Metabolism in Culture," R. J. Robins and M. J. C. Rhodes (eds.), Cambridge University Press, Cambridge, U.K., pp. 239-245.

M. A. Hjortso and H. E. Flores. 1987. Root Culture as an Experimental System for the Production of Plant Chemicals. Paper presented at AICHE Annual Meeting, New York.

N. G. Hurren, S. M. Zeadan, and R. D. MacLeod. 1988. Lateral Root Development in Attached and Excised Roots of *Zea mays* L. Cultivated in the Presence or Absence of Indol-3-yl Acetic Acid. *Ann. of Bot. 61*, 573-579.

M. B. Jackson and M. C. Drew. 1984. Effects of Flooding on Growth and Metabolism of Herbaceous Plants. In "Flooding and Plant Growth," T. T. Kozlowski (ed.), Academic Press, New York.

M. B. Jackson and M. J. Gross. 1978. Can Stress Experienced by Root Systems be Alleviated by Exogenous Growth Regulators? *Br. Crop Prot. Council Monogr.*, No. 21, 95-103.

L. Jouanin, P. Guerche, N. Pamboukdjian, C. Tourneur, F. Casse-Delbart, and J. Tourneur. 1987. Structure of T-DNA in Plants Regenerated from Roots Transformed by *Agrobacterium rhizogenes* Strain A4. *Mol. Gen. Genet. 206*, 387-392.

G. Jung and D. Tepfer. 1987. Use of Genetic Transformation by the R: T-DNA of *Agrobacterium rhizogenes* to Stimulate Biomass and Tropane Alkaloid Production in *Atropa belladonna* and *Calystegia sepium* Roots Grown *in vitro. Plant Sci. 50*, 145-151.

H. Kamada, N. Okamura, M. Satake, H. Harada and K. Shimomura. 1986. Alkaloid Production by Hairy Root Cultures in *Atropa belladonna. Plant Cell Rep. 5*, 239-242.

O. Kandler. 1951. As cited by Butcher and Street (1964).

R. Kato, H. Kamada, and M. Asashima. 1989. Effects of High and Very Low Magnetic Fields on the Growth of Hairy Roots of *Daucus carota* and *Atropa belladonna. Plant Cell Physiol. 30*, 605-608.

K. Kawaguchi, M. Hirotani, T. Yoshikawa, and T. Furuya. 1990. Studies on Plant Tissue Cultures. 66. Biotransformation of Digitoxigenin by Ginseng Hairy Root Cultures. *Phytochemistry 29*, 837-843.

P.J. King. 1980. Plant Tissue Culture and the Cell Cycle. *Adv. Biochem. Eng. 18*, 1-38.

E. Knopp, A. Strauss, and W. Wehrli. 1988. Root Induction on Several *Solanaceae* Species by *Agrobacterium rhizogenes* and the Determination of Root Tropane Alkaloid Content. *Plant Cell Rep.* 7, 590-593.

O. Kondo, H. Honda, M. Taya, and T. Kabayashi. 1989. Comparison of Growth Properties of Carrot Hairy Root in Various Bioreactors. *Appl. Microbiol. Biotechnol. 32*, 291-294.

H. Konings and M. B. Jackson. 1979. A Relationship between Rates of Ethylene Producion by Roots and the Promoting or Inhibiting Effects of Exogenous Ethylene and Water on Root Elongation. *Z. Pflanzenphysiol. Bd. 92*, 385-397.

A. Linkins, L. Lewis, and R. Palmer. 1973. Hormonally Induced Changes in the Stem and Petiole Anatomy and Cellulase Enzyme Patterns in *Phaseolus vulgaris* L. *Plant Physiol. 52*, 554-560.

H. Lundegardh and G. Stenlid. 1944. On the Oxidation of Nucleotides and Flavonone from Living Roots. *Ark. Bot. 31A*, 1-27.

R. J. Luxmoore, L. H. Stolzy, and J. Letey. 1970. Oxygen Diffusion in the Soil-Plant System. II. Respiration Rate, Permeability, and Porosity of Consecutive Excised Segments of Maize and Rice Roots. *Agron. J. 62*, 322-324.

T. E. Mallory, S. H. Chiang, E. G. Cutter, and E. M. Gifford, Jr. 1970. Sequence and Pattern of Lateral Root Formation in Five Selected Species. *Am. J. Bot.* 57, 800-809.

Y. Mano, H. Ohkawa, and Y. Yamada. 1989. Production of Tropane Alkaloids by Hairy Root Cultures of *Duboisia leichhardtii* Transformed by *Agrobacterium rhizogenes. Plant Sci. 59*, 191-201.

G. J. Martin. 1957. "Biological Antagonism." The Blackiston Co., New York.

R. Matsuki, H. Onodera, T. Yamauchi, and H. Uchimiya. 1989. Tissue-Specific Expression of the *rolC* Promotor of the Ri Plasmid in Transgenic Rice Plants. *Mol. Gen. Genet. 220*, 12-16.

E. McInnes, M. R. Davey, B. J. Mulligan, K. Davies, A. W. Sargent, and A. J. Morgan. 1989. Use of a Disarmed Ri Plasmid Vector in Analysis of Transformed Root Induction. *J. Exp. Bot. 40*, 1135-1144.

F. M. Melhuish. 1962. Aspects of the Amino Acid Metabolism of Roots. Ph.D. Thesis, University of Wales.

G. C. Mitra. 1968. *In Vitro* Growth of Excised Roots of *Rauvolfia serpentina* in Continuous Culture. *Indian J. Exp. Bot. 6*, 230-231.

G. C. Mitra. 1972. Atropine Production and Growth of Excised Belladonna Root Cultures. *Indian J. Exp. Biol. 10*, 217-218.

C. Moncousin, J. Favre, and T. Gaspar. 1989. Early Changes in Auxin and Ethylene Production in Vine Cuttings before Adventitious Rooting. *Plant Cell Tissue Organ Cult. 19*, 235-242.

K. W. Mudge. 1988. Effect of Ethylene on Rooting. In "Adventitious Root Formation in Cuttings." T. D. Davis, B. E. Hassig, and N. Sankhla (eds.), Dioscorides Press, Portland, Oregon, pp. 150-164.

J. Mugnier. 1988. Establishment of New Axenic Hairy Root Lines by Inoculation with *Agrobacterium rhizogenes. Plant Cell Rep.* 7, 9-12.

T. J. Mulkey, K. M. Kuzmanoff, and M. L. Evans. 1982. Promotion of Growth and Shift in the Auxin Dose/Response Relationship in Maize Roots Treated with the Ethylene Biosynthesis Inhibitors Aminoethoxyvinylglycine and Cobalt. *Plant Sci. Lett. 25*, 43-48.

A. Nir and L. M. Pismen. 1977. Simultaneous Intraparticle Forced Convection, Diffusion and Reaction in a Porous Catalyst. *Chem. Eng. Sci. 32*, 35-41.

R. A. Norton and G. H. N. Towers. 1986. Factors Affecting Synthesis of Polyacetylenes in Root Cultures of *Bidens alba. J. Plant Physiol. 122*, 41-53.

H. Ohkawa, H. Kamada, H. Harada, and H. Sudo. 1989. Effects of Gibberellic Acid on Hairy Root Growth in *Daturia innoxia. J. Plant Physiol. 134*, 633-636.

K. M. Oksman-Caldentey, O. Parkkinen, E. Joki, and R. Hiltunen. 1989. Increased Production of Tropane Alkaloids by Conventional and Transformed Root Cultures of Hyoscyamus-muticus. *Planta Medica 55*, 682-683.

A. J. Parr and J. D. Hamill. 1987. Relationship Between *Agrobacterium rhizogenes* Transformed Hairy Roots and Intact, Uninfected *Nicotiana* Plants. *Phytochemistry 26*, 3241-3245.

A. J. Parr, A. C. J. Peerles, J. D. Hamill, N. J. Walton, R. J. Robins, and M. J. C. Rhodes. 1988. Alkaloid Production by Transformed Root Cultures of *Catharanthus roseus. Plant Cell Rep.* 7, 309-312.

S. G. Peterson, B. M. Stumman, P. Olesen, and K. W. Henningsen. 1989. Structure and Function of Root - Including (Ri) Plasmids and Their Relation to Tumor-Inducing (Ti) Plasmids. *Physiol. Plant.* 77, 427-435.

A. Petit, C. David, G. Dahl, J. Ellis, P. Guyon, F. Casse-Delbart, and J. Tempe. 1983. Further Extension of the Opine Concept: Plasmids in *Agrobacterium rhizogenes* Cooperate for Opine Degradation. *Mol. and Gen. Genet. 190*, 204-214.

V. J. Philip and S. A. Z. Nainar. 1988. *In Vitro* Transformation of Root Meristem to Shoot and Plantlets in *Vanilla planifolia*. *Ann. of Bot. 61*, 193-199.

C. L. Prince. 1991. Production of Plant Secondary Metabolites from Root Cultures: 1) Kinetics and Reactor Design and 2) Flavor Production in *Allium* species. Ph. D. Thesis, Cornell University, Ithaca, New York.

C. L. Prince, V. Bringi, and M. L. Shuler. 1991. Convective Mass Transfer in Large Porous Biocatalysts: Plant Organ Cultures. *Biotechnol. Prog.* (Accepted).

J. W. Radin and R. S. Loomis. 1969. Ethylene and Carbon Dioxide in the Growth and Development of Cultured Radish Roots. *Plant Physiol. 44*, 1584-1589.

J. Rajnchapel-Messai. 1989. "Racines...Un Nouveau Scenario". *Biofutur.* 77, 18-23.

W. E. Rauser and R. F. Horton. 1975. Rapid Effects of Indoleacetic Acid and Ethylene on the Growth of Intact Pea Roots. *Plant Physiol. 55*, 443-447.

M. S. Reid. 1987. E1. Ethylene in Plant Growth, Development, and Senescence. In "Plant Hormones and Their Role in Plant Growth and Development," P. J. Davies (ed.), Martinus Nijhoff Publishers, Dordrecht, The Netherlands, pp. 257-279.

D. M. Reinecke and R. S. Bandurski. 1987. B. Hormone Synthesis and Metabolism. In "Plant Hormones and Their Role in Plant Growth and Development," P. J. Davies (ed.), Martinus Nijhoff Publishers, Dordrecht, The Netherlands. pp. 24-42.

M. J. C. Rhodes, M. Hilton, A. J. Parr, J. D. Hamill, and R. J. Robins. 1986. Nicotine Production by "Hairy Root" Cultures of *Nicotiana rustica:* Fermentation and Product Recovery. *Biotechnol. Lett. 8*, 415-420.

W. J. Robbins. 1922. Cultivation of Excised Root Tips and Stem Tips under Sterile Conditions. *Bot. Gaz. 73*, 376-390.

W. J. Robbins and M. Stebbins. 1949. An Additional Report on the Growth of Excised Tomato Roots. *Bull. Torrey Bot. Cl. 76*, 136-140.

A. E. Rodrigues and J. M. Orfao. 1984. Intraparticle Convection, Diffusion and Zero Order Reaction in Porous Catalysts. *Chem. Eng. Commun. 27*, 327-337.

L. A. Rupp and K. W. Mudge. 1985. Ethephon and Auxin Induce Mycorrhiza-Like Changes in the Morphology of Root Organ Cultures of Mugo Pine. *Physiol. Plant. 64*, 316-322.

P. H. Saglio, M. Rancillac, F. Bruzan, and A. Pradet. 1984. Critical Oxygen Pressure for Growth and Respiration of Excised and Intact Roots. *Plant Physiol. 76*, 151-154.

K. Saito, M. Yamazaki, K. Yamakawa, S. Fujisawa, S. Takamatsu, A. Kawaguchi, and I. Murakoshi. 1989. Lupin Alkaloids in Tissue Culture of *Sophora flavescens* var. *angustifolia:* Greening Induced Production of Matrine. *Chem. Pharm. Bull. 37*, 3001-3004.

F. B. Salisbury and C. W. Ross. 1985. "Plant Physiology," 3rd ed., Wadsworth Inc., Belmont, California.

T. Schmulling, J. Schell, and A. Spena. 1988. Single Genes from *Agrobacterium rhizogenes* Influence Plant Development. *EMBO J.* 7, 2621-2629.

T. Schmulling, J. Schell, and A. Spena. 1989. Promoters of the *rolA*, *B*, and *C* Genes of *Agrobacterium rhizogenes* Are Differentially Regulated in Transgenic Plants. *The Plant Cell 1*, 665-670.

R. Siebertz, P. Proksch, V. Wray, and L. Witte. 1989. Accumulation and Biosynthesis of Benzofurans in Root Cultures of *Eupatorium cannabinum*. *Phytochemistry 28*, 789-793.

M. W. Signs and H. E. Flores. 1990. The Biosynthetic Potential of Plant Roots. *BioEssays 12*, 7-13.

L. Spano, K. Mariotti, M. Cardarelli, C. Branca, and P. Constantino. 1988. Morphogenesis and Auxin Sensitivity of Transgenic Tobacco with Different Complements of Ri T-DNA. *Plant Physiol. 87*, 479-483.

A. Spena, T. Schmulling, C. Koncz and J. S. Schell. 1987. Independent and Synergistic Activity of *rol A*, *B*, and *C* loci in Stimulationg Abnormal Growth in Plants. *EMBO J. 6*, 3891-3899.

G. Stephanopoulos and K. Tsiveriotis. 1989. The Effect of Intraparticle Convection on Nutrient Transport in Porous Biological Pellets. *Chem. Eng. Sci. 44*, 2031-2039.

H. E. Street. 1957. Excised Root Culture. *Biol. Rev. 32*, 117-155.

H. E. Street and M. P. Mc Gonagle. 1953. Factors Controlling Meristematic Activity in Excised Roots. IV. Habituation of the Main Axis Meristem of Excised Tomato Roots to Repeated Subculture. *Physiol. Plant. 6*, 707-722.

Y. Sugimoto, Y. Sugimura, and Y. Yamada. 1988. Effects of Culture Conditions on Bisbenzylisoquinoline Alkaloid Production in Cultured Roots of *Stephania cepharantha*. *Agric. Biol. Chem. 52*, 1495-1498.

M. Taya, A. Yoyama, O. Kondo, T. Kobayashi, and C. Matsui. 1989a. Growth Characteristics of Plant Hairy Roots and Their Cultures in Bioreactors. *J. Chem. Eng. Japan 22*, 84-89.

M. Taya, A. Yoyama, R. Nomura, O. Kondo, C. Matsui, and T. Kobayashi. 1989b. Production of Peroxidase with Horseradish Hairy Root Cells in a Two Step Culture System. *J. Ferment. Bioeng.* 67, 31-34.

T. H. Teeri, H. Lehvaslaiho, M. Frank, J. Votila, P. Heino, E.T. Palva, M. Van Montagu, and L. Herrera-Estrella. 1989. Regulated Expression of lac Z Gene Fusions in Tobacco. In "Primary and Secondary Metabolism of Plant Cell Cultures II," W. G. W. Kurz (ed.), Springer-Verlag, New York, pp. 274-280.

D. Tepfer. 1989. Ri T-DNA from *Agrobacterium rhizogenes:* A Source of Genes Having Applications in Rhizoshere Biology and Plant Development, Ecology and Evolution. In "Plant-Microbe Interactions, 3," T. Kosuge and E. Nester (eds.), McGraw-Hill, New York.

C. J. Thomson, B. J. Atwell, and H. Greenway. 1989a. Response of Wheat Seedlings to Low O2 Concentrations in Nutrient Solution. *J. Exp. Bot. 40*, 985-991.

C. J. Thomson, B. J. Atwell, and H. Greenway. 1989b. Response of Wheat Seedlings to Low Oxygen Concentrations. II. K+/Na+ Selectivity of Root Segments of Different Age. *J. Exp. Bot. 40*, 993-999.

U. Thron, L. Maresch, R. Beiderbeck, and J. Reichling. 1989. Accumulation of Unusual Phenylpropanoids in Transformed and Non-Transformed Root Cultures of *Coreopsis tinctoria*. *Z. Naturforsch. 44C*, 573-577.

G. Toppel, L. Witte, B. Riebesehl, K. v. Borstel, and T. Hartmann. 1987. Alkaloid Patterns and Biosynthetic Capacity of Root Cultures from some Pyrrolizidine Alkaloid Producing *Senecio* species. *Plant Cell Reports 6*, 466-469.

J. G. Torrey. 1956. Physiological Bases of Organization and Development in the Root. *Annu. Rev. Plant Physiol.* 7, 237-266.

J. G. Torrey. 1976. Root Hormones and Plant Growth. *Annu. Rev. Plant Physiol.* 27, 435-459.

J. G. Torrey. 1988. Biotechnolgy Applied to the Improvement of Underground Systems of Woody Plants. In "Genetic Manipulation of Woody Plants," C. M. Wilson and G. Kuny (eds), Plenum, New York.

K. Ushiyama. 1989. Paper presented at 1989 International Chemical Congress of Pacific Basin Societies, Hawaii.

C. Violon, D. Dekegel, and A. Vercruysse. 1984. Relation between Valepotriate Content and Differentiation Level in Various Tissues from Valerianeae. *J. Nat. Prod.* 47, 934-940.

N. J. Walton and N. J. Belshaw. 1988. The Effect of Cadaverine on the Formation of Anabasine from Lysine in Hairy Root Cultures of *Nicotiana hesperis*. *Plant Cell Rep.* 7, 115-118.

N. J. Walton, R. J. Robins, N. J. Belshaw, A. J. Parr, J. D. Hamill, and M. J. C. Rhodes. 1988. Perturbation of Alkaloid Biosynthesis in Transformed Root Cultures of Solanaceous Plants by Metabolic Precursors and Their Analogues. In "Manipulating Secondary Metabolism in Culture," R. J. Robins and M. J. C. Rhodes (eds.), Cambridge University Press, Cambridge, U.K., pp. 73-78.

P. J. Weathers, A. DiIorio and R. D. Cheetham. 1989. A Bioreactor for Differentiated Plant Tissues. In "Proceedings of the Biotech USA Conference", San Francisco, California, pp. 247-256.

K. Weissing, J. Schell, and G. Kahl. 1988. Foreign Genes in Plants: Transfer Structure Expression and Applications. *Annu. Rev. Genet. 22*, 421-477.

M. C. Whalen and L. J. Feldman. 1988. The Effect of Ethylene on Root Growth of Zea mays Seedlings. *Can. J. Bot. 66*, 719-723.

J.R. Whitaker. 1976. Development of Flavor, Odor, and Pungency in Onion and Garlic. *Adv. Food Res. 22*, 73-133.

F. F. White and V. P. Sinkar. 1987. Molecular Analysis of Root Induction by *Agrobacterium rhizogenes*. In "Plant DNA Infectious Agents," T. Hohn and J. Schell (eds.), Springer-Verlag, New York, pp. 149-177.

P. R. White. 1936. Plant Tissue Cultures. *Bot. Rev. 2*, 419-437.

P. R. White. 1946. Plant Tissue Cultures II. *Bot. Rev. 12*, 521-529.

F. Wightman, E. Schneider, and K.V. Thimann. 1980. Hormonal Facors Controlling the Initiation and Development of Lateral Roots. *Physiol. Plant. 49*, 304-314.

R. Wijnsma, R. Verpoorte, P. A. A. Harkes, F. van Iren, and H. J. G. ten Hoopen. 1988. Conditioning of Media: An Elaborate Method of Optimizing Initial Growth Hormone Concentration. In "Plant Cell Biotechnology," M. S. S. Paris et al. (eds.), Springer-Verlag, Berlin, pp. 297-303.

P. D. G. Wilson, M. G. Hilton, R. J. Robins, and M. J. C. Rhodes. 1987. Fermentation Studies of Transformed Root Cultures. In "Bioreactors and Biotransformations," G. W. Moody and P. B. Baker (eds.), Elsevier, London, pp. 38-51.

M. M. Yeoman, K. Lindsey, M. B. Miedzybrodzka, and W. R. Mc Lauchlan. 1982. Accumulation of Secondary Products as a Facet of Differentiation in Plant Cell and Tissue Cultures. In "Differentiation in vitro," M. M. Yeoman and D. E. S. Truman (eds.), Cambridge University Press, New York.

T. Yoshikawa and T. Furuya. 1987. Saponin Production by Cultures of *Panax ginseng* Transformed with *Agrobacterium rhizogenes. Plant Cell Rep. 6*, 449-453.

T. Yoshioka, H. Yamagata, A. Ithoh, H. Deno, Y. Fujita, and Y. Yamada. 1989. Effects of Exogenous Polyamines on Tropane Alkaloid Production by a Root Culture of *Duboisia myoporoides. Planta Med. 55*, 523-524.

P. Zambryski, J. Schell, and J. Tempe. 1989. Transfer and Function of T-DNA Genes from *Agrobacterium* Ti-Plasmid and Ri-Plasmid in Plants. *Cell 56*, 193-201.

R. W. Zobel. 1973. Some Physiological Characteristics of the Ethylene-Requiring Mutant, *diageotropica. Plant Physiol. 52*, 385-389.

CHAPTER 9

SHOOT CULTURES

A. INTRODUCTION

The proliferation of shoots *in vitro* without intervening roots or undifferentiated tissue is referred to as a shoot culture.

Shoot cultures are potential sources of plant secondary metabolites that are not produced by undifferentiated cultures or that are not easily available from whole plants in nature (Table 9.1). Shoot cultures are also used for the clonal propagation of many horticultural plants. However, because of the labor-intensive nature of the current solidified-media-based shoot multiplication, there is interest in automated, large-scale, liquid-based culture for micropropagation purposes (Hussey 1986). Finally, shoot cultures are also used for the production and long-term preservation of virus-free germplasm (Kartha 1981; Withers 1985).

This chapter describes the growth and biosynthetic characteristics of shoot cultures, with the primary focus on secondary metabolite production. The latter part of the chapter deals with large-scale shoot culture in liquid medium and may be more relevant to those interested in shoot cultures for micropropagation purposes.

Shoot cultures obviously differ from root cultures, but they do possess some similarities based on their common differences with undifferentiated cultures, for example, enhanced genetic stability. The other similarity, namely, their gross morphological characteristics, calls for a similar approach when it comes to reactor design. We will allude to the more comprehensive discussions in the root culture chapter when such common issues arise, without repeating them at length in this chapter.

Table 9.1. Examples of cases in which shoot differentiation is necessary for enhanced secondary metabolism.

Plant	Compound	Reference
Catharanthus roseus	Vindoline, vinblastine	Constabel et al. (1982)
Chrysanthemum spp.	Pyrethrins	Staba et al. (1984)
Digitalis spp.	Cardenolides	Hagimori et al. (1982a)
Heimia salicifolia	Quinolizidine alkaloids	Pelosi et al. (1985)
Lupinus spp.	Quinolizidine alkaloids	Wink and Hartmann (1980)
Papaver somniferum	Morphinan alkaloids	Constabel (1985)
Withania somnifera	Withanolides	Heble (1985)
Pelargonium spp.	Essential oils	Charlwood et al. (1989)
Mentha spp.	Essential oils	Hirata et al. (1990)
Citrus spp.	Limonin, naringin	Barthe et al. (1987)

B. GROWTH CHARACTERISTICS OF SHOOT CULTURES

B.1. Initiation and Maintenance of Shoot Cultures

A shoot culture is typically initiated from a sterile germinated seedling or a whole plant by dissecting the shoot apical meristem, or a leaf, or a stem section onto solidified medium. Since many shoot cultures have been initiated with a view toward micropropagation, a large bank of data exists on their initiation and maintenance for a wide range of plants, including medicinal and aromatic plants (Bajaj 1988, 1989).

Shoot cultures proliferate by continuously forming new meristems. The meristems may be axillary, that is, those arising at the junction of a leaf and stem, or adventitious, those arising from anywhere on the leaf or stem. The greater the rate of formation of new meristems per unit mass of tissue, the greater will be the overall growth rate. Figure 9.1 shows the anatomy of a typical shoot apical meristem. Shoot growth and differentiation are complex processes and are not as well understood as root growth. The reader unfamiliar with shoot anatomy is urged to consult any plant biology text for further information (e.g., Esau 1977; Raven et al. 1981).

The important hormones for growth are cytokinins, alone or in combination with auxin. Cytokinins stimulate shoot bud growth by relieving apical dominance. Growth *in vitro* is usually heterotrophic, that is, with an exogenously supplied carbon source, though photoautotrophic growth is possible (Hagimori et al. 1984a). Though certain shoot cultures are capable of rapid growth in darkness (Hagimori et al. 1982a), light

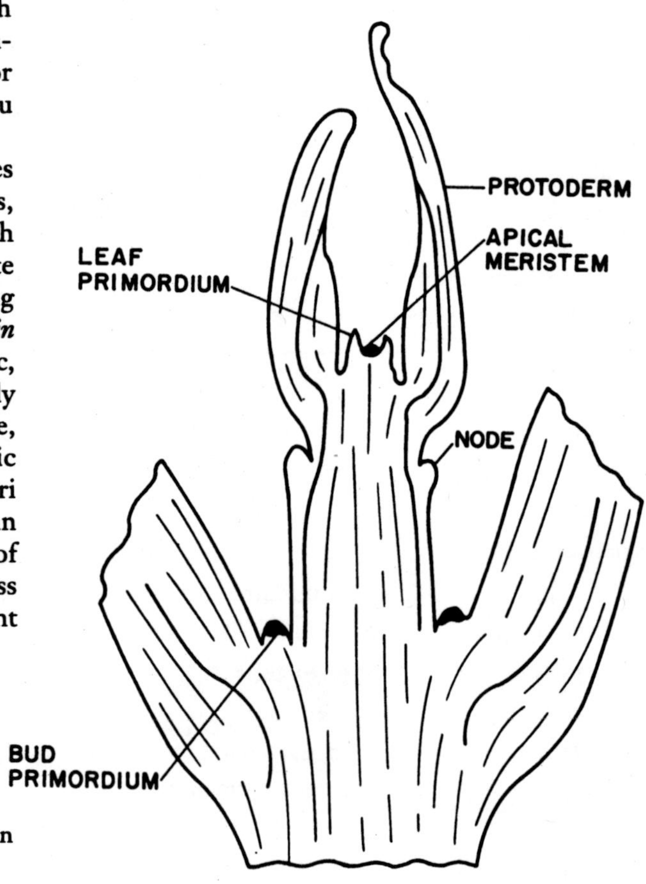

Figure 9.1. A shoot apical meristem of *Coleus*. After Raven et al. (1981), with permission.

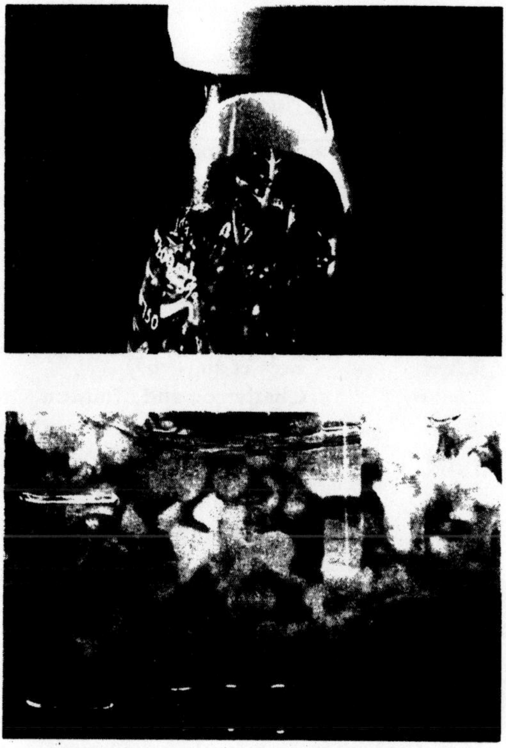

(A)

(B)

Figure 9.2. Shoot cultures of *Catharanthus roseus* (A) cultivated in a shake flask and *Nicotiana tabacum* (B) cultivated in a bioreactor.

is usually required for normal greening and leaf development. Figure 9.2 shows photographs of a tobacco and a periwinkle shoot culture, both grown in the presence of light. Just as infection with *Agrobacterium rhizogenes* yields hairy roots, infection with certain strains of *A. tumefaciens* and *A. rubi* also gives rise to "shooty teratomas" which synthesize their own hormones (Ondrej et al. 1987). The use of shooty teratomas for studies on secondary biosynthesis has been reported by Eilert (1989), Saito et al. (1989a), and Spencer, et al. (1990). The differences, if any, between the growth and biosynthetic capabilities of shooty teratomas and that of normal shoot cultures have yet to be studied.

The increase in weight over time in a batch vessel follows a sigmoidal curve as for undifferentiated cultures. Table 9.2 lists the doubling times of a few shoot cultures. Certain species are capable of very rapid growth. The growth rate of the *Digitalis* shoot culture of Hagimori et al. (1982a) compares very favorably with that of a suspension culture (t_D = ca. 3.2 days for both). However, further experiments with culture conditions are clearly required to improve the growth rates of the slower-growing cultures. The technological challenge of reproducing and sustaining high growth rates on a larger scale will be discussed in Section F.

Table 9.2. Growth rates of shoot cultures.

Species	Doubling time (days)	Average[a] or maximum?	Reference
Nicotiana tabacum	ca. 3	Max	Bringi (1991)
Withania somnifera	ca. 3	Ave	Heble (1985)
Digitalis purpurea	3.2	Max	Hagimori et al. (1982a)
Artemisia annua	4.3	Max	Park et al. (1989)
Begonia spp.	5.4	Ave	Takayama and Misawa (1981)
Digitalis lanata	6.6	Ave	Lui and Staba (1979)
Rauwolfia serpentina	8.5	Ave	Roja et al. (1985)
Pelargonium tomentosum	10	Max	Charlwood and Moustou (1988)
Cinchona spp.	10	Ave	Staba and Chung (1981)
Atropa belladonna	10	Ave	Benjamin et al. (1987)
Catharanthus roseus	10	Ave	Krueger et al (1982)
Catharanthus roseus	11.2	Ave	Hirata et al. (1987)

[a] Maximum doubling times were estimated by drawing a tangent to the batch growth curve at the maximum rate of biomass increase. Average doubling times were estimated from the reported growth index (final/initial weight) by assuming exponential growth throughout the growth period. Since average doubling times do not account for the lag and stationary phases, the maximum doubling time may be considerably lower than the average doubling time.

B.2. Genetic Stability of Shoot Cultures

The stability of cell lines is an important issue for propagation as well as for secondary metabolite production. Undifferentiated cultures of many plant species are notorious for their genetic and/or epigenetic instabilities (e.g., see Gould 1986). From the considerable research conducted on the genetic stability of micropropagated species, the following generalizations can be made (for a further discussion, see Hussey 1986). Organized meristems are generally more resistant to genetic alterations than are undifferentiated cultures. Furthermore, the genetic fidelity of the shoot culture depends on whether it is axillary or adventitious. Axillary shoot cultures are the most stable. Since adventitious shoots may arise from a single cell, they are more prone to genetic alterations. If any proliferation of undifferentiated tissue (callus) takes place before new adventitious shoot initials arise, then the culture is prone to variability.

Though thorough investigations into the long-term stability of secondary metabolite-producing shoot cultures have not been reported, anecdotal observations from various publications attest to the general stability of shoot cultures even in the absence of a constant selection pressure. Roja et al. (1987) reported the consistently high production of alkaloids in shoot cultures of *Rauwolfia serpentina* without major changes in the alkaloid profile and levels for five years. Miura et al. (1988) reported the stable maintenance of a relatively fast-growing, high-producing *Catharanthus roseus* shoot culture for 30 months with little change in the growth rates, morphological characteristics, or alkaloid productivity.

Whereas genetic fidelity is desirable once a high-producing culture is selected, genetic variability is desirable during screening and selection programs. Screening may be conducted using different shoot culture clones or by screening undifferentiated cultures followed by shoot redifferentiation. Further descriptions of selection and screening are found in Chapters 2, 8 and 11 and in Dougall (1987), Widholm (1987), and Yamada (1984).

C. BIOSYNTHETIC CHARACTERISTICS OF SHOOT CULTURES

This section is divided into three subsections. First, we will examine the possible reasons for the biosynthetic differences between shoot cultures and undifferentiated cultures. Second, we will answer the question, "If the leaf of a plant accumulates a particular secondary metabolite, will shoot cultures necessarily do so?" Finally, we will summarize attempts made at enhancing secondary biosynthesis in shoot cultures.

C.1. Biosynthetic Differences between Shoot Cultures and Undifferentiated Cultures

As shown in Table 9.1, shoot differentiation enables the expression of several biosynthetic pathways that are expressed at very low levels, if at all, in undifferentiated cultures. The technology for large-scale culture of undifferentiated cultures is well established, whereas for shoot cultures it is not. Therefore, much initial effort is usually spent in confirming that shoot differentiation is really necessary for secondary metabolism and that undifferentiated cultures cannot be easily coaxed to produce the secondary metabolite. In other systems, further efforts have been made to understand the biochemical basis for the differential expression. The pinpointing of precisely which biochemical step or steps along a biosynthetic pathway are blocked is a first step that could lead to the use of genetic engineering techniques to constitutively express the blocked genes in undifferentiated cultures. Alternatively, if the coordinated expression of many enzymes is necessary, or if organelle or cell specialization is necessary for storage, then genetic manipulation may perhaps not be as successful an approach.

C.1.1. Need for Specialized Cells Many secondary metabolites are synthesized and stored in specialized cells in the plant. Since leaf formation allows cell specialization, it is perhaps not surprising that shoot cultures allow the expression of certain pathways not normally expressed in undifferentiated cultures. For example, key enzymes in the monoterpene biosynthetic pathway have been localized in the glandular trichomes of spearmint (*Mentha spicata*) leaves (Gershenzon et al. 1989). Charlwood et al. (1989) were able to correlate essential oil accumulation with glandular hair density in shoot cultures of *Pelargonium fragrans*. The comprehensive review by Mulder-Krieger et al. (1988) lists many cases in which essential oils are accumulated in structurally differentiated structures. However, though specialized cells may be necessary in certain systems, their presence may not be sufficient to confer biosynthetic competence. For example, though artemisinin is accumulated in glandular trichomes in *Artemisia annua* leaves, shoot cultures possessing trichomes were still unable to synthesize appreciable levels of artemisinin (Martinez and Staba 1988).

C.1.2. Chloroplasts and Developmental Regulation Many studies have reported a positive correlation between light, chlorophyll content, and secondary metabolite production in shoot culture systems (Hagimori et al. 1982a; Pelosi et al. 1985; De Luca et al. 1988; Saito et al. 1989b), suggesting that certain key steps in the pathway may be chloroplast-associated. In some of these cases, chloroplast development is sufficient for expression of secondary metabolism, and shoot differentiation is not necessary, as evidenced by the ability of chlorophyllous suspensions (i.e., suspensions containing chloroplasts) to produce the secondary metabolite. In other cases, shoot differentiation is necessary, and chlorophyllous suspension cultures are incapable of producing the secondary metabolite (summarized in Table 9.3).

Examples of enhanced secondary biosynthesis by chlorophyllous undifferentiated cultures have been summarized by Husemann et al. (1989). Photoautotrophic cultures of *Morinda lucida* were capable of lipoquinone biosynthesis whereas heterotrophic cultures were not capable of forming this class of compounds (Igbavboa et al. 1985). Photosynthetically active suspension cultures of *Ruta graveolens* were capable of ac-

Table 9.3. Summary of effects of light, chloroplasts, and shoot differentiation on secondary metabolism. ***(See text for details.)***

Plant	Light or chloroplasts		Shoot differentiation		Reference
	Necessary	Sufficient	Necessary	Sufficient	
Sophora flavescens	Yes	Yes	No	No	Saito et al. (1989b)
Lupinus sp.	Yes	Traces[a]	Yes	No	Wink and Hartmann (1980)
Heimia salicifolia	Yes	No	Yes	No	Pelosi et al. (1985)
Digitalis purpurea	No[b]	No	Yes	Yes	Hagimori et al. (1982a)
Catharanthus roseus	Yes	No	Yes	No	De Luca et al. (1989)

[a] Light did increase accumulation, but levels were still two to three orders of magnitude below that of shoots.

[b] Though light was not necessary for biosynthesis, accumulation in shoot cultures increased by a factor of 3 in the presence of light.

cumulating a more complex spectrum of volatile components than were heterotrophic cultures (Drawert et al. 1984; Jordan et al. 1986). Similarly, the quinolizidine alkaloid matrine is produced by undifferentiated callus as well as by shoot cultures of the Chinese medicinal plant, *Sophora flavescens* (Saito, et al.,1989b). Light was found to be necessary, and the alkaloid content correlated well with chlorophyll content in differentiated and undifferentiated tissues. Though shoot differentiation was not necessary for matrine production, it was necessary for the production of a more complex spectrum of lupin alkaloids.

In the case of quinolizidine alkaloids production by *Lupinus* spp., shoot differentiation was found to be necessary for enhanced expression, and the presence of light and chloroplasts was required as well. In this system, the amino acid lysine undergoes transformations to yield the alkaloid lupanine, which in turn leads to a diverse spectrum of compounds. The biosynthesis of lysine, as well as of two enzymes involved in lupanine biosynthesis, has been localized in the chloroplasts (Wink and Hartmann 1982). Photomixotrophic suspension cultures of *Lupinus* spp., and other quinolizidine alkaloid-producing species, showed a 4- to 10-fold increase in alkaloid accumulation compared to heterotrophic cultures. This accumulation, however, was still two to three orders of magnitude lower than the levels accumulated by leaves in the whole plant or by shoot cultures. Furthermore, whereas lupanine was the main alkaloid accumulated by the chlorophyllous suspensions, shoot cultures accumulated a more complex spectrum of alkaloids typical of the whole plant (Wink and Hartmann, 1980). Similarly, quinolizidine alkaloids were accumulated by green shoot cultures of *Heimia salicifolia*, but not by chlorophyllous suspensions (Pelosi et al. 1985, 1986).

The production of digitoxin by *Digitalis purpurea* shoot cultures provides an example of a case in which shoot differentiation is necessary and sufficient for expression of secondary metabolism, but in which light enhances productivity. Hagimori et al. (1982a) found that heterotrophic and chlorophyllous suspensions of *Digitalis purpurea* were incapable of digitoxin accumulation whereas light-grown shoot cultures accumulated significant amounts of digitoxin. However, dark-grown shoot cultures were still capable of producing a third of the digitoxin, suggesting that active chloroplasts were not necessary for digitoxin biosynthesis and that shoot differentiation was both necessary and sufficient for secondary metabolite production. Light may have therefore favored both digitoxin accumulation as well as chloroplast development independently.

C.1.3. Case Study: *Catharanthus roseus* Shoot Cultures The production of indole alkaloids by *Catharanthus roseus* shoot cultures is examined in detail since it provides a good example of the biochemical differences between structurally undifferentiated and differentiated tissues. The inability of undifferentiated cultures of *Catharanthus roseus* to accumulate the commercially important, anticancer, dimeric indole alkaloids vincristine and vinblastine has long been recognized. Of the two monomeric precursors, catharanthine and vindoline, undifferentiated cultures are capable of producing catharanthine, but not vindoline. The coupling of the two

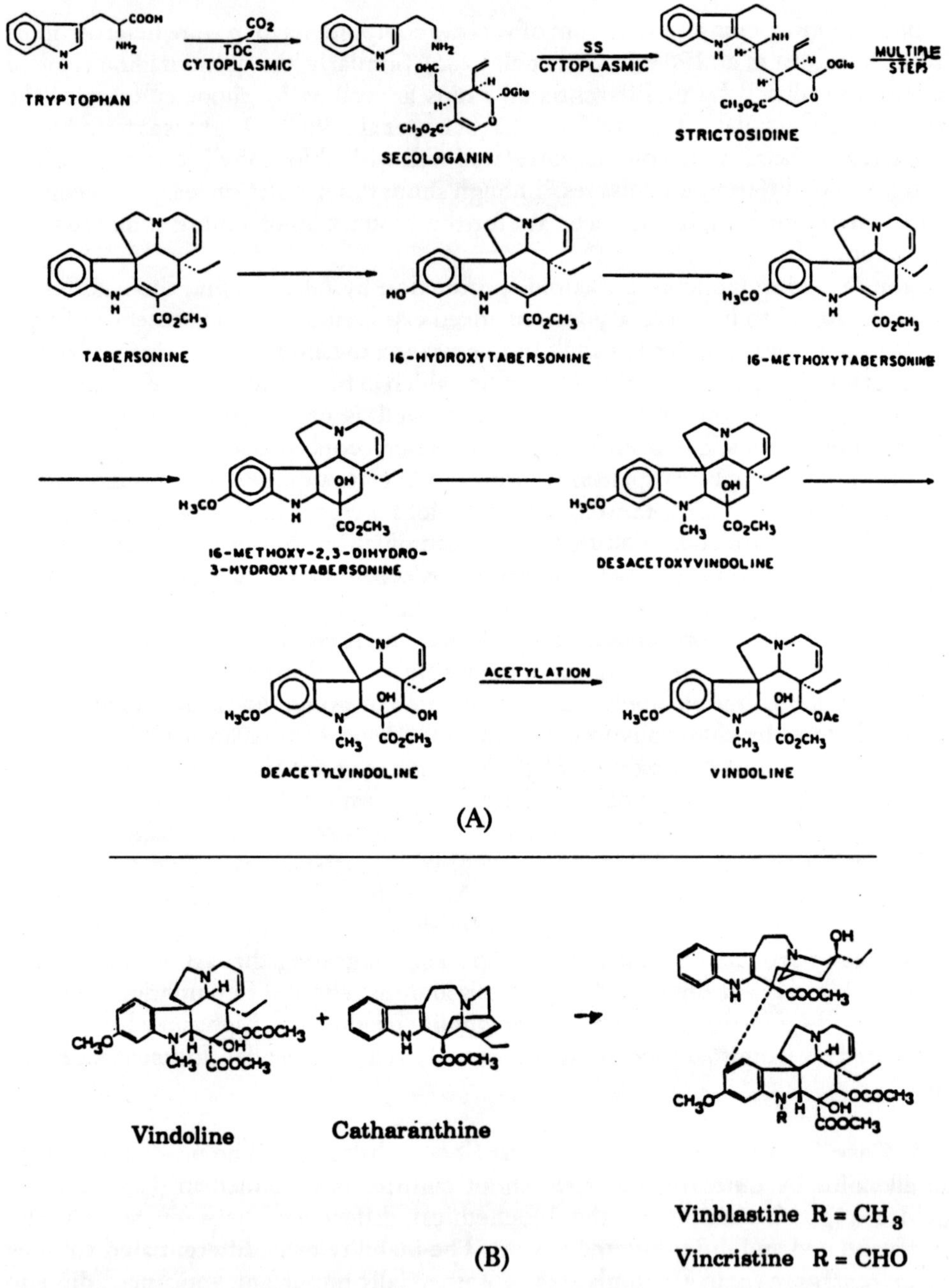

Figure 9.3. (A) Proposed pathway for vindoline biosynthesis in *Catharanthus roseus* after De Luca et al. (1989). (B) Formation of dimeric indole alkaloids from vindoline and catharanthine.

monomers to form the dimeric alkaloids is a relatively simple step and can be catalyzed by peroxidases present in *Catharanthus roseus* suspensions, as well as by other peroxidase systems (Endo et al. 1988; Goodbody et al. 1988). The production of vindoline has thus been the limiting step in the commercial production of vincristine and vinblastine by tissue culture and has long been a pursuit of plant biotechnologists.

Constabel et al. (1982) showed that callus initiated from hypocotyls of *Catharanthus roseus* did not produce vindoline. Regeneration of shoots from the callus was accompanied by the capacity for vindoline accumulation, which was lost again when a callus was induced from the regenerated shoot. These results clearly showed that undifferentiated cultures did possess the genetic ability to produce vindoline, but expressed this ability only upon shoot differentiation.

The investigations of De Luca and co-workers at the Plant Biotechnology Institute in Saskatoon, Canada, have helped in understanding the link between biochemical and structural differentiation at the enzymological level (summarized in De Luca et al. 1989). The formation of the intermediate tabersonine from tryptophan involves the cytoplasmic enzymes tryptophan decarboxylase (TDC), strictosidine synthase (SS), and a series of other steps. The suggested pathway for vindoline biosynthesis via tabersonine is shown in figure 9.3. The three enzymes involved in the final steps, namely, N-methyl transferase (NMT), an oxoglutarate-dependent dioxygenase, and deacetylvindoline acetyl-transferase (DAT), have been characterized. Whereas undifferentiated cultures showed TDC and SS activities upon transfer to production medium or upon exposure to microbial elicitors (De Luca et al. 1987; Eilert et al. 1987), NMT or DAT activities were not detectable.

Subcellular localization studies with seedlings showed that TDC, SS, the dioxygenase, and DAT were cytosolic enzymes, whereas NMT was probably associated with the thylakoid of chloroplasts (De Luca and Cutler 1987; De Luca, et al. 1989). Furthermore, whereas TDC, SS, and NMT were not affected by light, DAT was strongly activated when etiolated seedlings were exposed to light (De Luca et al. 1988). The association of NMT with chloroplasts and the activation by DAT by light prompted De Luca et al. (1987) to investigate whether photoautotrophic suspension cultures were capable of expressing NMT and DAT activities. The inability of the cultures to do so demonstrated that shoot differentiation was indeed necessary and that the presence of light and chloroplasts was not sufficient for vindoline biosynthesis.

Accordingly, several groups have established shoot cultures of *Catharanthus roseus* and have demonstrated accumulation of vindoline as well as of dimeric indole alkaloids in some cases (Krueger et al. 1982; Hirata et al. 1987; Endo et al. 1987; Miura et al. 1988).

C.2. Differences Between Shoot Cultures and Whole Plants

It may often be incorrect to assume that a shoot culture will produce all the compounds that are present in the leaves of a whole plant. The site of synthesis may be different from the site of storage (Wink 1987). For example, the commercially important purine and tropane alkaloids are accumulated in the leaves of *Nicotiana* and *Duboisia*, respectively, but are synthesized by the roots. Accordingly, root cultures of both species are capable of alkaloid biosynthesis (Rhodes et al. 1986; Yoshioka et al. 1989), whereas shoot cultures of both species do not accumulate alkaloids. Both shoot cultures do, however, take up exogenously added alkaloids and perform certain biotransformations (Saito et al. 1989a; Hashimoto and Yamada 1988).

Other factors affecting the whole plant such as leaf maturity, juvenility, flowering, and adaptation to climatic stresses and seasonal variations no doubt affect secondary metabolite production. Since shoot cultures are not subjected to the same factors *in vitro*, their secondary metabolite profile may vary qualitatively and quantitatively from that of the leaves of the whole plant. Shoot cultures of *Catharanthus roseus*, for example, accumulate significantly lower levels of the dimeric indole alkaloids than do leaves of a whole plant (Endo et al. 1987; Miura et al. 1988). Whereas the essential oil component profile of shoot cultures of *Pelargonium fragrans* was markedly different from that of the corresponding leaf oil with respect to the oxygenated monoterpenes, the essential oil components of the leaf oil and the shoot culture of *Pelargonium tomentosum* (Charlwood and Moustou 1988) and of *Mentha tricata* (Spencer et al. 1990) were very similar. The significant differences in the withanolide (a group of cytotoxic steroids) profiles between shoot cultures and whole plants of the ancient Indian medicinal plant *Withania somnifera* prompted Heble (1985) to suggest that shoot cultures may provide a source for obtaining novel secondary metabolites.

C.3. Issues Pertaining to Secondary Metabolite Productivity

Here we briefly summarize the few attempts made at using productivity enhancement strategies with shoot cultures. The efficacy of these strategies has been well documented with undifferentiated cultures, and they promise to be similarly successful with shoot cultures as well. Some broader issues pertaining to these strategies are discussed in Chapter 11. The discussion on selection and screening is very similar to that for root cultures and can be found in Section G.4.1 of Chapter 8.

C.3.1. Kinetics of Product Accumulation The design and operation of a production process requires a knowledge of the kinetics of product formation in relation to the growth rate. If the product is intracellularly accumulated, and if rapid growth and product formation occur concomitantly, then a process favoring rapid growth of tissue is the obvious choice. If product formation occurs only after growth ceases in a batch, then rapid growth of tissue to its final attainable density, followed by exposure to conditions favoring secondary metabolite formation, may be considered. This is analogous to the two-stage batch process used for shikonin production. If the tissue is capable of product secretion with prolonged retention of high biosynthetic rates and viability, then a continuous system with biomass reuse and product removal, much like an immobilized cell system, may be an alternative.

The kinetics of product formation are usually obtained from batch shake-flask experiments. As elaborated in Section G.2 of Chapter 8, such experiments should be designed carefully, and the data must be interpreted with caution. Generally, the increase in total biomass is monitored over time. However, the biomass itself comprises several different tissues, for example, stems, young leaves, and mature leaves. If the secondary metabolite of interest accumulated preferentially only in certain regions of the shoot (e.g., see Wink 1987; Zobel and Brown 1989a, b), say, in mature leaves, then the relationship between product formation and total biomass increase with time would be complex. However, in the few cases in which batch time courses of biomass increase and product formation have been reported, the complexities mentioned above have not been encountered, and the product has been growth-associated (Hagimori et al.1982a; Charlwood and Moustou, 1988; Pauthe-Dayde et al. 1990).

C.3.2. Product Secretion and *In Situ* Removal If the product of interest is growth-associated and completely intracellular, then the shoot culture simply provides an alternative to whole plants as a source of material for product extraction. If part of the secondary metabolite is secreted and can be produced by nongrowing tissue, then continuous operation with biomass reuse, much like an immobilized cell process, is possible. Digitoxin (Hagimori et al. 1984b) and vindoline (Krueger et al. 1982) have not been observed in the culture medium, whereas up to 35-40% of the total essential oil accumulated by *Pelargonium tomentosum* has been observed in the medium (Charlwood and Moustou 1988). Note that for a successful immobilized cell process, prolonged viability and sustained biosynthetic capacities are required (Section B.1.3 of Chapter 7). Therefore, the presence of product in the extracellular medium must be due to true secretion, and not, for example, simply due to lysis of the glandular hairs in which the essential oil accumulates.

Many shoots do possess specialized secretory apparatuses since many secondary metabolites such as resins are naturally secreted in nature. It is reasonable to expect partial product secretion in such systems if the shoots are suitably challenged, for example, by an elicitor (see Section C.3.5). The use of ethylene may be a promising approach in this regard, since ethylene does stimulate secretory processes as a wound response in whole plants. In *Hevea*, the rubber tree, the application of ethylene via ethephon is used commercially to maintain the flow of latex from the trees (Reid 1987). The various methods used to induce product secretion (listed in Chapter 7, Section B.2) in undifferentiated cultures have not been attempted as yet with shoot cultures.

The potential for the use of structurally differentiated tissue in a continuous immobilized cell process is illustrated by the results of Jacob John et al. (1990). The complex, oleo-resin gum "guggul" from *Commiphora wightii* contains the clinically interesting antichlolesterolemic compounds guggulsterone Z and E. The site of synthesis of the resin is localized in the highly specialized glandular epithelial cells of the gum resin duct. Jacob John et al. (1990) isolated tissue containing these specialized cells from stem sections of a plant and demonstrated the stable production of guggulsterones *in vitro*. Moreover, these compounds were selectively secreted into the medium, and high productivities were obtained with prolonged viability of the tissue.

As discussed in Chapter 11, if partial secretion does take place, then *in situ* product removal may increase the rate, selectivity, and extent of product secretion and may prevent loss of secondary metabolites due to degradation or volatility. It is likely that these advantages will apply to shoot culture processes as well.

C.3.3. Medium Optimization Most studies with shoot cultures have been conducted so far primarily with the intention of demonstrating unique biosynthetic capabilities. The optimization of medium components for high secondary metabolite productivity has been performed for few systems. As for undifferentiated cultures and root cultures, it is likely that an empirical approach will be necessary and that each system will have to be dealt with independently.

A systematic study of the effect of medium components on growth and digitoxin biosynthesis in *Digitalis purpurea* shoot cultures has been conducted by Hagimori and co-workers (Hagimori et al. 1982a,b; 1983). White light was necessary for high rates of biosynthesis, but the effects of different wavelengths were not studied. The independent effect of hormones on growth and secondary metabolism was clearly demonstrated. A readjustment of calcium, phosphate, and magnesium levels resulted in a threefold increase in product accumulation with no suppression of growth, showing that optimization of medium conditions can contribute to significant increases in productivity.

C.3.4. Use of Biosynthetic Precursors The successful use of biosynthetic precursors to increase secondary metabolite production by shoot cultures has been demonstrated in a variety of systems (Table 9.4). As discussed in Chapter 11, factors such as the level, rate, and timing of precursor addition, stoichiometry, and precursor uptake are likely to be important with shoot cultures as well.

Table 9.4. Summary of successful attempts at enhancing product accumulation in shoot cultures by precursor feeding.

Plant	Product	Precursor	% increase in specific content	Negative effect on growth (%)
Atropa belladonna (Benjamin et al.1987)	Tropane alkaloids	Phenylalanine + ornithine	400	30
Cinchona ledgeriana (Chung and Staba 1988)	Quinoline alkaloids	Tryptophan	66	8
		Secologanin	42	20
Digitalis lanata (Lui and Staba 1979)	Digoxin	Glycocholate	ca. 300	Not reported
		Cholesteryl acetate	ca. 300	Not reported
		Progesterone	ca. 250	Not reported
Digitalis purpurea (Hagimori et al. 1983)	Digitoxin	Progesterone	80	No effect

C.3.5. Elicitor Treatments Eilert (1989) studied the responses of suspension cultures, shoot cultures, and hydroponically grown whole plants of *Ruta graveolens* to a yeast elicitor. *R. graveolens* accumulates a wide spectrum of secondary metabolites, and levels of three classes of compounds were monitored; viz., acridone alkaloids, furoquinoline alkaloids, and furanocoumarins. Suspensions, shoot cultures, and whole plants all accumulated increased levels of acridone alkaloids in response to the elicitor. Suspensions and shoot cultures also accumulated higher levels of furoquinolines and furanocoumarins but showed qualitative and quantitative differences in the extent of the response. Interestingly, shoots from the whole plants did not alter their constitutive accumulation of these classes of compounds in response to the elicitor. It may thus be difficult to predict how a shoot culture will respond to elicitor treatment based on a knowledge of the response of undifferentiated cultures and whole plants to the same elicitor.

An interesting observation was that, of the five-to sixfold increase in the dictamnine (a furoquinoline) content in response to the elicitor, 60-80% of the total alkaloid was extracellular. A similar secretion effect was observed in the suspension cultures. The use of elicitor treatments as a productivity-enhancing strategy seems to be a promising avenue for further research.

D. PHYSIOLOGICAL EFFECTS OF LIGHT

Thus far, we have examined the growth and biosynthetic characteristics of shoot cultures. Before discussing the factors affecting scale-up, the physiological effects of light are briefly described. Light, besides being the fundamental source of energy, also profoundly affects almost every phase of plant growth and development. More information on the importance of light in plant biology can be found in standard texts (e.g., Kendrick and Kronenberg 1986; Salisbury and Ross 1985; Leopold and Kriedeman 1975). We have seen in the previous section that light can profoundly affect secondary metabolism, either by activating chloroplast development or by an independent mechanism. Though the optimum lighting strategy will have to be empirically determined for each system, a knowledge of the physiological basis of light action will aid in designing appropriate experimental strategies. The following variables may be expected to be of importance:

1. Quality of light - are particular wavelengths more effective?
2. Intensity of light
3. Periodic fluctuations in light supply

While perusing the literature, the reader unfamiliar with photomorphogenesis should note that studies usually involve two types of responses: (i) reversible responses to short-term light exposures, and (ii) responses to longer term exposures, known as the high-irradiation response ("HIR"). Though short-term exposure studies have provided valuable fundamental insights into the mechanisms underlying photomorphogenesis, the high-irradiance response is likely to be more important for secondary metabolite production.

D.1. Quality of Light

Photomorphogenesis, the control of morphogenesis by light, is mediated by several pigment molecules, some of whose structures and modes of action have yet to be elucidated. Phytochrome is an important photomorphogenic agent which is present in all organs. It is a protein to which a chromophore is attached, and its precise biochemical mode of action is still under study. The transformations of phytochrome in the presence of light of various wavelengths have been well documented (e.g., Salisbury and Ross 1985). Red light (666 nm) mediates the conversion of the inactive form of phytochrome, P_r, to the active form, P_{fr}, which in turn stimulates various responses. P_{fr} reverts to P_r upon exposure to far-red light (730 nm), or, more slowly, by dark reversion. Alternatively, P_{fr} can be degraded irreversibly. Other morphogenic responses involving other pigments are mediated by violet and blue light, and yet other responses may be activated by other wavelengths.

The effect of light on enzyme activation has been reviewed by Zucker (1972), Schopfer (1977), and, more recently, by Simpson and Herrera-Estrella (1990). The direct activation of enzymatic steps for anthocyanin biosynthesis *via* the photomorphogenic pigments has been well documented (e.g., see Mancinelli 1985) The light activation of the acetyltransferase step (DAT) in vindoline biosynthesis described in in Section C.1.3 may occur by a similar mechanism. The effects of light on secondary metabolism in tissue cultures have been reviewed (e.g., Seibert and Kadkade 1980; Mantell and Smith 1983; Collin, 1987).

We have seen that development of chloroplasts may be necessary for secondary metabolism. The triggering action of P_{fr} has been shown to initiate chlorophyll biosynthesis and chloroplast development. The main light-absorbing components of the chloroplasts are the chlorophylls a and b, with absorption maxima in the blue and red regions (Hoober 1984). Other light-absorbing molecules are the carotenes and the xanthophylls, with absorption spectra in the 390- to 510-nm range.

D.2. Light Intensity

There is evidence that very high light intensities may actually reduce the number of chloroplasts per cell, since excess absorbed light energy causes cellular damage. However, if the rate-limiting enzyme is thylakoid-associated, then it is reasonable to expect increased productivities with an increase in thylakoid protein content, which in turn may be directly proportional to the light intensity. Thus, the optimum light intensity may in fact be a compromise. It should be recognized that once light is supplied, then another non-light-regulated step could become rate limiting, in which case further light supply would not increase the overall rate of biosynthesis.

If light affects secondary metabolism by direct enzyme activation and not via chloroplast formation, then the effect of varying light intensities has to be empirically determined for each system. In the case of anthocyanin biosynthesis, the dependence of anthocyanin production on the intensity of illumination (also called the fluence rate) is different for different wavelengths of activation (Mancinelli 1985).

D.3. Fluctuations in Light Supply

From the bioreactor design point of view, it would be simpler to subject tissues to intermittent, rather than to constant, light. For example, if light were only supplied from the outside of a stirred vessel, then tissue would be exposed to light whenever it happened to be close to the walls of the vessel. There is evidence that rates of photosynthesis may be enhanced when cells are exposed to intermittent, rather than to constant, light (see Leopold and Kriedeman 1975). The response is, however, dependent on the intermittency, that is, the period of darkness between successive light exposures. The effects of intermittent light supply on anthocyanin biosynthesis have been investigated, and results indicate that intermittent light supports significant rates of secondary metabolite production (Mancinelli 1985).

The effect of diurnal variations, that is, day/night cycles, must also be studied independently (e.g., see Staba et al. 1984). The diurnal time scale is, of course, much different from the time scale for fluid mixing.

E. BIOREACTOR DESIGN CONSIDERATIONS FOR SHOOT CULTURES

Once a shoot culture has been developed to produce a secondary metabolite, and the appropriate environmental conditions for maximizing productivity have been established, the next step is process scale-up. The kinetics of product formation primarily determines the mode of operation to be used; for example, should the process be one- or two-stage batch or continuous?. Since the strategy used for process selection is substantially similar to that used for suspension cultures, readers are urged to consult Chapter 6 for further discussion of reactor operating modes. The considerations for reactor design, however, are more similar to those for root cultures than to those for undifferentiated cultures and are more comprehensively discussed in Section H of Chapter 8.

The shoot culture bioreactor should serve two purposes. First, it should provide an environment that supports rapid growth. The growth rate achieved in a bioreactor should approach or, better still, exceed that achievable in a shake flask. Moreover, the maximum growth rate should be sustained over several doublings, and the final density of shoots should be maximized.

If the product is non-growth-associated, and continuous operation with continuous product removal is desired, then the bioreactor should be designed to sustain the viability and productivity of tissue at high densities for prolonged periods. The constraints for doing so may be even more stringent than for providing rapid growth. For example, higher oxygen concentrations or light intensities may be required during secondary metabolite production than during growth. Most studies, so far, have focused on bioreactor design for growth of shoot cultures.

Before discussing bioreactor studies with shoot cultures, we first consider how the following issues will affect bioreactor design:

- Shoot morphology
- Effect of submersion
- Light requirements

E.1. Shoot Morphology

Shoot morphology determines the susceptibility of tissue to mechanical stresses associated with fluid flow, aeration, and agitation. Thus far, no attempts have been made to quantify the susceptibility of shoots to damage due to the stresses encountered in a bioreactor environment.

Shoot morphology also influences nutrient mass transfer. The problem of oxygen transfer to the interior of clumps or mats of roots was described in detail in Section H of Chapter 8. Shoot cultures, too, tend to form large clumps, and it is reasonable to expect that mass transfer into shoot clumps may be problematic as well if the clumps are very dense.

Shoot morphology varies considerably from species to species and also from clone to clone. It may be characterized by the tendency to clump, the clump dimensions, and by the clump's hydraulic permeability. Exogenous hormones may affect the shoot culture morphology. For example Bringi (1988, unpublished observations) found that benzyladenine levels profoundly affected the number of new leaf primordia formed in a tobacco shoot culture, whereas kinetin affected leaf expansion and leaf size. Addition of gibberellic acid caused elongation of leaves and stems (a more "stringy" appearance) and, no doubt, affected the hydraulic permeability of the shoot culture. Selection of clones that do not grow as dense clumps or the use of hormonal manipulations to alter morphology favorably may be factors to consider in choosing cultures capable of sustaining rapid growth rates in bioreactors.

E.2. Effect of Submersion

E.2.1. Growth under Submerged Conditions Shoots are aerial parts of the plant and have evolved to grow and develop in air. For example, aerial leaves possess stomata, which are cells specialized for regulation of gas exchange. Leaves of aquatic plants do not possess stomata and also have altered morphologies. A reasonable question can be raised as to whether aerial shoots can grow and develop at high rates under submerged conditions.

It has been recently demonstrated that shoot cultures of some species may indeed be able to thrive in submerged conditions (Levin et al. 1988; Bringi 1991), and grow at rates equal to those achieved in shake-flasks. Bringi (1991) used a recycle system, shown in figure 9.4, to provide a completely submerged environment for a tobacco shoot culture. Aeration was accomplished by sparging in the recycle vessel so that the shoots would not be subjected to the ensuing stresses. The aerated liquid medium was rapidly recirculated through the growth chamber to supply oxygen. Gentle agitation was provided with a magnetic stirrer in the growth chamber for adequate flow distribution, and the shoots were protected from the stirrer by a screen. Doubling times comparable to that observed in shake flasks (3±0.3 days) were observed in the reactor

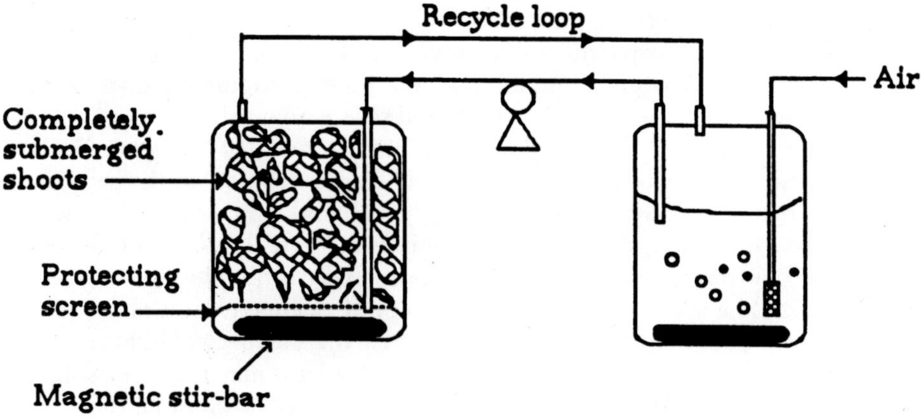

Figure 9.4. A recycle reactor for shoot growth under submerged conditions. Aeration is achieved in the recycle vessel, and the medium is rapidly recirculated. Details in Bringi (1991).

(average t_D = 2.9±0.3 days), showing conclusively that tobacco shoots are capable of growth under completely submerged conditions provided adequate nutrient supply is ensured. The successes of aeroponic mist cultures and of sequential exposure of shoots to liquid and air seem to indicate that shoots are capable of thriving in a range of environments, from full submersion to aerial conditions.

E.2.2. Vitrification Though shoot cultures of some species seem to thrive under submerged conditions, shoot cultures of other species often vitrify; that is, leaves and stems become turgid and distorted, and necrosis often results. Though vitrification has been reported mostly in the context of micropropagation, it is a potential problem for secondary metabolite-producing shoot cultures. Completely submerged shoots may be more susceptible to vitrification than shoots exposed to air. Debergh et al. (1981) and Kevers et al. (1984) have evaluated the various hypotheses regarding the factors responsible for vitrification. Factors that have been implicated include the type and concentration of cytokinin in the medium, the relative humidity in the air space, the water potential, the concentration of ammonium ions, and buildup of ethylene. No consensus seems to have been reached regarding the causes of the problem, nor is a universal solution available in the literature. Levin et al. (1988) have reported the use of an undisclosed, proprietary procedure that circumvented vitrification and allowed a number of species to be cultivated under submerged conditions.

E.3. Light Requirements

The light requirements of shoot cultures, as well as of other tissue cultures, provide a unique challenge for bioreactor design and scale-up. The physiological effects of light have been discussed in Section D. The mechanical constraints for adequate illumination are extreme. For external lighting, the surface area for light supply per unit volume should be sufficient and transparent reactor walls are required. Since light is attenuated by absorption and passage through tissue, the highest attainable tissue density may be constrained. Hagimori et al. (1984b) found severe nonuniformities of chlorophyll distribution in *Digitalis purpurea* shoots grown to a density of 15.4 g dry wt./L in a 3-L jar fermenter with external light supply. The accumulation of the secondary metabolite digitoxin was reduced correspondingly. When the final shoot density was reduced to ca. 10 g dry wt./L, slow mechanical agitation was still sufficient to keep the shoots suspended and mixed, and more uniform chlorophyll distributions were obtained. Frolich et al. (1983) have suggested that the concept of an effectiveness factor can be used on the reactor level to quantitatively characterize the adequacy of light supply in a photobioreactor. The use of light sources inside the reactor may provide a solution to some of the problems of meeting the light requirements of shoot cultures (e.g., Pohl et al. 1987).

F. BIOREACTOR STUDIES WITH SHOOT CULTURES

This section summarizes the few attempts that have been made at scaling up shoot cultures in bioreactors. The final use of the shoot cultures may determine the constraints upon bioreactor design and operation. For secondary metabolite production, rapid growth rates and high volumetric productivities are important and developmental nonuniformities may be acceptable. For micropropagation purposes, the quality of the "propagules" is of primary interest, and a slower growth rate may be acceptable. Here we will review earlier work with the bias of using shoot tissue for secondary metabolite production.

Since the performance of bioreactors is often compared to that of shake flasks, it should be noted that shake flasks provide a rather unique environment for shoot growth. Generally, a portion of the shoots remains submerged in the liquid medium, whereas the rest protrudes into the air space in the flask. The relative proportion of submerged tissue changes with growth. Submerged shoots are aerated by air dissolved in the liquid medium, whereas the rest of the shoots are aerated directly by exposure to gaseous air. The transfer of secondary metabolite productivity of a shoot culture from shake flask to bioreactor must be confirmed for each system.

We will classify bioreactor designs based on the mode of aeration as (i) conventional designs, where the shoots are submerged and aeration is via dissolved oxygen, or (ii) alternative designs, where aeration is achieved by direct exposure to air either by (a) alternate exposure of shoots to liquid and air or (b) constant exposure of shoots to nutrient mists, that is, to liquid-in-air dispersions.

F.1. Conventional Designs

Takayama and Misawa (1981) used a 3-L bubble column and a 10-L jar fermenter in order to scale up growth of *Begonia* shoot cultures. The growth rate achieved in a shake flask was not achievable in the larger scale reactors. The authors proposed that the injury caused by mechanical stresses associated with gas sparging was primarily responsible for the reduced growth rates.

The potential for achieving high growth rates in a sparged jar fermenter was unequivocally demonstrated by Hagimori et al. (1984b). The maximum growth rate of shoot cultures of *Digitalis purpurea* in a 3-L bioreactor (t_D = 3.2 days) actually exceeded that in a shake-flask. A final tissue density of 15.4 g dry wt./L was achieved. At these high densities, however, nonuniform light exposure was a problem, with chlorophyll and digitoxin contents varying with the position of shoots in the reactor. Lowering the final tissue density resulted in more uniform greening of the shoots.

Park et al. (1989) reported the growth of *Artemisia* shoot cultures in a sparged rectangular tank with no mechanical agitation. The maximum growth rate in shake flasks (t_D = 4.3-5.4 days) was superior to that in the bioreactor (t_D = 5.9 days). Shoots were successfully rooted in the bioreactor.

Bringi (1991) used a 1-L sparged vessel with mechanical agitation for cultivating tobacco shoot cultures. A stainless steel screen protected the shoots from the magnetic stirrer as shown in figure 9.5. Average doubling times of 3.9±0.3 days were achieved, which corresponds to slower growth than the same cultures are capable of in

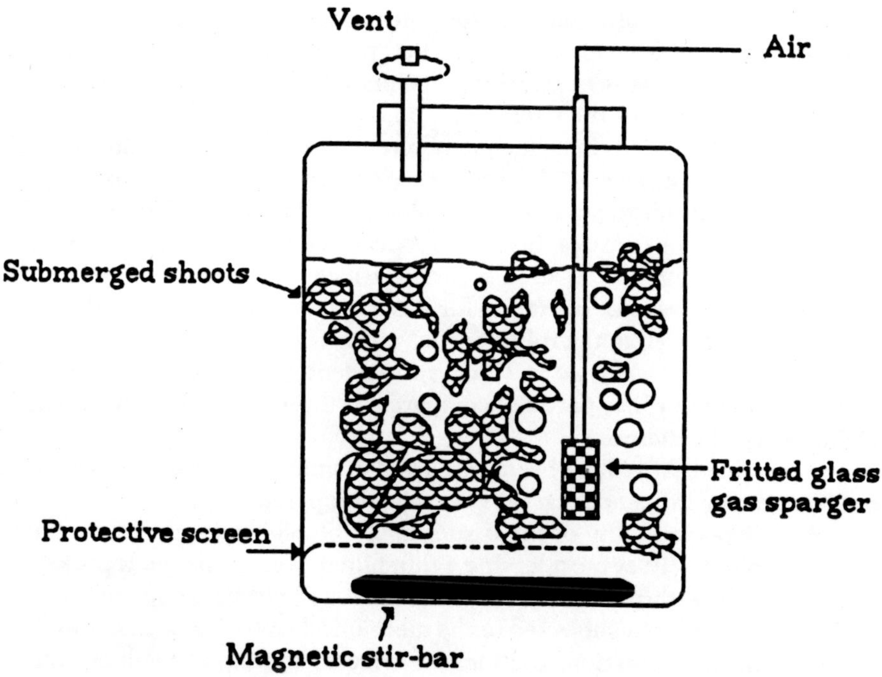

Figure 9.5. An air-sparged reactor for shoot growth. Liquid mixing is aided by a magnetic stirrer, which is separated from the tissue by a protective stainless steel screen. Details in Bringi (1991).

the recycle reactor of figure 9.4 (2.9±0.3 days). Bringi (1991) has argued that the probable cause of the slower growth in the sparged vessel was reduced oxygen supply to the interior of the large shoot clumps.

Whether a particular hydrodynamic reactor environment supports shoot growth can depend on the shoot morphology, that is, the tendency to clump, clump sizes, and hudraulic permeability. It is thus not surprising that shoot cultures of different species, with presumably differing morphologies, show differences in their ability to grow as fast in reactors as in shake flasks. As suggested in Chapter 8, reactor designs that allow for forced liquid flow through shoots may, in general, be more effective and scaleable than conventional sparged vessels.

A company in Israel, Plant Biotechnology Industries (no longer in existence), has reported the use of bioreactors for growth of submerged shoot cultures primarily for micropropagation purposes. The details of the reactor design and operating conditions have not been disclosed (Levin et al. 1988). The authors claim to have successfully used their bioreactor and automated "bioprocessing" scheme for propagating cultures of onion, potato, and other species.

F.2. Alternative Bioreactor Designs

F.2.1. Aeration by Alternate Exposure to Liquid Medium and Air The direct exposure of shoots to air is an alternative to sparging air into the liquid medium. Since the diffusion coefficient of oxygen in air is ca. 7000 times that in water, direct supply of oxygen from the gas phase is considerably facilitated. Dissolved components are supplied by periodic bathing of the tissue in liquid medium.

Kobayashi et al. (1988) used the principle of alternate exposure for culturing gel-immobilized cells of *Thalictrum minus* for berberine production. They found that varying the time of exposure to air affected productivity. Taya et al. (1989) also used the same principle for root cultures, but the performance was far inferior to that of shake flask cultures. Bringi (1991) used a simple roller bottle to demonstrate the feasibility of this mode of reactor operation for shoot cultures (figure 9.6). A doubling time of 3.5 days was achieved in the roller bottle, which compares favorably with that achieved in a small shake flask (t_D = 3±0.3 days). Incidentally, higher tissue densities have been observed in the roller bottle as compared to those for shoots grown under completely submerged conditions.

The studies described above have been conducted in small-scale reactors. It appears that several factors may influence reactor scale-up and operation:

(i) The hydraulic permeability of the tissue should be high enough for the liquid medium to drain completely, at most leaving a thin film to wet the tissue. If pockets of stagnant liquid remain within the tissue, then aeration will not be complete.

(ii) Though the tissue is not subjected to the mechanical stresses associated with air sparging and mechanical agitation, the inertial force of the liquid medium during filling and emptying must be considered, as should the weight of the tissue itself.

(iii) Constraints imposed on the times of exposure to liquid and air will likely affect

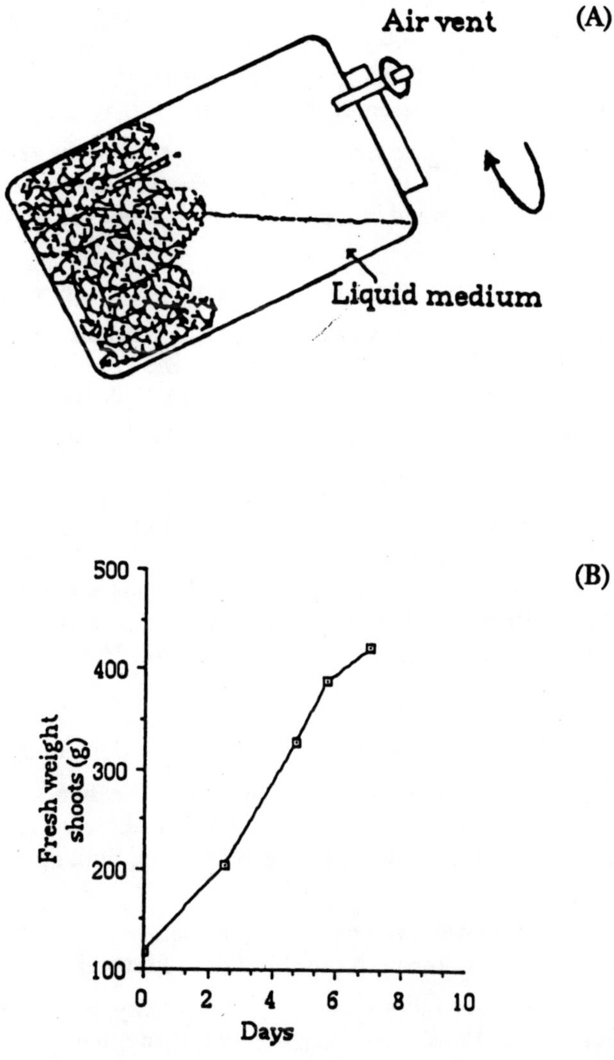

Figure 9.6. (A) A tilted rotating bottle for shoot growth in which shoots are alternately exposed to liquid medium and air. (B) Growth curve for ***Nicotiana tabacum*** shoots cultivated in a 0.7-L working volume jar rotating at 2 rpm. Details in Bringi (1991).

the scalability of this design. Estimates of these times can be obtained from the simple quantitative arguments based on a characteristic time analysis.

The characteristic time analysis is a useful engineering tool which can provide estimates of the time needed for transient processes to approach some arbitrary fraction of their steady-state values. Its use to estimate the times of exposure to liquid and air is illustrated here to reinforce some of the concepts introduced in Chapter 4.

Consider a representative "control volume" of 1 cm^3 in the reactor. Assume that half the volume is filled with 0.5 g fresh weight of shoots and that the remaining volume is void. The first question of relevance is:

(a) What is the maximum time that the shoots can be left submerged in the medium (initially saturated with air) without further aeration?

The answer may be obtained by a simple mass balance for oxygen over the control volume:

Amount of oxygen initially present in the medium = Uptake rate x Time

The amount of oxygen initially present in the air-saturated medium is given by the volume of medium in the control volume times the concentration of oxygen at air saturation, which is equal to $(0.5\ cm^3) \times (8 \times 10^{-3}\ mg\ O_2/cm^3) = 4 \times 10^{-3}\ mg\ O_2$. In the absence of uptake rate data for shoot cultures, we may assume a value typical for a suspension culture (say, 0.4 mg O_2 per g fresh weight per hour). We shall also assume for simplicity that the reaction is zero order; that is, it is independent of the oxygen concentration. Under these circumstances, the uptake rate may be estimated as: $(0.4/60)$ mg O_2/min/g fresh weight x 0.5 g fresh weight = $(0.2/60)$ mg O_2/min.

From the mass balance, the maximum time that the shoots can be left submerged is just over a minute. Note that this time depends on the tissue density in the reactor and also on the oxygen uptake rate of the shoot culture. If the oxygen uptake rate for shoots were half that for a suspension culture, then the shoots could remain submerged for about two minutes before the oxygen was depleted. The estimated time is a maximum since it may not be desirable to completely deplete the medium of oxygen.

(b) How long will it take for oxygen to penetrate through a liquid film to the leaf surface when a leaf initially starved for oxygen is suddenly exposed to air?

The resistances to the transport of oxygen from the bulk gas to the leaf surface are as follows:

1. The oxygen has to cross the gas-liquid barrier.
2. Oxygen must diffuse through the thin liquid film wetting the leaf.

Let us lump both resistances together for simplicity and assume that they have a collective stagnant film thickness of, say, 0.5 mm, through which oxygen has to diffuse before it reaches the leaf surface. The characteristic time for this diffusion process would be L^2/D, where L is the film thickness, assuming flat geometry, and D ($2 \times 10^{-5}\ cm^2/s$) is the diffusivity of oxygen through the film (see Appendix 1 of Chapter 4 for details).

The characteristic time for this process is about two minutes, suggesting that oxygen transport upon exposure to air is not instantaneous and that the shoots should be exposed to air for at least a few minutes before being submerged in liquid medium again. Note that the estimate of this characteristic time is very sensitive to the value of the assumed film thickness.

Indeed, experimental results seem to bear out these rough predictions. Kobayashi et al. (1988) found that exposure of Ca-alginate beads for 2 min to air followed by 30 s in liquid medium allowed successful operation of the immobilized cell reactor for 60 days. The failure of Taya et al. (1989) to culture roots by medium cycling may be attributed to the fact that the roots were submerged for ca. 15 min before exposure to air [see question (a) above]. Kondo et al. (1989) were able to culture horseradish roots successfully in a rotating drum reactor operating at 5 rpm, that is, alternate exposure every 6 s. Bringi (1991) exposed shoots alternately every 30 s. Note that when individual exposure times are short as in the latter two cases, steady-state conditions prevail, as opposed to alternating transients. The feasibility of this design for larger scale operation remains to be assessed.

F.2.2 Exposure of Shoots to Nutrient Mists The continuous exposure of shoots to liquid-in-air dispersions such as sprays or aerosols is another way to supply oxygen by direct exposure to air. Dissolved nutrients are supplied when the medium coalesces on the tissue, forming a thin liquid layer which drains off the tissue and can be continuously recirculated. Weathers and Giles (1988) have used the nutrient mist bioreactor for the culture of shoots, roots, and other plant tissues. Shoots grown in nutrient mists compared favorably with those grown on agar with respect to growth rates, uniformity, and quality of shoots for micropropagation. Importantly, vitrification may be circumvented in such an environment. With respect to secondary metabolite production, however, a more relevant comparison would be between nutrient mists and shake flasks or air-sparged bioreactors. Such comparisons have not yet been reported. Further description of this reactor and scale-up considerations are provided in Section H of Chapter 8.

G. CONCLUSIONS

The level of structural differentiation in a shoot culture is close to that in a plant. Consequently, the secondary biosynthetic capabilities of shoot cultures also reflect those of a whole plant more closely than do those of undifferentiated cultures such as suspensions and immobilized cells.

The scale-up of shoot cultures in bioreactors poses several challenges, largely stemming from unique morphological characteristics, susceptibility to mechanical damage, possible vitrification, and light requirements.

This chapter has attempted to summarize the issues involved in the design of a shoot culture process for secondary metabolite production.

References

Y. P. S. Bajaj (ed.). 1988. "Biotechnology in Agriculture and Forestry," Vol. 4, "Medicinal and Aromatic Plants I." Springer-Verlag, Berlin.

Y. P. S. Bajaj (ed.). 1989. "Biotechnology in Agriculture and Forestry," Vol. 7, "Medicinal and Aromatic Plants II." Springer-Verlag, Berlin.

G. A. Barthe, P. S. Jourdan, C. A. McIntosh, and R. L. Mansell. 1987. Naringin and Limonin Production in Callus Cultures and Regenerated Shoots from *Citrus* spp. *J. Plant Physiol. 127*, 55-65.

B. D. Benjamin, P. C. Roja, M. R. Heble, and M. S. Chadha. 1987. Multiple Shoot Cultures of *Atropa belladonna*: Effect of Physico-Chemical Factors on Growth and Alkaloid Formation. *J. Plant Physiol. 129*, 129-135.

V. Bringi. 1991. Differentiation in Plant Cultures: (1) Tracheary Element Induction in Immobilized Tobacco Cells; (2) Bioreactor Considerations for Shoot Culture Cultivation. Ph.D. Thesis. Cornell University, Ithaca, New York.

B. V. Charlwood and C. Moustou. 1988. Essential Oil Accumulation in Shoot-Proliferation Cultures of *Pelargonium* spp. In "Manipulating Secondary Metabolism in Culture," R. J. Robins and M. J. C. Rhodes (eds.), Cambridge University Press, Cambridge, U.K., pp. 187-194.

B. V. Charlwood, C. Moustou, J. T. Brown, P. K. Hegarty, and K. A. Charlwood. 1989. The Regulation of Accumulation of Lower Isoprenoids in Plant Cell Cultures. In "Primary and Secondary Metabolism of Plant Cell Cultures," W. G. W. Kurz (ed.), Springer-Verlag, Berlin, pp. 73-84.

C.-T. A. Chung and E. J. Staba. 1987. Effects of Age and Growth Regulators on Growth and Alkaloid Production in *Cinchona ledgeriana* Leaf-Shoot Organ Cultures. *Planta Med. 53*, 206-210.

C.-T. A. Chung and E. J. Staba. 1988. Effects of Precursors on Growth and Alkaloid Production in *Cinchona ledgeriana* Leaf-Shoot Organ Cultures. *Planta Med. 54*, 206-210.

H. A. Collin. 1987. Determinants of Yield in Plant Tissue Cultures. *Adv. Bot. Res. 13*, 145-187.

F. Constabel. 1985. Morphinan Alkaloids from Plant Cell Cultures. In "The Chemistry and Biology of Isoquinoline Alkaloids," J. D. Phillipson, M. F. Roberts, and M. H. Zenk (eds.), Springer-Verlag, Berlin, pp. 257-264.

F. Constabel, P. Gaudet-La Prairie, W. G. W. Kurz, and J. P. Kutney. 1982. Alkaloid Production in *Catharanthus roseus.* XII. Biosynthetic Capacity of Callus from Original Explants and Regenerated Shoots. *Plant Cell Rep. 1*, 139-142.

P. Debergh, Y. Harbaoui and R. Lemeur. 1981. Mass Propagation of Globe Artichoke (*Cynara scolymus*): Evaluation of Different Hypotheses to Overcome Vitrification with Special Reference to Water Potential. *Physiol. Plant. 53*, 181-187.

V. De Luca and A. J. Cutler. 1987. Subcellular Localization of Enzymes Involved in Indole Alkoid Biosynthesis in *Catharanthus roseus. Plant Physiol. 85*, 1099-1102.

V. De Luca, J. Balsevich, R. T. Tyler and W. G. W. Kurz. 1987. Characterization of a novel N-methyltransferase (NMT) from *Catharanthus roseus* plants. *Plant Cell Rep. 6*, 458-461.

V. De Luca, J. A. Fernandez, D. Campbell, and W. G. W. Kurz. 1988. Developmental Regulation of Enzymes of Indole Alkaloid Biosynthesis in *Catharanthus roseus. Plant Physiol. 86*, 447-450.

V. De Luca, N. Brisson, J. Balsevich, and W. G. W. Kurz. 1989. Regulation of Vindoline Biosynthesis in *Catharanthus roseus*: Molucular Cloning of the First and Last Steps in Biosynthesis. In "Primary and Secondary Metabolism of Plant Cell Cultures," W. G. W. Kurz (ed.), Springer-Verlag, Berlin, pp. 154-161.

D. K. Dougall. 1987. Cell Cloning and the Selection of High Yielding Strains. In "Cell Culture and Somatic Cell Genetics of Plants," Vol. 4, F. Constabel and I. K. Vasil (eds.), Academic Press, New York, pp. 117-124.

F. Drawert, R. G. Berger, R. Godelmann, S. Collin, and W. Barz. 1984. Characterization of Volatile Constituents from Photomixotrophic Cell Suspension Cultures of *Ruta graveolens*. *Z. Naturforsch. 39C*, 525-530.

U. Eilert. 1989. Elicitor Induction of Secondary Metabolism in Dedifferentiated and Differentiated *In Vitro* Systems of *Ruta graveolens*. In "Primary and Secondary Metabolism of Plant Cell Cultures," W. G. W. Kurz (ed.), Springer-Verlag, Berlin, pp. 219-228.

U. Eilert, V. De Luca, F. Constabel, and W. G. W. Kurz. 1987. Elicitor-Mediated Induction of Tryptophan Decarboxylase and Strictosidine Synthase Activities in Cell Suspension Cultures of *Catharanthus roseus*. *Arch. Biochem. Biophys. 254*, 491-497.

T. Endo, A. Goodbody, and M. Misawa. 1987. Alkaloid Production in Root and Shoot Cultures of *Catharanthus roseus*. *Planta Med. 53*, 479-482.

T. Endo, A. Goodbody, J. Vukovic and M. Misawa. 1988. Enzymes from *Catharanthus roseus* Cell Suspension Cultures That Couple Vindoline and Catharanthine to Form 3', 4'-Anhydrovinblastine. *Phytochemistry* 7, 2147-2149.

K. Esau. 1977. "Anatomy of Seed Plants," John Wiley, New York.

B. T. Frohlich, I. A. Webster, M. M. Ataai, and M. L. Shuler. 1983. Photobioreactors: Models for Interaction of Light Intensity, Reactor Design, and Algal Physiology. *Biotechnol. Bioeng. Symp.*, No. 13, 331-350.

J. Gershenzon, M. Maffei, and R. Croteau. 1989. Biochemical and Histochemical Localization of Monoterpene Biosynthesis in the Glandular Trichomes of Spearmint (*Mentha spicata*). *Plant Physiol. 89*, 1351-1357.

A. Goodbody, T. Endo, J. Vukovic, and M. Misawa. 1988. The Coupling of Catharanthine and Vindoline to Form 3',4'-Anhydrovinblastine by Haemoproteins and Haemin. *Planta Med. 54*, 210-214.

A. R. Gould. 1986. Factors Controlling Generation of Variability in Vitro. In "Cell Culture and Somatic Cell Genetics of Plants," Vol. 3, I. K. Vasil (ed.), Academic Press, New York, pp. 549-567.

M. Hagimori, T. Matsumoto, and Y. Obi. 1982a. Studies on the Production of *Digitalis purpurea* Cardenolides by Plant Tissue Culture. *Plant Physiol. 69*, 653-656.

M. Hagimori, T. Matsumoto, and Y. Obi. 1982b. Studies on the Production of *Digitalis purpurea* Cardenolides by Plant Tissue Culture-III. Effects of Nutrients on Digitoxin Formation by Shoot-Forming Cultures of *Digitalis purpurea* Grown in Liquid Media. *Plant Cell Physiol. 23*, 1205-1211.

M. Hagimori, T. Matsumoto, and Y. Obi. 1983. Effects of Mineral Salts, Initial pH, and Precursors on Digitoxin Formation by Shoot-Forming Cultures of *Digitalis purpurea* Grown in Liquid Media. *Agric. Biol. Chem. 47*, 565-571.

M. Hagimori, T. Matsumoto, and Y. Mikami. 1984 a. Photoautotrophic Culture of Undifferentiated Cells and Shoot-Forming Cultures of *Digitalis Purpurea* L. *Plant and Cell Physiol. 25*, 1099-1102.

M. Hagimori, T. Matsumoto and Y. Mikami. 1984 b. Jar Fermenter Culture of Shoot-Forming Cultures of *Digitalis purpurea* L. Using a Revised Medium. *Agric. Biol. Chem. 48*, 965-970.

T. Hashimoto and Y. Yamada. 1988. Biosynthesis of Scopolamine from (7-β-H-2)6-β-Hydroxyhyoscyamine in *Duboisia* Shoot Cultures. *Agric. Biol. Chem. 53*, 863-864.

M. R. Heble. 1985. Multiple Shoot Cultures: A Viable Alternative In Vitro System for the Production of Known and New Biologically Active Plant Constituents. In "Primary and Secondary Metabolism of Plant Cell Cultures," K. -H. Neumann, W. Barz, and E. Reinhard (eds.), Springer-Verlag, Berlin, pp.281-289.

K. Hirata, A. Yamanaka, N. Kurano, K, Miyamoto, and Y. Miura. 1987. Production of Indole Alkaloids in Multiple Shoot Culture of *Catharanthus roseus* (L.) G. Don. *Agric. Biol. Chem. 51*, 1311-1317.

T. Hirata, S. Murakami, K. Ogihara, and T. Suga. 1990. Volatile Monoterpenoid Constituents of the Plantlets of *Mentha spicata* Produced by Shoot Tip Culture. *Phytochemistry 29*, 493-495.

K. J. Hoober. 1984. "Chloroplasts." Plenum Press, New York.

W. Husemann, K. Fischer, I, Mittelbach, S. Hubner, G. Richter, and W. Barz. 1989. Photoautotrophic Plant Cell Cultures for Studies on Primary and Secondary Metabolism. In "Primary and Secondary Metabolism of Plant Cell Cultures," W. G. W. Kurz (ed.), Springer-Verlag, Berlin, pp. 35-46.

G. Hussey. 1986. Problems and Prospects in the *In Vitro* Propagation of Herbaceous Plants. In "Plant Tissue Culture and Its Agricultural Applicatons," L. A. Withers and P. Alderson (eds.), Butterworths, Boston, pp. 69-84.

U. Igbavboa, H. J. Sieweke, E. Leistner, E. Rower, W. Huseman, and W. Barz. 1985. Alternative Formation of Anthroquinones and Lipoquinones in Heterotrophic and Photoautotrophic Cell Suspension Cultures of *Morinda lucida*. *Planta 166*, 537-544.

S. Jacob John, J. Subramani, P. N. Bhatt, and A. R. Mehta. 1990. A Novel Approach for the Production of Guggulipid using Immobilized Differentiated Cells of *Commiphora wightii*. In "Physiology of Immobilized Cells," J. A. M. de Bont, J. Visser, B. Mattiasson, and J. Tramper (eds.), Elsevier, Amsterdam, pp. 467-474.

M. Jordan, C. H. Rolfs, W. Barz, R. G. Berger, H. Kohlmannsberger, and F. Drawert. 1986. Characterization of Volatile Constituents from Heterotrophic Cell Suspension Cultures of *Ruta graveolens*. *Z. Naturforsch. 41C*, 809-812.

K. K. Kartha. 1981. Meristem Culture and Cryopreservation-Methods and Applications. In "Plant Tissue Culture-Methods and Applications in Agriculture," T. A. Thorpe (ed.), Academic Press, New York, pp. 181-211.

R. E. Kendrick and G. H. M. Kronenberg (eds.). 1986. "Photomorphogenesis in Plants." Martinus Nijhoff Publishers, Dordrecht, The Netherlands.

C. Kevers, M. Coumans, M.-F. Coumans-Gilles, and Th. Gaspar. 1984. Physiological and Biochemical Events Leading to Vitrification of Plants Cultured *In Vitro*. *Physiol. Plant. 61*, 69-74.

Y. Kobayashi, H. Fukui, and M. Tabata. 1988. Berberine Production by Batch and Semi-Continuous Cultures of Immobilized *Thalictrum* Cells in an Improved Bioreactor. *Plant Cell Rep.*, 7, 249-252.

O. Kondo, H. Honda, M. Taya, and T. Kobayashi. 1989. Comparison of Growth Properties of Carrot Hairy Roots in Various Bioreactors. *Appl. Microbiol. Biotechnol. 32*, 291-294.

R. J. Krueger, D. P. Carew, J. H. C. Lui, and E. J. Staba. 1982. Initiation, Maintenance and Alkaloid Content of *Catharanthus roseus* Leaf Organ Cultures. *Planta Med. 45*, 56-57.

A. C. Leopold and P. E. Kriedeman. 1975. "Plant Growth and Development," 2nd ed., McGraw-Hill, New York.

R. Levin, V. Gaba, B. Tal, S. Hirsch, D. DeNola, and I. K. Vasil. 1988. Automated Plant Tissue Culture for Mass Propagation. *Bio/Technology 6*, 1035-1040.

J. H. C. Lui and E. J. Staba. 1979. Effects of Precursors on Serially Propagated *Digitalis lanata* Leaf and Root Cultures. *Phytochemistry 18*, 1913-1916.

A. L. Mancinelli. 1985. Light-Dependent Anthocyanin Synthesis: A Model System for the Study of Plant Photomorphogenesis. *Bot. Rev. 51*, 107-157.

S. H. Mantell and H. Smith. 1983. Cultural Factors that Influence Secondary Metabolite Accumulations in Plant Cell and Tissue Cultures. In "Plant Biotechnology," S. H. Mantell and H. Smith (eds.), Cambridge University Press, Cambridge, U.K., pp. 75-108.

B. C. Martinez and E. J. Staba. 1988. The Production of Artemisinin in *Artemisia annua* L. Tissue Cultures. *Adv. Cell Culture 6*, 69-87.

Y. Miura, K. Hirata, N. Kurano, K. Miyamoto, and K. Uchida. 1988. Formation of Vinblastine in Multiple Shoot Culture of *Catharanthus roseus*. *Planta Med. 54*, 18-20.

Th. Mulder-Krieger, R. Verpoorte, A. Baerheim Svendsen, and J. J. C. Scheffer. 1988. Production of Essential Oils and Flavours in Plant Cell and Tissue Cultures. A Review. *Plant Cell Tissue Organ Culture 13*, 85-154.

M. Ondrej, J. Matousek, and J. Vlasak. 1987. Differentiation of Transformed Plants from Tumors Induced by *Agrobacterium rubi* ATCC 13335. *J. Plant Physiol. 126*, 397-407.

J. M. Park, W.-S. Hu, and E. J. Staba. 1989. Cultivaton of *Artemisia annua* L. Plantlets in a Bioreactor Containing a Single Carbon and a Single Nitrogen Source. *Biotechnol. Bioeng. 34*, 1209-1213.

D. Pauthe-Dayde, M. Rochd, and M. Henry. 1990. Triterpenoid Saponin Production in Callus and Multiple Shoot Cultures of *Gypsophila* spp. *Phytochemistry 29*, 483-487.

L. A. Pelosi, A. Rother, and J. M. Edwards. 1985. The Effect of Light on the Production of *Heimia* Alkaloids. *Phytochemistry 24*, 2215-2218.

L. A. Pelosi, A. Rother, and J. M. Edwards. 1986. Lysine Decarboxylase Activity and Alkaloid Production in *Heimia salicifolia* Cultures. *Phytochemistry 25*, 2315-2319.

P. Pohl, M. Kohlase, and M. Martin. 1987. Photobioreactors for the Axenic Mass Cultivation of Microalgae. In "Algal Biotechnology," T. Stadler, J. Mollion, M.-C. Verdus, Y. Karamanos, H. Morvan, and D. Christaen (eds.), Elsevier Applied Science, London, pp.209-217.

P. H. Raven, R. F. Evert, and H. Curtis. 1981. "Biology of Plants," 3rd ed., Worth Publishers, New York.

M. S. Reid. 1987. Ethylene in Plant Growth, Development and Senescence. In "Plant Hormones and Their Role in Plant Growth and Development," P. J. Davies (ed.), Martinus Nijhoff, Dordrecht, The Netherlands, pp. 257-279.

M. J. C. Rhodes, M. Hilton, A. J. Parr, J. D. Hamill, and R. J. Robins. 1986. Nicotine Production by "Hairy Root" Cultures of *Nicotiana rustica*: Fermentation and Product Recovery. *Biotechnol. Lett. 8*, 415-420.

P. C. Roja, B. D. Benjamin, M. R. Heble, and M. S. Chadha. 1985. Indole Alkaloids from Multiple Shoot Cultures of *Rauwolfia serpentina*. *Planta Med. 51*, 73-74.

P. C. Roja, A. T. Sipahimalani, M. R. Heble and M. S. Chadha. 1987. Multiple Shoot Cultures of *Rauwolfia serpentina*. *J. Nat. Prod. 50*, 872-875.

K. Saito, I. Murakoshi, D. Inze, and M. Van Montagu. 1989a. Biotransformation of Nicotine Alkaloids by Tobacco Shooty Teratomas Induced by a Ti Plasmid Mutant. *Plant Cell Rep.* 7, 607-610.

K. Saito, M. Yamazaki, K. Yamakawa, S. Fujisawa, S. Takamatsu, A. Kawaguchi, and I. Murakoshi. 1989b. Lupin Alkaloids in Tissue Culture of *Sophora flavescens* var. *angustifolia*: Greening Induced Production of Matrine. *Chem. Pharm. Bull.* 37, 3001-3004.

F. B. Salisbury and C. W. Ross. 1985. "Plant Physiology," 3rd ed. Wadsworth, Belmont, California.

P. Schopfer. 1977. Phytochrome Control of Enzymes. *Annu. Rev. Plant Physiol. 28*, 223-252.

M. Seibert and P. G. Kadkade. 1980. Enviromental Factors: A. Light. In "Plant Tissue Culture as a Source of Biochemicals," E. J. Staba (ed.), CRC Press, Boca Raton, Florida, pp. 123-142.

J. Simpson and L. Herrera-Estrella. 1990. Light-Regulated Gene Expression. *Crit. Rev. Plant Sci. 9*, 95-109.

A. Spencer, J. D. Hamill, and M. J. C. Rhodes. 1990. Production of Terpenes by Differentiated Shoot Cultures of *Mentha citrata* Transformed with *Agrobacterium tumefaciens* T37. *Plant Cell Rep. 8*, 601-604.

E. J. Staba and A. C. Chung. 1981. Quinine and Quinidine Production by *Cinchona* Leaf, Root and Unorganized Cultures. *Phytochemistry 20*, 2495-2498.

E. J. Staba, B. G. Nygaard, and S. W. Zito. 1984. Light Effects on *Pyrethrum* Shoot Cultures. *Plant Cell Tissue Organ Culture 3*, 211-214.

S. Takayama and M. Misawa. 1981. Mass Propagation of *Begonia* X *hiemalis* Plantlets by Shake Culture. *Plant Cell Physiol.* 22, 461-467.

M. Taya, A. Yoyama, O. Kondo, T. Kobayashi, and C. Matsui. 1989. Growth Characteristics of Plant Hairy Roots and Their Cultures in Bioreactors. *J. Chem. Eng. Japan.* 22, 84-89.

P. J. Weathers and K. Giles. 1988. Regeneration of Plants Using Nutrient Mist Culture. *In Vitro 24*, 727-732.

J. M. Widholm. 1987. Selection of Mutants Which Accumulate Desirable Secondary Metabolites. In "Cell Culture and Somatic Cell Genetics of Plants," Vol. 4, F. Constabel and I. K. Vasil (eds.), Academic Press, New York, pp. 125-137.

M. Wink. 1987. Physiology of the Accumulation of Secondary Metabolites with Special Reference to Alkaloids. In "Cell Culture and Somatic Cell Genetics of Plants," Vol. 4, F. Constabel and I. K. Vasil (eds.), Academic Press, New York, pp. 17-42.

M. Wink and T. Hartmann. 1980. Production of Quinolizidine Alkaloids by Photomixotrophic Cell Suspension Cultures: Biochemical and Biogenetic Aspects. *Planta Med. 40*, 149-155.

M. Wink and T. Hartmann. 1982. Localization of the Enzymes of Quinolizidine Alkaloid Biosynthesis in Leaf Chloroplasts of *Lupinus polyphyllus*. *Plant Physiol. 70*, 74-77.

L. A. Withers. 1985. Cryopreservation of Cultured Cells and Meristems. In "Cell Culture and Somatic Cell Genetics of Plants," Vol. 2, I. K. Vasil (ed.), Academic Press, New York, pp. 253-316.

Y. Yamada. 1984. Selection of Cell Lines for High Yields of Secondary Metabolites. In "Cell Culture and Somatic Cell Genetics of Plants," Vol. 1, I. K. Vasil (ed.), Academic Press, New York, pp. 629-636.

T. Yoshioka, H. Yamagata, A. Ithoh, H. Deno, Y. Fujita, and Y. Yamada. 1989. Effects of Exogenous Polyamines on Tropane Alkaloid Production by a Root Culture of *Duboisia myoporoides*. *Planta Med. 55*, 523-524.

A. M. Zobel and S. A. Brown. 1989a. Histological Localization of Furanocoumarins in *Ruta graveolens* Shoots. *Can. J. Bot. 67*, 915-921.

A. M. Zobel and S. A. Brown. 1989b. Localization of Daphnetin and Umbelliferone in Different Tissues of *Daphne mezereum* Shoots. *Can. J. Bot. 67*, 1456-1459.

M. Zucker. 1972. Light and Enzymes. *Annu. Rev. Plant Physiol. 23*, 133-156.

CHAPTER 10

SOMATIC EMBRYOGENESIS

A. INTRODUCTION

In this chapter, we will consider somatic embryogenesis -especially focusing on the formation of embryos from previously unorganized cells. Reviews on the cultivation of embryos obtained from reproductive tissue can be found in Rangan (1982) and Raghavan (1980), while reviews describing the various techniques and cell lines used for embryogenesis can be found in Evans et al. (1981), Ammirato (1983), and Dodds and Roberts (1985). We will first describe biological aspects of somatic embryogenesis and then briefly consider possible applications.

B. SOMATIC EMBRYOGENESIS

Steward et al. (1958) and Reinert (1959) observed that when carrot cells are grown in suspension or callus cultures, they are able to form structurally distinct units which develop in a similar manner as the zygotic embryo. Although their origin was uncertain, these somatic embryos morphologically developed through the globular, heart, and torpedo stages shown in Figure 10.1. Schiavone and Cooke (1985) recently suggested that it may be helpful to consider an intermediate developmental state between the globular and heart stages. These authors suggested that the transition

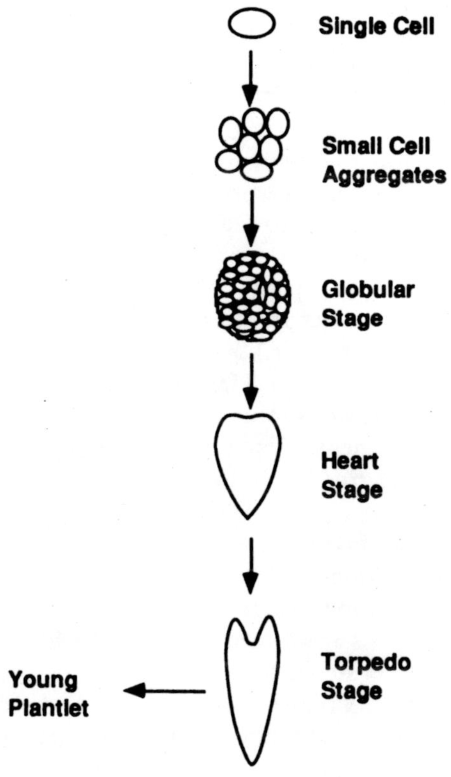

Figure 10.1. The various stages of somatic embryo development. Adapted from Dodds and Roberts (1985).

Table 10.1. The differing stages of somatic embryos were separated using sieves of various sizes. Compiled from Giuliano et al. (1983).

Stage of embryo development	Sieve size which embryo passed through (µm)	Sieve size which retained embryo (µm)
Globular	120	-
Heart	200	170
Torpedo	350	300

from the isodiametric expansion characteristic of the globular embryo to a longitudinal extension results in an "oblong" embryo. The emergence of cotyledons then could be used to distinguish between the oblong and the heart stages. Al-Abta and Collin (1978) reported that at the globular and heart stages, cells in celery embryos were densely cytoplasmic and showed no signs of differentiation. At the torpedo stage, however, cell differentiation occurred and a "more formal internal organization of tissue" was observed. Small cytoplasmically rich cells were observed in the root meristem and the cotyledon node, while neighboring cells in the embryo were larger, vacuolated and oriented in the direction of growth. Also, xylem elements were observed in torpedo-stage embryos (Al-Abta and Collin 1978).

Initially, it was observed that these embryos originated from aggregates; however, McWilliam et al. (1974) reported that the "young embryoids have always a clearly defined outline indicating discontinuity with the surrounding cells." (It should be mentioned that our use of the term "aggregate" does not imply a mechanism by which the cells come together - rather it is believed that "aggregates" arise because cells fail to separate upon division. To distinguish this latter mechanism from aggregation, some authors prefer to use the term cell clump or cluster.) In later research, individual embryos at varying developmental stages were studied (Halperin and Jensen 1967; Fujimura and Komamine 1980). Table 10.1 provides some indication of the sizes of the embryos at the various stages of development. This table was compiled from results of Giuliano et al. (1983), who sieved aggregates to isolate embryos at various developmental stages.

After the initial observations of somatic embryogenesis, there were considerable efforts devoted to defining the medium conditions which permitted embryogenesis and characterizing the morphological and developmental events leading to embryo formation. Embryogenesis is often thought of as a two-step process. The first step is the induction of embryogenesis, and the second step is the actual development of the embryo through the stages shown in Figure 10.1 (Christianson 1985). The requirements for embryo induction and embryo development differ, and thus separate media are often used for each step.

B.1. Medium Studies

Media and species used for somatic embryogenesis are reviewed by Evans et al. (1981), Ammirato (1983), and Dodds and Roberts (1985) and will be only briefly discussed here. Most media studies have focused on the role of the growth regulators. Auxins, especially the synthetic auxin 2,4-D, appear to be required for embryo induction but adversely affect embryo development (Fujimura and Komamine 1975; Kamada and Harada 1979a; Borkird et al. 1986). Cytokinins, except zeatin (Fujimura and Komamine 1975), suppress embryogenesis (Kamada and Harada 1979a). Ethylene may also suppress embryogenesis (Tisserat and Murashige 1977). Abscisic acid (ABA) was reported to suppress the abnormal development of embryos (Ammirato 1974) whereas gibberellic acid and zeatin resulted in an increased frequency of abnormal embryo development (Ammirato 1977). Later work examined endogenous levels of these growth regulators (Tianran and Neumann 1985). A mutant carrot cell line which accumulates the auxin indole-3-acetic acid (IAA) was observed to have a reduced regenerative capacity (Sung 1979). ABA levels were observed to increase during embryo development (Kamada and Harada 1981), and Rajasekaran et al. (1982) suggested that endogenous ABA may play a role in embryo dormancy.

The nitrogen source also appears to play an important role in embryogenesis (Wetherell and Dougall 1976). Although not strictly required for embryo induction, a reduced form of nitrogen (e.g., ammonium or amino acids) has been observed to be required for embryo development (Kamada and Harada 1979b).

A few studies of the carbohydrate source have also been reported. Verma and Dougall (1977) reported that the ability of various carbohydrates to support embryogenesis was directly related to their ability to support carrot cell growth. Thus, it was reasoned that the carbohydrates are probably metabolized via a common intermediate and that observed differences between these carbohydrates are probably due to general metabolic effects rather than effects specific to embryogenesis. In contrast, Strickland et al. (1987) reported maltose to be the best carbohydrate for alfalfa somatic embryogenesis.

Interesting work by Kessell and co-workers (Kessell and Carr 1972; Kessell et al. 1977) has suggested that dissolved oxygen can play a critical role in differentiation. Below a dissolved oxygen level of 16% they observed carrot cell differentiation via embryogenesis, while above 16% dissolved oxygen, differentiation occurred via rhizogenesis. Stuart et al. (1987) observed that higher dissolved oxygen levels were required for alfalfa embryogenesis. Thus, the effects of dissolved oxygen on embryogenesis warrant further study.

B.2. Embryo Development

In addition to the medium studies, a considerable effort was devoted to understanding how these embryos originated. In one of the initial reports, Steward et al. (1958) stated that a certain developmental state "does not form until the cell aggregate or colony reaches a size that the inner cells of the mass behave differently from the outer cells." These authors then postulated that the outer cells of the culture shielded the internal cells from stimuli that suppressed organized development.

Although aggregates seem to be a prerequisite, embryos appear to originate from single, or a small number of, cells within the aggregate (McWilliam et al. 1974). These cells are typically small and cytoplasmically rich (McWilliam et al. 1974; Fujimura and Komamine 1979). Later work demonstrated that the minimum aggregate size in which embryos would develop was 31-47 μm and contained 3-10 cells (Fujimura and Komamine 1979). Nomura and Komamine (1985) further reported that these embryos could be derived from individual cells if these single cells were first cultured in a 2,4-D-containing medium which permitted cell division and resulted in the formation of an embryogenic aggregate. Subsequent transfer of these aggregates to a 2,4-D-deficient medium resulted in embryo development. Similar findings were reported by Kim and Janick (1989a) for callus-derived embryos.

The common observation that embryos are derived from larger pieces of tissue has been considered in several ways. Initially, Steward et al. (1958) speculated that the outer cells shielded the embryo from medium-derived inhibitors - presumably associated with the coconut milk (Steward et al. 1964), which was a common medium component in the early days of plant cell tissue culture. Further support for the role of inhibitors is provided by the facts that in some cases activated carbon has improved embryo development (Fridborg et al. 1978; see Ammirato 1983) and that dilution of carrot cell line W001C into fresh medium could result in globular embryo formation even in the presence of 2,4-D (Sung and Okimoto 1981). These observations may suggest that limitations in the mass transfer of these inhibitors, either from the medium to the embryo or from the embryo to the medium, may be important for obtaining embryogenesis in culture. If true, this suggestion would predict that an optimal aggregate size may be required for somatic embryogenesis.

In addition to the above observations which suggest that inhibitors may be important, other, somewhat contradictory, observations can be cited which suggest that endogenous conditioning agents may be required for embryogenesis. These agents could become limiting, especially when small aggregates are studied or when embryogenic aggregates are diluted. Halperin (1967) showed that carrot embryogenesis could be suppressed by diluting small embryogenically competent aggregates into fresh medium. This author hypothesized that a critical metabolite was diluted from the aggregate into the liquid. De Vries et al. (1988) reported that embryo development was accelerated by a high-molecular-weight, heat-labile fraction from an established embryogenic culture. In addition to the possible role of mass transfer limitations, the above observations suggest that specific cell-to-cell interactions may be required for the formation of microenvironments suitable for embryogenesis. Thus, embryo induction and development may be directly analogous to tracheary element formation discussed in Chapter 7.

B.3. Embryogenesis as a Model of Developmental Regulation

Once the culturing techniques were established, somatic embryogenesis became a model for studying developmental regulation of gene expression and function. However, "the acquisition of embryogenic potential and subsequent formation of somatic embryos is accompanied by only minor changes in gene expression as visualized by two-dimensional polyacrylamide gel electrophoresis (PAGE) analysis of in vitro translation products and hybridization analysis with a cloned complementary (c)DNA probe" (de Vries et al. 1988). It has been estimated that only about 1-2% of the proteins show quantitative differences between embryos and unorganized cells (see Choi and Sung 1989). Although large changes in gene expression between unorganized tissue and developing embryos have not been observed, Sung and Okimoto (1981, 1983) have identified both callus-specific and embryo-specific proteins. The embryo- specific proteins (Sung and Okimoto 1981), designated E1 and E2, were observed four hours after culture transfer whereas embryo formation could not be detected for several days. The observation that embryo-specific genes are expressed very early relative to the observation of morphologically recognizable structures is in agreement with results of Wilde et al. (1988), who observed that "many of the molecular processes of embryogenesis are already established in PEMs [proembryogenic masses] in the presence of auxin."

A final example of developmentally regulated gene expression is provided by Satoh et al. (1986), who observed that a glycoprotein (GP65) was released into the medium when embryogenic carrot cell aggregates were transferred into a 2,4-D-free medium and embryo development was initiated. A second glycoprotein (GP57) was released into the medium when embryogenic aggregates were cultured in the presence of 2,4-D or when a nonembryogenic cell line was cultured. These authors also observed that cultivation of embryogenic aggregates in the presence of increasing levels of 2,4-D resulted in reduced embryogenesis and reduced levels of the embryo-specific GP65, whereas increasing levels of GP57 were observed. This study illustrates that although embryogenesis can often be correlated with other morphological, physiological, or developmental processes (e.g., in this case glycoprotein release), it has been difficult to establish causal relationships between these processes.

In addition to characterizing gene products, there have been efforts aimed at identifying physiological and functional differences between embryos and unorganized cells. Sengupta and Raghavan (1980a, b) observed rapid increases in the rates of protein and RNA synthesis after transfer of embryogenically competent carrot cells to an auxin-free medium. Consistent with this observation is the report by Ashihara et al. (1981) that carrot somatic embryos had increased enzyme activities for the synthesis of pyrimidine nucleotides. Polyamine synthesis and the potential role of polyamines in embryogenesis have also been studied (Montague et al. 1978, 1979; Feirer et al. 1984). Feinberg et al. (1984) correlated polyamines with embryogenesis by showing that (1) embryogenic cultures had higher levels of polyamines; (2) inhibitors of polyamine biosynthesis could suppress embryogenesis whereas the addition of the polyamine spermidine could reverse this inhibition; and (3) levels of enzymes for polyamine synthesis, arginine decarboxylase and S-adenosylmethionine decarboxylase, were higher under embryogenic conditions. A final example of a developmentally related function is provided by Sung et al. (1981), who reported that the inactivation of the antibiotic cycloheximide was expressed by differentiated, and not undifferentiated, tissues.

C. APPLICATIONS OF SOMATIC EMBRYOGENESIS

C.1. Artificial Seeds

Since the last stage of embryo development shown in Figure 10.1 results in a regenerable unit, somatic embryogenesis can be used for the regeneration of plants from tissue culture. Thus, the potential exists for performing specific genetic manipulations on the single-cell level and ultimately regenerating improved plants from these cells. In comparison to other means of regeneration, embryogenesis has several advantages. Because both meristematic regions are present in the somatic embryo, regeneration requires less manual manipulation than other methods of plant regeneration. Also, since many of these manipulations are performed on embryos that are small (less than one millimeter) and less sensitive to physical damage, there exists a significant potential for developing automated processes in which embryos are produced in liquid culture and delivered into an appropriate environment for plantlet formation. Such automation could significantly enhance the economics of plant improvement through tissue culture. Because of the similarities of the somatic and zygotic embryos, it has been suggested that somatic embryogenesis may provide a means of producing "artificial" or "synthetic" seeds. In the following section, we will briefly discuss the major obstacles which currently limit the use of artificial seeds. Further discussion of artificial seeds can be found in Redenbaugh et al. (1986, 1988), Gupta and Durzan (1986, 1987), Durzan and Gupta (1988), Lutz et al. (1985), and Gray (1987a).

Figure 10.2 outlines the steps which are required for the large-scale production and use of artificial seeds. Although this sequence of operations seems most reasonable at this time, the reader should be cautioned that since there are significant scientific and technological gaps in our knowledge, this sequence is speculative. For our discussion, we will consider only the initial steps in which cells and aggregates are cultured in liquid phase medium. If somatic embryos are to be induced from unorganized tissue (e.g., the upper left in Figure 10.2), then it is likely that, first, undifferentiated cells will be proliferated in suspension culture. Following undifferentiated cell growth, medium exchange will be required to provide the stimuli for embryo development. Styer (1985) and Ammirato and Styer (1985) reported that a spin filter bioreactor could be used for medium exchange. In the spin filter bioreactor, a membrane filter is used in conjunction with the agitator. After growing the cells, the growth or embryo induction medium can be filtered off while the cells are retained within the reactor. The second medium, which permits embryo development, can then be added. Using different bioreactor systems for embryo development, Stuart et al. (1987) reported that the embryo yield was similar if an airlift or an agitated bioreactor with a propellor was used (embryo yields appear to be about 100 embryos/mL).

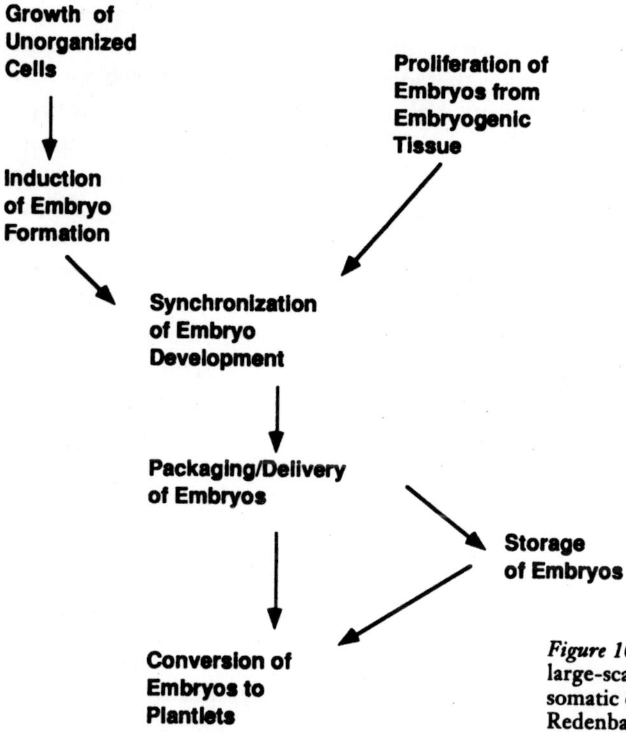

Figure 10.2. Proposed steps required for the large-scale production of synthetic seeds via somatic embryogenesis. Adapted from Redenbaugh et al. (1987).

The upper right-hand side of Figure 10.2 shows that embryos can also be formed directly from tissue without an intermediate formation of callus or unorganized cell growth. Examples include the somatic polyembryogenesis of oil palm (Jones 1983), sugar pine (Gupta and Durzan 1986), and loblolly pine (Gupta and Durzan 1987).

Once the embryos are induced, it will probably be necessary that the embryos develop in a synchronous fashion. Most methods to synchronize embryo development have relied on physical methods such as sieving or density gradient centrifugation (Warren and Fowler 1977; Fujimura and Komamine 1979; Giuliano et al. 1983) to isolate embryos at specific developmental states. Subsequent cultivation of embryos isolated in specific developmental states often results in synchronous development. Due to the operational difficulties and concerns about asepsis with these physical methods for synchronization, chemical methods to synchronize embryo development have also been suggested (Lutz et al. 1985; Kamada and Harada 1981; Bradley et al. 1984). More recently, Cazzulino et al. 1990) developed an image analysis system to statistically categorize embryos at various developmental states. If a means of embryo separation could be coupled to this analyzer, then it should be possible to automatically identify and separate embryos and thus synchronize further development.

The above discussion suggests that cultivation and induction of embryogenesis is straightforward. However, the development of culture methods for the generation of vigorous and mature embryos is a significant limitation to the development of artificial seeds (Gray 1987a). The first problem is that, over extended periods of subculture, cell lines often lose their regenerative ability. Although the causes of lose of embryogenic potential are uncertain, Smith and Street (1974) suggested that both genetic and physiological factors are involved. A second problem is that, once embryos are induced, it may be difficult to provide conditions in a bioreactor which support further embryo development. This is illustrated by work of Stuart et al. (1987), who reported that the conversion efficiency of plantlets from embryos was less than 30% for embryos cultured in flasks and only 2-3% for bioreactor-cultured embryos. This compares to a conversion efficiency of 70 - 90% for embryos cultured on a semisolid agar surface. The reasons for these low conversion efficiencies are unknown and represent an intriguing research problem.

Once the embryos have developed in the bioreactor, they must be either delivered to an environment for further plantlet development or stored in a resting state. Research efforts to induce a resting state while maintaining embryo viability (Gray 1987b; Kim and Janick 1989b) and encapsulating embryos in protective gels (Redenbaugh et al. 1987) will not be discussed here.

C.2. Natural Product Synthesis

Somatic embryogenesis may provide a means for producing chemicals normally associated with the zygotic embryo or seed. Examples include characteristic celery flavors (Al-Abta et al. 1979) and cacao seed lipids (Janick et al. 1982). In both these cases, the compounds are not produced by unorganized cells but are produced by embryos. In this section, we will summarize studies involving embryo culture for chemical synthesis.

As mentioned previously, Al-Abta and Collin (1978) reported that celery cells in early-stage embryos (e.g., globular and heart) were undifferentiated whereas cells in the torpedo stages began to show signs of cellular differentiation. This group also observed that characteristic celery flavors were only produced by torpedo, or further developed, tissue (Al-Abta et al. 1979). Since oil ducts, which are known to accumulate celery flavors, were not observed in these torpedo-stage embryos (Al-Abta and Collin 1978), Collin and Watts (1983) concluded that production could not be correlated to the formation of these specialized tissues. These authors did however suggest a correlation between flavor production and tissue greening.

Cardenolides (e.g., digitoxigenin derivatives) also appear to be produced in association with organ differentiation (see reviews by Luckner and Diettrich 1985, 1988). In plants, cardenolide levels are higher in the shoot apex and the leaves than in the stems or roots. Although shoot culture has been considered for cardenolide production, we will only discuss the use of embryos in this chapter. In cell culture, Nover et al. (1980) and Garve et al. (1980) observed that cardenolide levels were low in unorganized cells, however when embryoid formation was triggered, cardenolide levels were enhanced. Kuberski et al. (1984) examined hormonal and nutritional requirements for *Digitalis* growth, embryo formation and cardenolide accumulation. In addition to the nutritional issues, light has also been observed to be important for cardenolide accumulation. Although not a strict requirement for prodution, light has been observed to enhance cardenolide accumulation (Hagimori et al. 1982; Ohlsson et al. 1983; Kuberski et al. 1984). Thus, chloroplasts do not appear to be strictly required for production (Hagimori et al. 1982), but rather the conditions which promote greening appear to also promote cardenolide production (Ohlsson et al. 1983). With respect to the commercial potential, Luckner and Diettrich (1988) concluded that significant improvements in productivities are necessary before plant cell processes for cardenolide production would be practical.

Embryo culture has also been considered for the synthesis of cacao seed lipids for cocoa butter (for a review, see Janick 1986). Pence et al. (1980) observed that somatic embryos of *Theobroma cacao* could be initiated from zygotic embryos without the intermediary formation of callus. In contrast, Tsai and Kinsella (1981) did not observe organogenesis or embryogenesis from calli or suspension cultures of *Theobroma cacao*. Thus, for this case, somatic embryos have not been induced to form from unorganized cells. As the somatic embryos developed, they were observed to accumulate lipids (Janick et al. 1982), anthocyanins (Pence et al. 1981a), and alkaloids (Paiva and Janick 1983) characteristic of the *Theobroma* zygotic embryo. Paiva and Janick (1983) also observed that the relative anthocyanin and alkaloid levels of various somatic embryos were similar to the levels of the zygotic embryos from which they were derived. This observation suggests that favorable cell lines could be derived from the zygotic embryos of high-quality plants. Pence et al. (1981b) observed that when the sucrose concentration was increased to 27%, the fatty acid composition of the somatic embryos was similar to that of the zygotic embryo. This latter point is important since characteristic flavors often result from a mixture and not from a single compound.

A final interesting observation by Ozeki and Komamine (1981, 1985) is that a carrot cell culture could be induced to form embryos and induced to produce anthocyanins under the same conditions (i.e., transfer to an auxin-free medium). However, when this culture was subjected to sieving and Ficoll density gradient centrifugation, it was observed that the higher density fraction (>14% Ficoll) formed embryos while the lower density fraction (<12% Ficoll) formed anthocyanins. Thus, although the conditions which induced these two differentiation processes were similar, the cells that produced anthocyanins were not the same as those that underwent embryogenesis.

D. CONCLUSION

We have focused our discussion on the culture of embryos in liquid systems with the goals of efficiently producing regenerable units or synthesizing zygote-specific natural products. Compared to vegetative propagation, the advantage of somatic embryogenesis for plant generation is that the regenerable unit (the embryo) is small, contains both meristems, and is less sensitive to shear damage. Thus, automated processes can be envisioned for the mass production, transport, and planting of these artificial seeds. With respect to secondary metabolite production, somatic embryos may provide a means for producing zygote-specific products. Difficulties in reproducibly inducing embryogenesis and ensuring appropriate embryo development need to be overcome.

References

S. Al-Abta and H. A. Collin. 1978. Cell Differentiation in Embryoids and Plantlets of Celery Tissue Cultures. *New Phytol. 80*, 517-521.

S. Al-Abta, I. J. Galpin, and H. A. Collin. 1979. Flavor Compounds in Tissue Cultures of Celery. *Plant Sci. Lett. 16*, 129-134.

P. V. Ammirato. 1974. The Effects of Abscisic Acid on the Development of Somatic Embryos from Cells of Caraway (*Carum carvi* L.). *Bot. Gaz. 135*, 328-337.

P. V. Ammirato. 1977. Hormonal Control of Somatic Embryo Development from Cultured Cells of Caraway: Interactions of Abscisic Acid, Zeatin, and Gibberellic Acid. *Plant Physiol. 59*, 579-586.

P. V. Ammirato. 1983. Embryogenesis. In "Handbook of Plant Cell Culture," Vol. 1, D. A. Evans, W. R. Sharp, P. V. Ammirato, and Y. Yamada (eds.), Macmillan, New York, pp. 82-123.

P. V. Ammirato and D. J. Styer. 1985. Strategies for Large-Scale Manipulation of Somatic Embryos in Suspension Culture. In "Biotechnology in Plant Science: Relevance to Agriculture in the Eighties," M. Zaitlin, P. Day, and A. Hollaender (eds.), Academic Press, New York, pp.161-178.

H. Ashihara, T. Fujimura, and A. Komamine. 1981. Pyrimidine Nucleotide Biosynthesis During Somatic Embryogenesis in a Carrot Cell Suspension Culture. *Z. Pflanzenphysiol. 104*, 129-137.

C. Borkird, J. H. Choi, and Z. R. Sung. 1986. Effect of 2,4-Dichlorophenoxyacetic Acid on the Expression of Embryogenic Program in Carrot. *Plant Physiol. 81*, 1143-1146.

P. M. Bradley, F. El-Fiki, and K. L. Giles. 1984. Polyamines and Arginine Affect Somatic Embryogenesis of *Daucus carota*. *Plant Sci. Lett. 34*, 397-401.

D. L. Cazzulino, H. Pederson and C.-K. Chin. 1990. Monitoring Somatic Embryo Development in Carrot Cell Cultures. *Dev. Ind. Microbiol. 31*, 285-292.

J. H. Choi and Z. R. Sung. 1989. Induction, Commitment, and Progression of Plant Embryogenesis. In "Plant Biotechnology," S.-D. Kung and C. J. Arntzen (eds.), Butterworths, Boston, pp. 141-159.

M. L. Christianson. 1985. An Embryogenic Culture of Soybean: Towards a General Theory of Somatic Embryogenesis. In "Tissue Culture in Forestry and Agriculture," R. R. Henke, K. W. Hughes, M. J. Constantin, and A. Hollaender (eds.) Plenum Press, New York, pp. 83-103.

H. A. Collin and M. Watts. 1983. Flavor Production in Culture. In "Handbook of Plant Cell Culture," Vol. 1, D. A. Evans, W. R. Sharp, P. V. Ammirato, and Y. Yamada (eds.), Macmillan, New York, pp. 729-747.

S. C. de Vries, H. Booij, P. Meyerink, G. Huisman, H. D. Wilde, T. L. Thomas, and A. van Kammen. 1988. Acquisition of Embryogenic Potential in Carrot Cell-Suspension Cultures. *Planta 176*, 196-204.

J. H. Dodds and L. W. Roberts. 1985. "Experiments in Plant Tissue Culture" 2nd ed., Cambridge University Press, New York, pp. 122-132.

D. J. Durzan and P. K. Gupta. 1988. Somatic Embryogenesis and Polyembryogenesis in Conifers. In "Biotechnology in Agriculture," A. Mizrahi (ed.), Alan R. Liss Inc., New York, pp. 53-81.

D. A. Evans, W. R. Sharp, and C. E. Flick. 1981. Growth Behavior of Cell Cultures: Embryogenesis and Organogenesis. In "Plant Tissue Culture: Methods and Applications in Agriculture," T. A. Thorpe (ed.), Academic Press, New York, pp. 45-113.

A. A. Feinberg, J. H. Choi, W. P. Lubich, and Z. R. Sung. 1984. Developmental Regulation of Polyamine Metabolism in Growth and Differentiation of Carrot Culture. *Planta 162*, 532-539.

R. P. Feirer, G. Mignon, and J. D. Litvay. 1984. Arginine Decarboxylase and Polyamines Required for Embryogenesis in Wild Carrot. *Science 223*, 1433-1435.

G. Fridborg, M. Pedersen, L.-E. Landstrom, and T. Eriksson. 1978. The Effect of Activated Charcoal on Tissue Cultures: Adsorption of Metabolites Inhibiting Morphogenesis. *Physiol. Plant. 43*, 104-106.

T. Fujimura and A. Komamine. 1975. Effects of Various Growth Regulators on the Embryogenesis in a Carrot Cell Suspension Culture. *Plant Sci. Lett. 5*, 359-364.

T. Fujimura and A. Komamine. 1979. Synchronization of Somatic Embryogenesis in a Carrot Cell Suspension Culture. *Plant Physiol. 64*, 162-164.

T. Fujimura and A. Komamine. 1980. The Serial Observation of Embryogenesis in a Carrot Cell Suspension Culture. *New Phytol. 86*, 213-218.

R. Garve, M. Luckner, E. Vogel, A. Tewes, and L. Nover. 1980. Growth, Morphogenesis and Cardenolide Formation in Long-Term Cultures of *Digitalis lanata. Planta Med. 40*, 92-103.

G. Giuliano, D. Rosellini, and M. Terzi. 1983. A New Method for the Purification of the Different Stages of Carrot Embryoids. *Plant Cell Rep. 2*, 216-218.

D. J. Gray. 1987a. Introduction to the Symposium. *HortScience 22*, 796-797.

D. J. Gray. 1987b. Quiescence in Monocotyledonous and Dicotyledonous Somatic Embryos Induced by Dehydration. *HortScience 22*, 810-814.

P. K. Gupta and D. J. Durzan. 1986. Somatic Polyembryogenesis from Callus of Mature Sugar Pine Embryos. *Biotechnology 4*, 643-645.

P. K. Gupta and D. J. Durzan. 1987. Biotechnology of Somatic Polyembryogenesis and Plantlet Regeneration in Loblolly Pine. *Biotechnology 5*, 147-151.

M. Hagimori, T. Matsumoto, and Y. Obi. 1982. Studies of the Production of *Digitalis* Cardenolides by Plant Tissue Culture. *Plant Physiol. 69*, 653-656.

W. Halperin. 1967. Population Density Effects on Embryogenesis in Carrot-Cell Cultures. *Exp. Cell Res. 48*, 170-173.

W. Halperin and W. A. Jensen. 1967. Ultrastructral Changes during Growth and Embryogenesis in Carrot Cell Cultures. *J. Ultrastruct. Res. 18*, 428-443.

J. Janick. 1986. Embryogenics: The Technology of Obtaining Useful Products from the Culture of Asexual Embryos. In "Biotechnology of Plants and Microorganisms," O. J. Crocomo, W. R. Sharp, D. A. Evans, J. E. Bravo, F. C. A. Tavares and E. F. Paddock, Ohio State University Press, Columbus, Ohio, pp. 97-117.

J. Janick, D. C. Wright, and P. M. Hasegawa. 1982. In Vitro Production of Cacao Seed Lipids. *J. Am. Soc. Hort. Sci. 107*, 919-922.

L. H. Jones. 1983. The Oil Palm and Its Clonal Propagation by Tissue Culture. *Biologist 30*, 181-188.

H. Kamada and H. Harada. 1979a. Studies on the Organogenesis in Carrot Tissue Cultures. I. Effects of Growth Regulators on Somatic Embryogenesis and Root Formation. *Z. Pflanzenphysiol. 91*, 255-266.

H. Kamada and H. Harada. 1979b. Studies on the Organogenesis in Carrot Tissue Cultures. II. Effects of Amino Acids and Inorganic Nitrogenous Compounds on Somatic Embryogenesis. *Z. Pflanzenphysiol. 91*, 453-463.

H. Kamada and H. Harada. 1981. Changes in the Endogenous Level and the Effects of Abscisic Acid During Somatic Embryogenesis of *Daucus carota* L. *Plant Cell Physiol. 22*, 1423-1429.

R. H. J. Kessell and A. H. Carr. 1972. The Effect of Dissolved Oxygen Concentration on Growth and Differentiation of Carrot (*Daucus carota*) Tissue. *J. Exp. Bot. 23*, 996-1007.

R. H. J. Kessell, C. Goodwin, J. Philp, and M. W. Fowler. 1977. The Relationship between Dissolved Oxygen Concentration, ATP and Embryogenesis in Carrot (*Daucus carota*) Tissue Cultures. *Plant Sci. Lett. 10*, 265-274.

Y.-H. Kim and J. Janick. 1989a. Origin of Somatic Embryos in Celery Tissue Culture. *HortScience 24*, 671-673.

Y.-H. Kim and J. Janick. 1989b. ABA and Polyox-Encapsulation or High Humidity Increases Survival of Desiccated Somatic Embryos of Celery. *HortScience 24*, 674-676.

C. Kuberski, M. Scheibner, C. Steup, B. Diettrich, and M. Luckner. 1984. Embryogenesis and Cardenolide Formation in Tissue Cultures of *Digitalis lanata*. *Phytochemistry 23*, 1407-1412.

M. Luckner and B. Diettrich. 1985. Formation of Cardenolides in Cell and Organ Cultures of *Digitalis lanata*. In "Primary and Secondary Metabolism of Plant Cell Cultures," K.-H. Neumann, W. Barz, and E. Reinhard (eds.), Springer-Verlag, Heidelberg, pp. 154-163.

M. Luckner and B. Diettrich. 1988. Cardenolides. In "Cell Culture and Somatic Cell Genetics of Plants," Vol. 5, F. Constabel and I. K. Vasil (eds.), Academic Press, New York, pp. 193-212.

J. D. Lutz, J. R. Wong, J. Rowe, D. M. Tricoli, and R. H. Lawrence. 1985. Somatic Embryogenesis for Mass Cloning of Crop Plants. In "Tissue Culture in Forestry and Agriculture," R. R. Henke, K. W. Hughes, M. J. Constantin, and A. Hollaender (eds.), Plenum Press, New York, pp. 105-116.

A. A. McWilliam, S. M. Smith, and H. E. Street. 1974. The Origin and Develpment of Embryoids in Suspension Cultures of Carrot (*Daucus carota*). *Ann. Bot. 38*, 243-250.

M. J. Montague, J. W. Koppenbrink, and E. G. Jaworski. 1978. Polyamine Metabolism in Embryogenic Cells of *Daucus carota*. I. Changes in Intracellular Content and Rates of Synthesis. *Plant Physiol. 62*, 430-433.

M. J. Montague, T. A. Armstrong, and E. G. Jaworski. 1979. Polyamine Metabolism in Embryogenic Cells of *Daucus carota*. II. Changes in Arginine Decarboxylase Activity. *Plant Physiol. 63*, 341-345.

K. Nomura and A. Komamine. 1985. Identification and Isolation of Single Cells That Produce Somatic Embryos at a High Frequency in a Carrot Suspension Culture. *Plant Physiol. 79*, 988-991.

L. Nover, M. Luckner, A. Tewes, R. Garve, and E. Vogel. 1980. Cell Specialization and Cardiac Glycoside Formation in Cell Cultures of *Digitalis* Species. *Acta Hort. 96*, 65-74.

A. B. Ohlsson, L. Bjork, and S. Gatenbeck. 1983. Effect of Light on Cardenolide Production by *Digitalis lanata* Tissue Cultures. *Phytochemistry 22*, 2447-2450.

Y. Ozeki and A. Komamine. 1981. Induction of Anthocyanin Synthesis in Relation to Embryogenesis in a Carrot Suspension Culture: Correlation of Metabolic Differentiation with Morphological Differentiation. *Physiol. Plant. 53*, 570-573.

Y. Ozeki and A. Komamine. 1985. Induction of Anthocyanin Synthesis in Relation to Embryogenesis in a Carrot Suspension Culture - A Model System for the Study of Expression and Repression of Secondary Metabolism. In "Primary and Secondary Metabolism of Plant Cell Cultures," K.-H. Neumann, W. Barz, and E. Reinhard (eds.), Springer-Verlag, Heidelberg, pp. 99-106.

M. Paiva and J. Janick. 1983. In Vivo and In Vitro Production of Alkaloids in *Theobroma cacao* L. *Acta Hort. 131*, 265-273.

V. C. Pence, P. M. Hasegawa, and J. Janick. 1980. Initiation and Development of Asexual Embryos of *Theobroma cacao* L. *In Vitro. Z. Pflanzenphysiol. 98*, 1-14.

V. C. Pence, P. M. Hasegawa, and J. Janick. 1981a. In Vitro Cotyledonary Development and Anthocyanin Synthesis in Zygotic and Asexual Embryos of *Theobroma cacao*. *J. Am. Soc. Hort. Sci. 106*, 381-385.

V. C. Pence, P. M. Hasegawa, and J. Janick. 1981b. Sucrose Mediated Regulation of Fatty Acid Composition in Asexual Embryos of *Theobroma cacao*. *Physiol. Plant. 53*, 378-384.

V. Raghavan. 1980. Embryo Culture. *Int. Rev. Cytol., Suppl. 11B*, 209-240.

K. Rajasekaran, J. Vine, and M. G. Mullins. 1982. Dormancy in Somatic Embryos and Seeds of *Vitis*: Changes in Endogenous Abscisic Acid during Embryogenesis and Germination. *Planta 154*, 139-144.

T. S. Rangan. 1982. Ovary, Ovule, and Nucellus Culture. In "Experimental Embryology of Vascular Plants," B. M. Johri (ed.), Springer-Verlag, Heidelberg, pp. 105-129.

K. Redenbaugh, B. D. Paasch, J. W. Nichol, M. E. Kossler, P. R. Viss, and K. A. Walker. 1986. Somatic Seeds: Encapsulation of Asexual Plant Embryos. *Biotechnology 4*, 797-801.

K. Redenbaugh, D. Slade, P. R. Viss, and J. A. Fujii. 1987. Encapsulation of Somatic Embryos in Synthetic Seed Coats. *HortScience 22*, 803-809.

K. Redenbaugh, J. A. Fujii, and D. Slade. 1988. Encapsulated Plant Embryos. In "Biotechnology in Agriculture," A. Mizrahi (ed.), Alan R. Liss Inc., New York, pp. 225-248.

J. Reinert. 1959. Uber die Kontrolle der Morphogenese und die Induktion von Adventiveembryonen an Gewebekulturen aus Karotten. *Planta 53*, 318-333.

S. Satoh, H. Kamada, H. Harada, and T. Fujii. 1986. Auxin-Controlled Glycoprotein Release into the Medium of Embryogenic Carrot Cells. *Plant Physiol. 81*, 931-933.

F. M. Schiavone and T. J. Cooke. 1985. A Geometric Analysis of Somatic Embryo Formation in Carrot Cell Cultures. *Can. J. Bot. 63*, 1573-1578.

C. Sengupta and V. Raghavan. 1980a. Somatic Embryogenesis in Carrot Cell Suspension. I. Pattern of Protein and Nucleic Acid Synthesis. *J. Exp. Bot. 31*, 247-258.

C. Sengupta and V. Raghavan. 1980b. Somatic Embryogenesis in Carrot Cell Suspension. II. Synthesis of Ribosomal RNA and Poly(A)+RNA. *J. Exp. Bot. 31*, 259-268.

S. M. Smith and H. E. Street. 1974. The Decline of Embryogenic Potential as Callus and Suspension Cultures of Carrot (*Daucus carota* L.) Are Serially Subcultured. *Ann. Bot. 38*, 223-241.

F. C. Steward, M. O. Mapes, and K. Mears. 1958. Growth and Organized Development of Cultured Cells. II Organization in Cultures Grown in Freely Suspended Cells. *Am. J. Bot. 45*, 705-708.

F. C. Steward, M. O. Mapes, A. E. Kent, and R. D. Holsten. 1964. Growth and Development of Cultured Plant Cells. *Science 143*, 20-27.

S. Strickland, J. W. Nichol, C. M. McCall, and D. A. Stuart. 1987. Effect of Carbohydrate Source on Alfalfa Somatic Embryogenesis. *Plant Sci. 48*, 113-121.

D. A. Stuart, S. G. Strickland and K. A. Walker. 1987. Bioreactor Production of Alfalfa Somatic Embryos. *HortScience 22*, 800-803.

D. J. Styer. 1985. Bioreactor Technology for Plant Propagation. In "Tissue Culture in Forestry and Agriculture," R. R. Henke, K. W. Hughes, M. J. Constantin, and A. Hollaender (eds.), Plenum Press, New York, pp. 117-130.

Z. R. Sung. 1979. Relationship of Indole-3-Acetic Acid and Tryptophan Concentrations in Normal and 5-Methyltryptophan-Resistant Cell Lines of Wild Carrot. *Planta 145*, 339-345.

Z. R. Sung and R. Okimoto. 1981. Embryonic Proteins in Somatic Embryos of Carrot. *Proc. Natl. Acad. Sci. USA 78*, 3683-3687.

Z. R. Sung and R. Okimoto. 1983. Coordinate Gene Expression during Somatic Embryogenesis in Carrots. *Proc. Natl. Acad. Sci. USA 80*, 2661-2665.

Z. R. Sung, G. B. Lazar, and D. Dudits. 1981. Cycloheximide Resistance in Carrot Culture: A Differentiated Function. *Plant Physiol. 68*, 261-264.

L. Tianran and K.-H. Neumann. 1985. Embryogenesis and Endogenous Hormone Content of Cell Cultures of Some Carrot Varieties (*Daucus carota* L.). *Ber. Dtsch. Bot. Ges. 98*, 227-235.

B. Tisserat and T. Murashige 1977. Effects of Ethephon, Ethylene, and 2,4-Dichlorophenoxyacetic Acid on Embryogenesis *in Vitro*. *Plant Physiol. 60*, 437-439.

C. H. Tsai and J. E. Kinsella. 1981. Initiation and Growth of Callus and Cell Suspensions of *Theobroma cacao* L. *Ann. Bot. 48*, 549-557.

D. C. Verma and D. D. Dougall. 1977. Influence of Carbohydrates on Quantitative Aspects of Growth and Embryo Formation in Wild Carrot Suspension Cultures. *Plant Physiol. 59*, 81-85.

G. S. Warren and M. W. Fowler. 1977. A Physical Method for the Separation of Various Stages in the Embryogenesis of Carrot Cell Cultures. *Plant Sci. Lett. 9*, 71-76.

D. F. Wetherell and D. K. Dougall. 1976. Sources of Nitrogen Supporting Growth and Embryogenesis in Cultured Wild Carrot. *Physiol. Plant. 37*, 97-103.

H. D. Wilde, W. S. Nelson, H. Booij, S. C. de Vries, and T. L. Thomas. 1988. Gene-Expression Programs in Embryogenic and Non-Embryogenic Carrot Cultures. *Planta 176*, 205-211.

CHAPTER 11

QUESTIONS AND STRATEGIES FOR PRODUCTIVITY IMPROVEMENTS

A. INTRODUCTION

Throughout this book, we have given many examples of how significant increases in cell-specific and reactor volumetric productivities have been obtained with suspension cultures, immobilized cells, and organized tissues. Because of the complex and incompletely understood nature of plant cells in *in vitro* culture, empirical, case-by-case studies have been used. In this chapter, we shall discuss approaches for productivity enhancement in terms of the key experimental parameters and questions that need to be asked, no matter what species and form of culture is being investigated. Although generalization of the answers appears impossible, generalization of questions is an achievable and worthwhile exercise.

A key to generalizable approaches to improved productivity is to recognize that all questions must be asked in a "context." It is well known in optimization that an optimal process cannot be found by sequentially optimizing each step of a process individually. The complex, nonlinear, interactive nature of most processes requires optimization to take place over the whole process, and usually the optimal process will consist of components that operate in a suboptimal manner.

The formal mathematical optimization of the process of developing an optimally productive process with plant cell culture is impossible. However, all approaches to productivity increases are potentially interactive (often synergistically), and an integrated and iterative approach to productivity enhancement is required. Thus, each question we ask will have answers which depend on the context in which it is asked.

B. SELECTION, SCREENING, AND VARIABILITY

The intrinsic variability of *in vitro* cultures can be extremely valuable if controlled. However, it is inconceivable that a large investment in a commercial-size plant would ever be made unless a stable production strain were available, so uncontrolled variation is unacceptable from a process perspective.

Within the early stages of process development, variability is a blessing in providing a rich source of plant material to be screened. Many plants, strains, and related species should be chosen to establish the initial callus cultures, and from the *in vitro* culturing many spontaneous variants arise (e.g., Gould 1986). Mutagens and radiation can be used to enhance this level of variation (see Widholm 1987), but this is usually not necessary. More predictable methods of generating desirable clones using the techniques of molecular biology will be available in the future.

Generating variants is much less of a problem than is screening them. For products that are colored (e.g., pigments), the selection process can be done by simple visual inspection. In other cases, fluorescence or some other property associated with a desired metabolite can be used to develop a screening procedure, often with the aid of a microscope. In still other cases, more advanced and time-consuming techniques such as radioimmunoassays may be required (e.g., Kemp and Morgan 1987).

The problems, however, are not so much in generating variability or even in screening, but in screening under those conditions most likely to identify the best producers. Should screening be done in growth or production medium? Should an elicitor be used? Will the screen detect both intracellular and extracellular product? In what phase of batch growth should the cultures be monitored?

Whether these questions are important depends on how a product is formed in that cell line. For example, in screening for berberine production, only one cell line excreted the product. If the screen was only for the intracellular product, the excreting cell line would have to be dismissed as unproductive. If cell line A responds to an elicitor to increase production 10- fold, whereas a second cell line, B, is "constitutive" or unresponsible to elicitors and produces at a level twofold higher than that of unelicited A, strain A would be eliminated on an initial screen without elicitor. If only end-point assays are used to screen for product, factors such as product dilution by growth, product degradation, and delayed synthesis all could lead to suboptimal selection of strains. Since the degree to which a product is growth-associated may change from strain to strain, the use of a growth- or a product formation-oriented medium could easily influence the choice of strain. The above factors are not the only ones that could influence the choice of optimal strain, but they do serve to illustrate the difficulties of finding the optimal strain when the screen is done under conditions which do not represent the final process.

Selection refers to the technique of repeatedly selecting, from within a cell line, high-producing cells, which, upon subculture, give rise to a new cell line of higher productivity. Several excellent sources of information are available on the various aspects of selection and screening (e.g., Yamada 1984; Berlin and Sasse 1985; Dougall 1987).

C. MEDIUM FORMULATION

If a product is primarily growth-associated, a single medium may be sufficient to achieve high-level production. A single medium simplifies process design and operation. In some cases, a metabolite may be formed in a largely, but not completely growth-associated, manner, and a "compromise" medium can be developed. Botta et al. (1989) have discussed a statistical approach to development of such a medium.

In many cases, however, the product has such a strong, non-growth-associated character that two media must be developed. How the second "production" medium is formulated must respond to questions of what type of bioreactor operation is envisioned. If a two-stage batch system with suspension cultures is envisioned, the constraints on medium formulation may differ from that for a continuous immobilized cell culture. Features that would be important are whether growth is acceptable (it would be in a suspension system but not in an immobilized system), whether product excretion is necessary (it would be for the immobilized but not the suspension systems), and whether sufficient nutrients are available to sustain the population for extended periods (this would be more crucial for immobilized cell systems, although potentially useful in both cases).

In comparing these two reactor options, the crucial parameter for optimization is volumetric productivity, not specific productivity. For the two-stage batch system, the minimum time to achieve an acceptable product concentration is critical. A 10% increase in concentration at the cost of a 40% increase in time would not be attractive. Further, the period of adaptation between growth and production media must give a very short lag phase if high volumetric activities are to be obtained (Berlin et al. 1987). For an immobilized cell system, where the cells may be used for many months, the adaptation lag period is of lesser importance.

A further implication of optimizing volumetric productivity is that a highly concentrated cell mass from the growth phase is desired. Production medium formulation would be best optimized by using a highly concentrated inoculum (e.g., 30 to 50 vol.% cells). The choice of inoculum level may alter the optimal concentration of medium components. If the component is not taken up by the cells to a large extent, then the concentration in the medium will be critical and independent of inoculum level. If, however, the product is taken up rapidly by the cells and stored intracellularly, the important factor in medium design will not be the concentration, but the ratio of component to initial cell mass. Many hormones and phosphate can be withdrawn from the medium rapidly and stored intracellularly.

The above experiment also should consider whether or not the concentration of dissolved gaseous species is controlled. Different inoculum levels in a shake flask will give rise to different concentrations of dissolved gaseous species (O_2, CO_2, and ethylene). Further, if light is a parameter, its penetration and utilization will depend on inoculum density.

Of course, other interactions such as strain choice and the use of elicitors can markedly alter what the optimal medium composition might be.

D. SECRETION AND SEQUESTRATION OF SECONDARY METABOLITES

The final location of a secondary metabolite is a critical parameter in process selection. Most products are stored in the central vacuole. If the metabolite is membrane-permeable, but is trapped within the vacuole by a mechanism such as ion-trapping (summarized in Guern et al. 1987), it should be possible to induce product secretion. If the metabolite undergoes an irreversible modification and is rendered impermeable upon entry into the vacuole (e.g., see Hartmann and Toppel 1987), then product secretion may be difficult to induce.

Although secretion is often desirable, its detection is not always simple. The presence of an extracellular component can be the result of true secretion or simply lysis. Lysis is undesirable since expensive biomass is lost and the release of the intracellular material is not controlled as it would be if cells were harvested directly for an intracellular product.

Lysis can be detected if the fraction of the desired product which is extracellular is the same as the extracellular fraction of a normally intracellular component. Total protein in the medium can be determined fairly simply, and a high fraction of the total protein being extracellular would suggest lysis. Further confirmation can be obtained from gel electrophoresis, where the protein pattern can be observed. If the patterns are identical for extracellular and intracellular measurements, then lysis would be essentially confirmed. Specific enzyme assays [e.g., isocitrate dehydrogenase (Brodelius and Nilsson 1983) and 5-phosphodiesterase (Tanaka et al.1985)] have also been used to detect lysis.

In a continuous system, the actual production and release of a product can be distinguished from the simple loss of intracellular material by lysis. If the total amount of a product recovered at the end of an experiment (extracellular plus intracellular) is the same as the total product in the inoculum, it is reasonable to presume that lysis was responsible for the extracellular portion. On the other hand, if the total amount that is extracellular is greater than the amount in the inoculum and also the amount expected to be made by any growth and lysis, then secretion would be probable.

Once a method is in place to distinguish between secretion and lysis, then methods for inducing secretion can be evaluated. Some methods, such as chemical permeabilization and electroporation, appear to be ineffective as general strategies, because of the difficulty of maintaining cell viability. The use of pH cycling appears to be effective in the case of ajmalicine, but only for newly synthesized ajmalicine (Bouyssou et al. 1987). Elicitation and immobilization (Chapter 7) often confer upon cells the ability to secrete products, although the mechanisms by which this occurs is not clear. Another strategy which has the possibility to enhance secretion is the use of *in situ* product removal.

E. *IN SITU* PRODUCT REMOVAL

The use of *in situ* product removal has a number of key potential advantages beyond promoting secretion. If the formation of a product is subject to feedback inhibition or intracellular degradation, the removal and sequestering of the product in a nonbiological compartment may increase total production. Further processes of recovery and purification are generally simplified, thus reducing overall production costs. Recovery is simplified if the process of *in situ* recovery concentrates the product. Downstream purification requirements are reduced if product removal from the fermentation broth and cells is selective.

Potential methods for *in situ* product recovery are reversible conversion to another species by reaction, volatilization, two-phase liquid extraction (usually into an organic phase), clathration by cyclodextrins, and binding to solid adsorbents. The solid adsorbents may be better suited to the requirements of plant cell culture. Liquid extractants, such as organics, must be phase immiscible with water. Thus, highly hydrophobic solvents must be used whereas more hydrophilic solvents, such as methanol, may be better extractants but are too soluble to be useful in practical processes. With solid adsorbents, a much wider range of chemistries are available, and the degree of hydrophobicity and hydrophilicity can be better designed to fit a particular product. Further, the biocompatibility of a liquid extractant is often a severe constraint. Biocompatibility is less of a problem with solid adsorbents because the cells can be physically separated from contact with the solid phase and the solubility of the solid is negligible.

F. ELICITORS

The use of elicitors (Eilert 1987) to stimulate product formation and secretion has become an important process strategy that integrates easily with strategies such as *in situ* product removal and immobilization. The use of elicitors has been very useful to reduce the process time necessary to reach a high product concentration, thereby increasing volumetric productivity. Further, elicitation may result in the formation of novel compounds leading to wholly new processes.

Important, but as yet unanswered, questions arise when the use of elicitors is considered. The mechanism of action of elicitors is poorly understood. However, there is significant interest in this topic, and the recent advances in molecular biology hold great promise for giving direct insight into these mechanisms. Elicitors ultimately must affect gene regulation, and a clear understanding of regulation of secondary metabolism in plants would allow a much more predictable, efficient approach to product synthesis to replace the current empirical approaches we rely upon.

However, empirical approaches are the only ones currently available, and key questions must be addressed in developing a strategy for elicitation. The first is simply what elicitor to use. Although a reasonable first choice is a known plant pathogen for the plant species in question, other elicitors often work better. Fragments from either funal or plant cell walls are possible elicitors. Nonbiological compounds (vandium salts) can also be effective. In all cases, the plant cell is activating secondary metabolite pathways in response to stress, or the perception of stress.

One challenge in the choice of elicitor is ensuring its purity. A homogenate of a fungus is likely to contain toxic metabolites as well as elicitors. Purification of the elicitor may result in retention of the eliciting effect with a reduction in undesirable side effects. Repeated elicitation could be an important process strategy (see Funk et al. 1987) but is only viable if inhibitory side effects are minimized.

Once a elicitor is identified, its use interacts with the other process strategies and variables we have discussed. Important considerations may be the ratio of elicitor to inoculum size and how highly selected the cell line is. A low-yielding cell line may show much more dramatic responses than a high-yielding cell line [see discussion in Berlin (1988)]. The "history" of the culture and timing of elicitation may be critical.

An underappreciated fact is that elicitation will only work when cells are given the correct environment. The potential interaction of elicitation with the composition of a production medium requires careful thought. For example, if a nutrient (e.g., SO_4^{2-}) is a necessary component of the final product and if that nutrient is absent or at very low levels, elicitation *cannot* result in any increases in productivity because the cell lacks the building blocks. Similarly, the use of light may be important to provide the intracellular generation of precursors that would feed into a pathway activated by elicitation. It is also interesting to note that oxygen consumption increases upon elicitation (Funk et al. 1987). Thus, a culture that was oxygen sufficient before elicitation could become oxygen starved after elicitation, reducing the level of response to elicitation. Thus, elicitation will work only when the proper conditions and building blocks are present.

G. PRECURSORS AND BIOTRANSFORMATION

An additional strategy to increase production is to feed specific precursors to a particular pathway to encourage formation of a desired product of higher value. Such transformation strategies can be used not only to increase product levels but also to alter a product spectrum (see Chapter 8). Further, the use of synthetic analogues to natural substrates can result in the production of completely novel compounds that could have important pharmacological uses.

The use of precursors presents a wide range of questions as well as interactions with other process strategies. The importance of stoichiometry must be appreciated; one cannot make product without the minimum amount of precursor dictated by stoichiometry. Providing the right amount of precursors in the medium does not guarantee that the precursor will be taken into the cell at the correct rate. Even if uptake is sufficiently rapid to be nonlimiting, the precursor must be sequestered with the biosynthetic pool of the precursor. Since multiple pools can exist in the cell, uptake does not guarantee availability to the appropriate enzyme system. Other factors such as potential toxicity and the timing of precursor addition can be critical.

Also, coupling precursor feeding with production medium development, elicitation, and *in situ* product removal should be done carefully to ensure optimal response.

H. SYNERGY AND SUMMARY

The process strategies we have discussed are interactive. As described in Chapter 7 (see Figure 7.4), the use of gel immobilization, *in situ* resin extraction, elicitation, and production medium results in a synergetic response for extracellular ajmalicine production from *Catharanthus roseus.* For plant cell cultures to become economically attractive, a hundredfold increase in productivity is often required. Since no single strategy usually gives that large an increase, a strategy of integrated process development is required. The interactive nature of process improvement strategies must be appreciated if efficient, rapid process development is to be expected. The combination of empirical treatments that can alter productivity is large, and rational process development is a formidable challenge. The purpose of this book has been to provide the reader with the tools to meet such a challenge and to succeed.

References

J. Berlin. 1988. Formation of Secondary Metabolites in Cultured Plant Cells and Its Impact on Pharmacy. In "Biotechnology in Agriculture and Forestry," Vol. 4, "Medicinal and Aromatic Plants I," Y. P. S. Bajaj (ed.), Springer-Verlag, Berlin, pp. 37-59.

J. Berlin and F. Sasse. 1985. Selection and Screening Techniques for Plant Cell Cultures. *Adv. Biochem. Eng. Biotechnol. 31*, 99-102.

J. Berlin, C. Mollenschott, and F. DiCosmo. 1987. Comparison of Various Strategies to Optimize Alkaloid Accumulation of a Cell Suspension Culture of *Catharanthus roseus. Z. Naturforsch. 42C*, 1101-1108.

B. Botta, G. Dall'Olio, F. Ferrari, B. Monacelli, G. Pasqua, R. Scurria, and G. Delle Monarche. 1989. Cell Suspension Cultures of *Cassia didymobotrya:* Optimization of Growth and Secondary Metabolite Production by Application of the Orthogonal Design Method. *J. Plant Physiol. 135*, 290-294.

H. Bouyssou, A. Pareilleux, and G. Marigo. 1987. The Role of pH Gradients across the Plasmalemma of *Catharanthus roseus* and Its Involvement in the Release of Alkaloids. *Plant Cell Tissue Organ Culture 10*, 91-99.

P. Brodelius and K. Nilsson. 1983. Permeabilization of Immobilized Plant Cells, Resulting in Release of Intracellularly Stored Products with Preserved Cell Viability. *Eur. J. Appl. Microbiol. Biotechnol. 17*, 275-280.

D.K. Dougall. 1987. Cell Cloning and the Selection of High Yielding Strains. In "Cell Culture and Somatic Cell Genetics of Plants," Vol. 4, F. Constabel and I. K. Vasil (eds.), Academic Press, New York, pp. 117-124.

U. Eilert. 1987. Elicitation: Methodology and Aspects of Application. In "Cell Culture and Somatic Cell Genetics of Plants," Vol. 4, F. Constabel and I. K. Vasil (eds.), Academic Press, New York, pp. 153-196.

C. Funk, K. Gugler, and P. Brodelius. 1987. Increased Secondary Product Formation in Plant Cell Suspension Cultures after Treatment with a Yeast Carbohydrate Preparation (Elicitor). *Phytochemistry 26*, 401-405.

A. R. Gould. 1986. Factors Controlling Generation of Variability In Vitro. In "Cell Culture and Somatic Cell Genetics of Plants," Vol. 3, I. K. Vasil (ed.), Academic Press, New York, pp. 549-567.

J. Guern, J. P. Renaudin, and S. C. Brown. 1987. The Compartmentation of Secondary Metabolites in Plant Cell Cultures. In "Cell Culture and Somatic Cell Genetics of Plants," Vol. 4, F. Constabel and I. K. Vasil (eds.), Academic Press, New York, pp. 43-76.

T. Hartmann and G. Toppel. 1987. Senecionine *N*-Oxide, the Primary Product of Pyrrolizidine Alkaloid Biosynthesis in Root Cultures of *Senecio vulgaris. Phytochemistry 26*, 1639-1643.

H. A. Kemp and M. R. A. Morgan. 1987. Use of Immunoassays in the Detection of Plant Cell Products. In "Cell Culture and Somatic Cell Genetics of Plants," Vol. 4, F. Constabel and I. K. Vasil (eds.), Academic Press, New York, pp. 287-302.

H. Tanaka, C. Hirao, H. Semba, Y. Tozawa, and S. Ohmomo. 1985. Release of Intracellularly Stored 5'-Phosphodiesterase with Preserved Plant Cell Viability. *Biotechnol. Bioeng. 27*, 890-892.

J. M. Widholm. 1987. Selection of Mutants Which Accumulate Desirable Secondary Metabolites. In "Cell Culture and Somatic Cell Genetics of Plants," Vol. 4, F. Constabel and I. K. Vasil (eds.), Academic Press, New York, pp. 125-137.

Y. Yamada. 1984. Selection of Cell Lines for High Yields of Secondary Metabolites. In "Cell Culture and Somatic Cell Genetics of Plants," Vol. 1, I. K. VAsil (ed.), Academic Press, New York, pp. 629-636.

INDEX